Environmental Chemistry for a Sustainable World

Volume 10

Other Publications by the Editors

Books
Scientific Writing for Impact Factor Journals
https://www.novapublishers.com/catalog/product_info.php?products_id=42242
http://fr.slideshare.net/lichtfouse/scientific-writing-for-impact-factor-journals

Environmental Chemistry
http://www.springer.com/978-3-540-22860-8

Organic Contaminants in Riverine and Groundwater Systems
http://www.springer.com/978-3-540-31169-0

Sustainable Agriculture
Volume 1: http://www.springer.com/978-90-481-2665-1
Volume 2: http://www.springer.com/978-94-007-0393-3

Book series
Environmental Chemistry for a Sustainable World
http://www.springer.com/series/11480

Sustainable Agriculture Reviews
http://www.springer.com/series/8380

Journals
Environmental Chemistry Letters
http://www.springer.com/10311

Agronomy for Sustainable Development
http://www.springer.com/13593

Publier La Science
https://listes.inra.fr/sympa/d_read/veillecaps
(in French and English)

More information about this series at http://www.springer.com/series/11480

Jayanta K. Saha • Rajendiran Selladurai
M. Vassanda Coumar • M.L. Dotaniya
Samaresh Kundu • Ashok K. Patra

Soil Pollution - An Emerging Threat to Agriculture

Springer

Jayanta K. Saha
Division of Environmental Soil Science
Indian Institute of Soil Science
Bhopal, Madhya Pradesh
India

Rajendiran Selladurai
Division of Environmental Soil Science
Indian Institute of Soil Science
Bhopal, Madhya Pradesh
India

M. Vassanda Coumar
Division of Environmental Soil Science
Indian Institute of Soil Science
Bhopal, Madhya Pradesh
India

M.L. Dotaniya
Division of Environmental Soil Science
Indian Institute of Soil Science
Bhopal, Madhya Pradesh
India

Samaresh Kundu
Division of Environmental Soil Science
Indian Institute of Soil Science
Bhopal, Madhya Pradesh
India

Ashok K. Patra
Indian Institute of Soil Science
Bhopal, Madhya Pradesh
India

ISSN 2213-7114 ISSN 2213-7122 (electronic)
Environmental Chemistry for a Sustainable World
ISBN 978-981-10-4273-7 ISBN 978-981-10-4274-4 (eBook)
DOI 10.1007/978-981-10-4274-4

Library of Congress Control Number: 2017934645

Printed on acid-free paper

This Springer imprint is published by Springer Nature
The registered company is Springer Nature Singapore Pte Ltd.
The registered company address is: 152 Beach Road, #21-01/04 Gateway East, Singapore 189721, Singapore

Foreword

त्रिलोचन महापात्र, पीएच.डी.
एफ एन ए, एफ एन ए एस सी, एफ एन ए ए एस
सचिव एवं महानिदेशक

TRILOCHAN MOHAPATRA, Ph.D.
FNA, FNASc, FNAAS
SECRETARY & DIRECTOR GENERAL

भारत सरकार
कृषि अनुसंधान और शिक्षा विभाग एवं
भारतीय कृषि अनुसंधान परिषद
कृषि एवं किसान कल्याण मंत्रालय, कृषि भवन, नई दिल्ली 110 001

GOVERNMENT OF INDIA
DEPARTMENT OF AGRICULTURAL RESEARCH & EDUCATION
AND
INDIAN COUNCIL OF AGRICULTURAL RESEARCH
MINISTRY OF AGRICULTURE AND FARMERS WELFARE
KRISHI BHAVAN, NEW DELHI 110 001
Tel.: 23382629; 23386711 Fax: 91-11-23384773
E-mail: dg.icar@nic.in

Globally agriculture has been recognized as an engine of economic development and is an integral part of any agenda for addressing global issues of 21st century. According to numerous studies, pollution in agricultural soils has become a growing concern in most of the developed and developing nations due to enhanced industrialization and urbanization. Intake of heavy metal via soil-crop system has been considered as the predominant pathway of human exposure to environmental heavy metals and organic pollutants in agricultural areas. The sources of heavy metals and other pollutants in the environment are mainly derived from anthropogenic sources that include mining, smelting, waste disposal, urban effluent, vehicle exhaust, sewage sludge, and pesticide and fertilizer application. Soils fulfill a wide variety of environmental services as a foundation for biomass production, a filter and buffer for water, an archive of natural and human history, and an important store of carbon, and these ecosystem services have become core issues of modern ecology. In India, agricultural soils may indeed be managed to enhance the productive capacity and to reduce the vulnerability to pollution at farm level but are often traded off against higher priority for short-term food and water security under the pressure of increasing production goal. The farm system is a complex social-ecological system wherein biophysical and socioeconomic concerns interact and sometimes counteract in complex ways. Finding solution to check soil pollution requires integrated approaches, and it has to be at the national scale where regulations are designed and implemented at the farm level.

I am sure that the book entitled *Soil Pollution: An Emerging Threat to Agriculture*, authored by a group of professional experts from Indian Institute of Soil Science, will serve as a useful reference material for researchers, developmental officials, policy makers, and others engaged in soil protection endeavors.

Dated the 8th March, 2017
New Delhi

(T. MOHAPATRA)

Foreword

Post-liberalization India has recorded considerable industrial growth accounting for more than 30% of its GDP. This impressive growth is, however, accompanied by an increasing threat to its environment from emission, discharge and disposal of pollutants from industries, higher consumerism, etc. Rapid urbanization in the country has also posed a serious threat to its environment for unscientific disposal of huge solid and liquid wastes to its precious water bodies and agricultural land. Soil is the principal recipient of such wastes generated from industrial and urban sectors either through direct discharge or through contaminated irrigation water. Several toxic metals and compounds generating out of the above anthropogenic activities are contaminating our precious natural resources which have evolved through millions of years of pedogenic processes and also threatening the very existence of different forms of life including human beings. Although India is blessed with vast area of fertile land, plenty of water resources and favourable climatic conditions for high agricultural growth, the above happenings are affecting her natural capital at an alarming rate. The degradation of soil resources due to faulty land management practices has been highlighted for long, but the same for industrial and urban activities is, by and large, ignored. We cannot take no notice on this for the fact that (i) though estimated land area affected with pollution is smaller compared to other types of degradation, it occurs in more fertile land near cities and towns and also (ii) reclamation of such land is very costly (even with low success) compared to the degraded land caused by faulty management practices. The pressure of increasing and sustaining agricultural productivity on every piece of land in order to feed the ever-increasing population is again forcing us to pay almost no attention towards this vital area of land degradation. To combat it and develop an effective action plan for remediation of such area, we need to have first-hand information on the status of soil pollution in the country.

Although sporadic information generated from some parts of the country indicated the build-up of pollutants in agricultural land, a systemic and comprehensive report on anthropogenic activities and the status of soil pollution is not available.

I am extremely happy that the Division of Environmental Soil Science at the Indian Institute of Soil Science has compiled both primary and secondary information on the topic and brought out a useful publication entitled 'Soil Pollution: An Emerging Threat to Agriculture' and congratulate its authors. I hope researchers and policy makers working on the subject will be highly benefitted from the contents of the publication.

Former Vice Chancellor and Professor (Soil Science)
Bidhan Chandra Krishi Viswavidyalaya, India
Kalyani, West Bengal, India

L.N. Mandal
(L. N. Mandal)

Preface

Due to the unique geographical location and climatological advantage, India is bestowed with fertile land capable of producing diversified crops and food for the second largest population of the world. Soil has been revered by mankind in the Indian subcontinent since ancient times as evidenced in the *Atharva Veda* (12-1-12) which solemnly declared the enduring filial allegiance of humans to Mother Earth 'माता भूमिः, पुत्रोऽहं पृथिव्याः'. In tune to that, soil was cared for maintaining its productivity through soil and water conservation and application of organic manure. However, due to greed for growing more food, the need for sustainable use of land has been ignored during the post-Green Revolution period. This, in conjunction with increased developmental activities (mining, infrastructure, industry, urbanization, etc.), has put considerable pressure on agricultural land causing its degradation. Since the last decade of the previous century, the need for assessment of soil quality was being felt to address the issue of declining partial factor productivity of inputs during growing of agricultural crops. Although considerable effort is being made to address the issues of soil conservation, nutrient mining and soil organic matter maintenance through research, extension and modification of the fertilizer policy, the issue of declining soil quality due to entry of harmful contaminants is by and large ignored in spite of their accelerated generation and spread in the environment. This is probably because soil is generally viewed mainly as a medium of plant growth and is considered a domain of agricultural scientists where enhancing food productivity from a fixed land area remains the major challenge. However, limited researches conducted by ICAR-Indian Institute of Soil Science gave enough indication that the area of soil pollution and its remediation needs to be addressed adequately in order to prevent permanent deterioration of soil quality in vulnerable areas. However, this cannot be done without sincere involvement of other stakeholders responsible for generation, monitoring and control of the contaminants. Also, this requires considerable awareness among different stakeholders like students, educationalists, environmentalists, industrialists, policy makers and farmers on the importance of soil quality maintenance and its protection from polluting activities so that right direction can be followed during the country's development in generations to come. This issue had been deliberated considerably in Research Advisory Committee meetings of ICAR-Indian Institute of Soil Science where the need for a publication on the status of soil pollution in the country and on adoptable remedial measures for its protection was felt. Several textbooks have been written in the perspective of imparting basic knowledge to graduate students focussing on remediation technologies for polluted lands. However, a comprehensive book dealing on the subject of 'soil pollution-pollutants, their sources, impact, assessment methodology, status of soil pollution in India and ways to address it in Indian context' is meagre. Keeping this in mind, this book has been written after analysis of vast research and other information generated in this subject, which will hopefully be suitable both as textbook for students and as a reference book for researchers in the field related to soil and environmental science, as well as for experts involved in Environmental Impact Assessment and policy making.

I, on behalf of all the authors, take this opportunity to express our gratitude to members of the Research Advisory Committee of ICAR-Indian Institute of Soil Science for their keen interest in bringing out this publication. I am also grateful to the scientists and other staff of the Division of Environmental Soil Science for extending their help during the preparation of the manuscript. It is hoped that the community of students, researchers, policy makers, planners, farmers, environmentalists and EIA experts will find this publication useful.

Bhopal, Madhya Pradesh
India

Jayanta K. Saha

Contents

About the Authors

Dr. Jayanta Kumar Saha is head and principal scientist of the Division of Environmental Soil Science, ICAR-Indian Institute of Soil Science, Bhopal, India. He did his M.Sc. and Ph.D. in agricultural chemistry and soil science from BCKV, Mohanpur (West Bengal), India. His major research areas are chemistry & analysis of plant nutrients and pollutants in soil, soil pollution impact assessment, food contamination, risk assessment on soil biological activities and toxicity amelioration. He has worked under various capacities in national and international forums. He has published several original research articles (52) in renowned national and international journals, technical bulletins and books (9) and book chapters (12) and serves as reviewer of several international journals. He is the recipient of S. N. Ranade Memorial Junior Scientist Award for Excellence in Micronutrient Research.

Dr. Rajendiran Selladurai is scientist in the Division of Environmental Soil Science, ICAR-Indian Institute of Soil Science, Bhopal, India. He did his M.Sc. and Ph.D. in agricultural chemistry and soil science from Chandra Shekhar Azad University of Agriculture and Technology, Kanpur, and ICAR-Indian Agricultural Research Institute, New Delhi, respectively. His specialized areas of work are environmental impact assessment specifically soil pollution, soil quality assessment and management and carbon sequestration. He has published many numbers of research articles (more than 20) in renowned national and international journals, book chapters (8) and technical bulletins. He is recipient of the Silver Medal for University Second Topper, Junior and Senior Research Fellowships, Young Scientist Awards, Best Poster and Paper Prizes, etc.

Dr. Mounisamy Vassanda Coumar is scientist in the Division of Environmental Soil Science, ICAR-Indian Institute of Soil Science, Bhopal, India. He did his M. Sc. and Ph.D. in agricultural chemistry and soil science from ICAR-Indian Agricultural Research Institute, New Delhi. His specialized areas of work are environmental impact assessment, soil quality and resilience and toxicity amelioration. He has published several research articles (more than 20), books, book chapters and bulletins in renowned national and international journals. He is the recipient of many awards like Junior and Senior Research Fellowships, Best Poster and Paper Prizes, etc.

Dr. Mohanlal Dotaniya is currently scientist at ICAR-Indian Institute of Soil Science, Bhopal, India. He obtained his M.Sc. in soil science from GBPUA&T, Pantnagar, and Ph.D. in soil science and agricultural chemistry from IARI, New Delhi. His specialized areas of work are heavy metal interaction in soil and plant and soil and water pollution. He has published more than 25 research papers/reviews in reputed national and international journals. He is recipient of Young Scientist Awards and the Excellence in Research Award. He serves as editorial member of national journals as well as reviewer of many international reputed journals.

Dr. Samaresh Kundu is former head and principal scientist in the Division of Environmental Soil Science, ICAR-Indian Institute of Soil Science, Bhopal, India. He received his M.Sc. and Ph.D. in soil science at IARI, New Delhi, and postdocs at the USSR and at Australia. His research areas are soil chemistry/fertility/soil microbiology/application of nanoscience and nanotechnology in soil science and plant nutrition research. He has published 130 research papers, 4 books, 8 research bulletins and 20 book chapters. He is a recipient of several awards/recognitions and fellowships, namely, the Krishak Bharti Barani Kheti Award, Dr. Rajendra Prasad Puraskar Award of ICAR, Meritorious Invention Award of NRDC, Hari Om Ashram Award of ICAR and ICAR Award for Outstanding Team Research, as well as fellow of the Indian Society of Soil Science and of the National Academy of Agricultural Sciences.

Dr. Ashok Kumar Patra is director of ICAR-Indian Institute of Soil Science, Bhopal, India, since 2014. He obtained his MSc and PhD in soil science and agricultural chemistry from ICAR-Indian Agricultural Research Institute (IARI), New Delhi. Earlier he had served in ICAR-Indian Grassland and Fodder Research Institute, Jhansi; ICAR-Central Institute of Fisheries Education, Mumbai; and IARI, New Delhi. He has made outstanding contribution on various aspects of nitrogen cycling and its ecology and biodiversity in different agroecosystems. He has contributed more than 200 research publications. He is a recipient of British Council TCT Award; DBT Overseas Associateship Award; FAI Dhiru Morarji Memorial Award; Bharat Jyoti Award; Rajiv Gandhi Excellence Award; ISSS Dr. G.S. Sekhon Memorial Lecture Award; Bioved Agri-Innovation Award; and Hooker Award of IARI, New Delhi. He was editor of *Range Management and Agroforestry* and *Journal of the Indian Society of Soil Science*. He is associate editor of the *European Journal of Soil Science* (2014–present); councillor of the ISSS (2005–2006); vice president, Indian Society of Soil Science (2016 & 2017); and president of Agriculture and Forestry Sciences Section of the Indian Science Congress Association (2016–2017); He served as editorial advisor for several international journals. He is a fellow of Indian Society of Soil Science, National Academy of Agricultural Sciences, and Range Management Society of India.

Chapter 1
Agriculture, Soil and Environment

Abstract Agriculture has always remained backbone of economy and sociopolitical stability of low and middle income countries, and employs largest work force in India. Maintenance of soil quality has been pointed out as one of the major contributing factors for sustainability of agricultural production. Rising contribution of industry to India's GDP has been speculated to cause adverse impact on the precious land resources; though the issue has received less attention among the policy makers. There is a complex relationship of Indian economy with industry and agriculture, both of which competing for same natural resources with differential impact on land resources. This chapter highlights the need for reviewing different developmental activities having direct or indirect impact on agricultural productivity and the need for accelarating soil research to investigate upon migration of pollutants from source of generation and their impact on soil so as to devise strategies for protection and remediation of precious non-renewable soil resources.

Keywords Agriculture • Industry • Economy • India • Land • Environmental Impact

1.1 Agriculture in Indian Economy

The economy of India is the seventh-largest in the world by nominal GDP with an average growth rate of approximately 7% over the last two decades. Sources contributing to Indian economy is normally categorized in three sectors—Agriculture (also includes livestock, forestry, fishing and related activities), Industry and Services. While contribution of agriculture and allied sector to GDP of the country was 51.81% in 1950–1951 at current prices, the same has declined gradually to 18.20% by the year 2013–2014 which is still higher than similar figures from most of the developed and developing countries (Table 1.1). Although contribution of agriculture & allied sector in Indian economy is smaller as compared to industry and service sector, it engages about 50% of the total workforce which is the highest among major developed and developing countries. This emphasizes the role of agriculture & allied sector in maintaining socioeconomic and sociopolitical stability of the country. Due to integrated efforts through genetic manipulation of crop varieties as well as nutrient, water and pest management during the last 30 years,

J.K. Saha et al., *Soil Pollution - An Emerging Threat to Agriculture*, Environmental Chemistry for a Sustainable World 10, DOI 10.1007/978-981-10-4274-4_1

Table 1.1 Status of India agriculture and economy among the major countries in the world in 2014

	GDP (billion US$)	Population density (per km^2)	Average annual rainfall (mm)	% contribution by agriculture in GDP	% irrigated land	Cereal yield (kg/ha)	Employment in agriculture (% of total employment)
USA[a]	17348.0	35	715	1.3	5.5	7341	–
China[a]	10354.8	145	645	9.2	10.5	5891	3
Germany	3868.3	232	700	0.7	–	7318	1
UK	2988.9	267	1220	0.7	–	6630	1
France	2829.2	121	867	1.7	–	7074	3
Brazil[a]	2346.1	25	1761	5.6	–	4826	15
Russia	1860.6	9	460	4.2	–	2240	7
Canada	1785.4	4	537	–	–	4170	2
Australia[a]	1454.7	3	534	2.4	0.6	1992	3
Netherlands	879.3	501	778	1.8	–	8653	2
Denmark	342.4	133	703	1.4	–	6211	2
Greece	235.6	85	652	3.8	16.9	4481	13
India[a]	2048.5	436	1083	17.8	36.3	2962	50

Source: World Bank data (http://data.worldbank.org)
[a]Megadiverse country

India's food grains production nearly doubled from 102 megatonne (Mt) in 1973 to nearly 200 Mt in 1999. In 2013–2014, India achieved a record food grains production of 264 Mt in spite of continuous shrink in cultivated area due to its diversion for non-agricultural use. However during the same period, Indian population increased by about 3.4 times and is projected to be growing at 1.25% rate (Economic Survey 2014–2015 of Union Budget, Government of India). The massive increase in population poses a challenge before the farmers and government to produce about 2.5 Mt of food grains extra each year; besides increasing supply of livestock, fish and horticultural products. Being a source of livelihood and food for a vast majority of low income, poor and vulnerable sections of society, performance of agriculture assumes greater significance in view of the government's target for ensuring food security. In the last two 5-year plans, it has been clearly indicated that for the economy to grow at the rate of 9%, it is important that agriculture should grow at least by 4% per annum. In spite of the considerable emphasis given by government in the agriculture sector, productivities of most of the crops are considerably lower as compared to major countries of the world. This indicates that there is still a considerable scope of increasing per-unit-area productivity in the country, provided soil and water quality is maintained for achieving the targeted potential.

In this context, important issue that needs to be addressed is whether soil resources of the country are adequate to meet the food demands for rapidly increasing human population. Due to development of high yielding fertilizer responsive cultivars and increased use of fertilizer nutrients and agrochemicals in the post green revolution period in tandem with the enhanced availability of irrigation water, agriculture witnessed major transformation in terms of increased crop intensity and productivity per unit area during post 80s. However, sharply declining partial factor productivity (yield response per unit quantity of NPK applied) of fertilizer over the years (Fig. 1.1) has been attributed to rising concerns for decreasing profitability and unsustainability of Indian agriculture, which is also one of the major reasons for migration of human force from agriculture to other sectors and causing the former a more vulnerable sector of Indian economy.

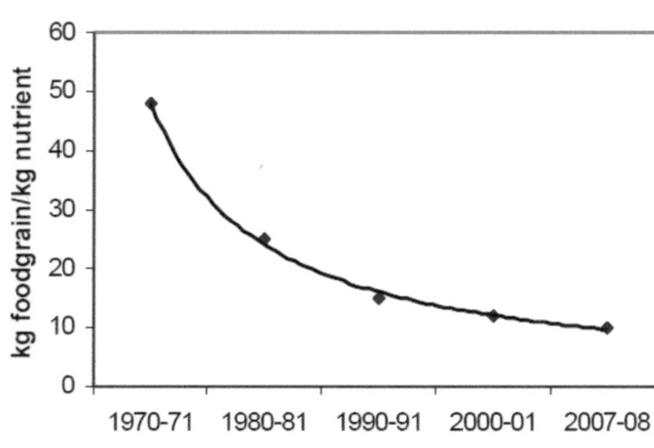

Fig. 1.1 Partial factor productivity of fertilizer NPK (Yadav et al. 1998)

1.2 Role of Soil in Sustainability of Agriculture in India

An earlier estimate indicated that 89.5% of land area in the country was considered threatened due to human activity (Eswaran et al. 1999). In broader context of environment management, Wackernagel et al. (2002) defined sustainability as living within the regenerative capacity of the biosphere. Within micro-environment also, regenerative capacity of the soil contributes towards sustainability of agro-ecosystem. Thus, enhancement and maintenance of soil regenerative capacity is essential for the sustainability of agriculture and for meeting basic needs of rising population of country. Wackernagel et al. (2002) also indicated that human demands for goods and services have exceeded the biosphere regenerative capacity since the 1980s and quantified this analysis by suggesting that humanity's load corresponded to 70% of the global biosphere in 1961 and this ballooned to 120% in 1999. Although quantitative estimate is not available, contribution of soil in agricultural production during pre-green revolution period was assumed to be higher than during later period when seed, water, fertilizers and agrochemicals played larger role (Yadav et al. 1998). As majority of essential nutrient elements are taken up by plant roots, quality of rhizospheric environment is important for higher fertilizer nutrient use efficiency and agricultural profitability. Optimum root proliferation, proper vertical air diffusion and absence of any adverse interactions among soil constituents are some of the soil criteria which determine rhizosphere quality for higher nutrient use efficiency. Beside several inappropriate agricultural management practices, excessive build-up of several elements/compounds has been attributed to declining soil quality in and around the root zone. Hence, resilience (i.e. regenerative capacity) of soil resources to the degrading forces is an emerging area of investigation due to emergence of such forces caused by increased anthropogenic activities.

1.3 Threat to Soil and Environment

India has been identified by World Conservation Monitoring Centre under UNEP in 1988 as one of the 17 megadiverse countries that harbor the majority of the Earth's species and high numbers of endemic species. Wide variation in climates, land topography, soil types are some of the major drivers responsible for such large biodiversity. Soil is by far the most biologically diverse part of Earth and has been evolved in the nature to perform several ecosystem services critical for life and performs as filter for water; acts as medium and nutrient reserve for plant growth; provides habitat for billions of organisms contributing to biodiversity; supplies most of the antibiotics used to fight diseases and regulates the Earth's temperature as well as many of the important greenhouse gases. However, with the progress of modern civilization, humans additionally use land as a dumping site for solid wastes, filter for wastewater, and foundation for our cities and towns. Except in forest ecosystem at far away from pollution activities, positive balances of heavy metals have been estimated during computation of trace elements budget (Kabata-Pendias and Pendias 1992). This is due to their higher rate of inputs through fertilizers, sludges, manures, residue recycling, atmospheric inputs etc. as compared to the removal through crop uptake and infiltrating water, which indicates gradual build-up of heavy metals in soils under agro-ecosystem.

In the quest of rapid growth in the national economy, India is increasingly pushing forth for rapid urbanization, industrialization and transportation; all of which are competing for agriculturally important natural resources like land and water. From 2001 to 2010, it was observed that each state lost about 1% of its total geographical area due to diversion of agricultural land for urban expansion and magnitude of such diversion has been increasing steadily (Pandey and Seto 2015). Large scale urbanization, industrialization and associated environmental alteration and their impact on soil quality pose a threat to biodiversity as evident from the extensive studies conducted at several developed countries. Although industry has several adverse impacts on agriculture, their influential & dependency relationship is many a times quite complex because of the unlimited human need. Different agricultural and industrial requirements and activities affecting environment and human life and their relationships are depicted below in Fig. 1.2.

Above complex relationships among different economic and social sectors and associated activities indicate that most of the anthropogenic activities are essential for meeting expectations and need of increasing population and socio-economic development of the country; and any such developmental activities leaves irreparable damage to the environment including its natural resources. This is a transition period for many developing economies like India to strike a balance between industrial development and environmental protection so as to feed the bulging population adequately with quality food as well as to improve the comfort and quality of lifestyle. Although considerable researches are being conducted by different institutes under Indian Council of Agricultural Research as well as by State Agricultural Universities to address the issues of declining soil quality due to faulty agricultural practices, insufficient research efforts have been put to address the issue of soil and water pollution arising from high inputs of contaminated fertilizers, soil amendments and agro-chemicals. Although long-term fertilizer experiments at different agro-ecosystems across the country have conclusively indicated that integrated and balanced fertilizer management systems in crop production do not leave any adverse effect on soil quality (Swarup 1998), a study conducted by ICAR-Indian Institute of Soil Science indicated groundwater contamination with nitrate beyond permissible level at several places in high fertilizer using districts of the country (Biswas et al. 2013).

In the early Anthropocene, rural population dominated but in the last 70–80 years, urban centers expanded rapidly with higher concentrations of people. These urban centers and their encompassing ecosystems initiated a new set of environmental problems mostly related to land surface sealing, water use, waste management, and an insatiable appetite for food, fiber and energy needs.

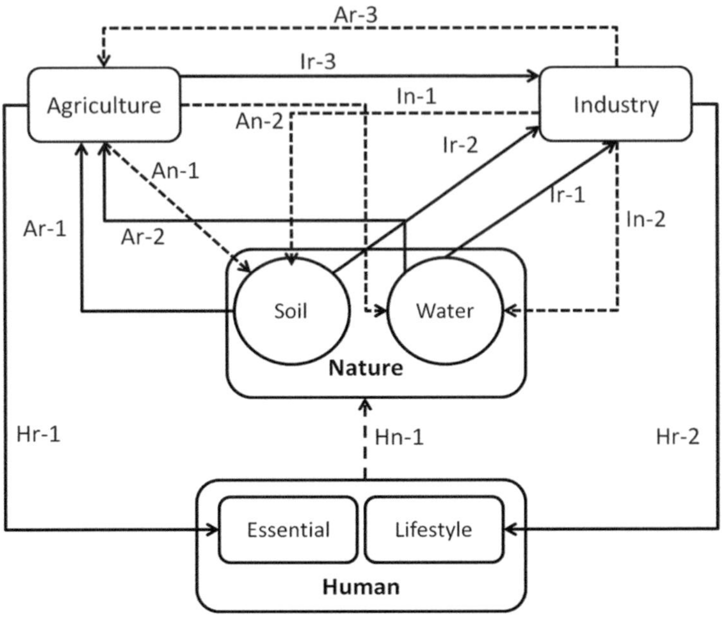

Legends for Figure

Sector	Requirement	Activities affecting nature
Agriculture	From nature: Good quality soil (**Ar-1**)	Frequent tilling of land leading to enhanced erosion (**An-1**)
	From nature: Good quality water (**Ar-2**)	Use of agrochemicals affecting food quality, soil biodiversity, water quality (**An-2**)
	From industry: Fertilizers, pesticides, mine-products like lime, rock phosphates etc. (**Ar-3**)	
Industry	From nature: Good quality water (**Ir-1**)	Release of effluents in water bodies (**In-1**)
	From nature: Fertile land, mine deposits (**Ir-2**)	Emission of pollutants to deposit on land; hazardous waste dumping (**In-2**)
	From agriculture: Raw materials like fibre, sugarcane, oil, starch (**Ir-3**)	
Human	From agriculture: Adequate quantity and quality of food (**Hr-1**)	Sewage generation, solid wastes (**Hn-1**)
	From industry: Housing material, clothing, comfort and lifestyle products etc. (**Hr-2**)	-

Fig. 1.2 Different agricultural and industrial requirements and activities affecting environment and human life and their relationships

Widespread development of infrastructure such as housing, roads and railways have contributed to environmental degradation by changing the general hydrology and energy fluxes of the area. Apart from the threat of land consumption by non-agricultural activities like urbanization, transport and industrialization, soil contamination by different pollutants has become a major cause for deterioration of its health; and in the long run the polluted soil may not be fit to grow food which in turn may affect Indian economy directly or indirectly through reduction in raw material supply to industries and emergence of several socio-economic problems. Urbanization and industrialization exploited natural resources and damaged the environment through their activities at every phases and processes of operation. It is evident from several environmental tragedies that pollution from various sources particularly from industries and mining had negative impact on environment which was mostly irreversible in nature, and cause extinction of species resulting in loss of unique genetic resources of valuable use. In agro-ecosystem, soil pollution is caused by the presence of xenobiotic chemicals or other alteration in the natural soil environment by industrial activity, agricultural chemicals, or improper disposal of waste. The most common chemicals involved are petroleum hydrocarbons, aromatic hydrocarbons, pharmaceutical products, solvents, pesticides, lead, and other heavy metals. Impact of pollution on soil is quite alarming and introduction of these hazardous chemicals into the soil environment results in huge disturbances to the ecological balance of soil and injures health of living creatures on earth.

1.4 Addressing the Issue of Soil Pollution

To address issues related to land degradation, different countries have formulated action plan and strategies to identify and combat soil pollution for the purpose of protecting environment and organisms. United States Environmental Protection Agency has taken up ‘Superfund’ program for cleaning up of some of the nation’s most contaminated land and responding to anticipated environmental emergencies, oil spills and natural disasters. The European Commission has established a framework based on common principles and objectives to address the different facets of soil degradation, where member countries identified the issues and location of problems and devised their own strategies to address those problems. In India, attention has been given to protect environment from different types of pollution in the country through environmental laws and act like ‘Environment Protection Act, 1986’, as well as through awareness program and education. However, almost all the emphasis has been given to air and water pollution, probably due to higher mobility of toxicants towards organisms through these matrices. Unlike many developed countries, the information and awareness on pollution of land resource and it impact is lacking in the country. In most of the Environmental Impact Assessment (EIA) of different industrial/mining/infrastructural projects, impact on land resources and agroecosystems are largely ignored.

Table 1.2 Degraded land areas of India

	Type of degradation	Area under arable land (million ha)	Area under open forest with <40% canopy (million ha)
1.	Water erosion (>10 t/ha)	73.27	9.30
2.	Wind erosion (Aeolian)	12.40	–
3.	Physical degradation		
	(a) Mining and industrial wastes	0.19	–
	(b) Permanent surface inundation	0.88	–
4.	Chemical degradation		–
	(a) Exclusively salt-affected soils	5.44	–
	(b) Salt-affected and water eroded soils	1.20	0.10
	(c) Exclusively acidic soils (pH <5.5)	5.09	–
	(d) Acidic (pH <5.5) and water eroded soils	5.72	7.13
	Total degraded area	104.19	16.53

NAAS (2010)

In India, information on extent of soil pollution, its impacts on other functional area, plant and human health is not complete. About 60% of the geographical area of the country is occupied by agricultural land, most of which is facing one or more kind degradation stresses. The data given by different agencies on soil degradation mainly focus on physical aspects of soil deterioration like erosion, waterlogging etc. and in case of chemical degradation not much attention was given except on soil salinity problems (Table 1.2).

In many developed countries like in North America and Western Europe, extent of contaminated land been estimated with many of these having a legal framework to identify and deal with the environmental problems. However, information on magnitude of different developmental activities like mining, industries, urbanization, transportation and others, and their impact on land/soil quality are meager in India. Instances of soil pollution or chemical degradation are cited to be very sporadic and region specific; and hence assessment of soil pollution, locating their sources and impact across the country is specifically required for devising suitable action plan for its protection and remediation. In India, each area is unique in respect of nature and extent of anthropogenic interventions, climate, land morphology, geology, hydrology and biodiversity. For precise EIA focusing land degradation or soil pollution, one needs to have comprehensive knowledge on the soil resources and its interaction with other environmental components as well as pollutants so that protective/remedial soil management can be planned effectively. To keep in view of the above, an attempt has been made to review the status of soil pollution in India under the background of ongoing anthropogenic activities with appropriate interpretation using scientific knowledges generated on the subject and to suggest future course of action.

References

Biswas AK, Srivastava S, Subba Rao A (2013) Delineation and mapping of nitrate contamination of ground water in some intensively cultivated districts. In: Kundu S, Manna MC et al (eds) IISS contribution in frontier areas of soil research. Indian Institute of Soil Science, Bhopal, pp 337–369

Eswaran H, Beinroth F, Reich P (1999) Global land resources and population supporting capacity. Am J Alter Agric 14:129–136

Kabata-Pendias A, Pendias H (1992) Trace elements in soils and plants. CRC Press, Baton Raton

NAAS (2010) Degraded and wastelands of India: Status and spatial distribution. National Academy of Agricultural Sciences. ICAR, New Delhi

Pandey B, Seto KC (2015) Urbanization and agricultural land loss in India: comparing satellite estimates with census data. J Environ Manag 148:53–66

Swarup A (1998) Emerging soil fertility management issues for sustainable crop productivity in irrigated system. In: Swarup A, Reddy DD, Prasad RN (eds) Proceeding of national workshop on long-term fertility management through integrated nutrient supply (2–4 April, 1998). Indian Institute of Soil Science, Bhopal, pp 54–58

Wackernagel M, Schulz B, Deumling D et al (2002) Tracking the ecological overshoot of the human economy. PNAS, Proc Nat Acad Sci 99:9266–9271

Yadav RL, Prasad K, Gangwar KS (1998) Prospects of Indian agriculture with special reference to nutrient management under irrigation system. In: Swarup A, Reddy DD, Prasad RN (eds) Proceedings of a national workshop on long-term soil fertility management through integrated plant nutrient supply. Indian Institute of Soil Science, Bhopal, pp 16–33

Chapter 2
Soil and Its Role in the Ecosystem

Abstract Soil is highly heterogeneous body in the terrestrial ecosystem that has evolved through thousands of years of natural processes and has remained habitat for enormous biodiversity. Due variability in seasonal temperature, rainfall, parent materials and vegetation, different types of soils have been found in India; and each of these soils have distinct mineralogical compositions, physical and chemical properties. Such heterogeneity has resulted wide variations in the response of soils to polluting activities which causes differential location specific impacts. To understand the interaction of pollutants with soil constituents and their impact on agroecosystems, basic knowledge on various aspects of soil resources and its functions are essential. This chapter describes in brief, the major soil types of India and their properties, role of soil constituents on its quality, different soil forming processes, inhabiting organisms and their role in different soil nutrient cycling processes affecting crop productivity.

Keywords Soil type • Ecosystem • Organisms • Pedogenic processes • Agroecological zone • India

Soil is a highly heterogeneous body in respect of physical, chemical and biological characteristics as well as constituents and such heterogeneity arises both at micro- and macro-scale in all three dimensions. Pollutants interact quite differently with each of its constituents and hence, impact of anthropogenic activities on ecosystem varies widely. This necessitates a comprehensive understanding on soil and its role in the ecosystem by the personnel's involved in environmental impact assessment. This chapter briefly describes the important aspects of soil science mainly in the perspective of crop productivity.

Soil is a resource on which every person's life, well-being and fulfillment depend. Poor management reveals itself in terms of social and economic costs along with political repercussions. No sustainable agriculture and sustainable development is possible without this awareness. Present societies show little interest in this regard, and soil is often considered as no more than a support for human activities. We abuse the land because we regard it as a commodity for use by us but when we see it as a part of community to which we belong, we start looking at it with respect. If soil is viewed as precious resource, its use should abide by certain rules and a number of conditions so that different ecosystem functions are protected

J.K. Saha et al., *Soil Pollution - An Emerging Threat to Agriculture*, Environmental Chemistry for a Sustainable World 10, DOI 10.1007/978-981-10-4274-4_2

and sustained. In sum, soil is the source of life for humankind. Through the intermediary of plants, animals, water and minerals, the soil shelters and nourishes the people and their well being and health are dependent upon it. Soil is thus essential for humanity, whether its uses are of an agricultural, forestry, industrial, urban or ecological nature. No society can develop without using soils. Having said this, all is a question of balance between functional and renovation capacities of soil and anthropological pressure; the future is finely poised between the realization of soil potential and the pressure of human activities. Indeed, the development of human activity is linked with the functioning and properties of soils throughout the world. Anthropogenic changes in the pedosphere influence the other spheres with which it is interpenetrated *viz.;* lithosphere, hydrosphere, atmosphere and biosphere. The soil cover is thus in a pivotal position in relation to whatever local, regional or global changes the earth is undergoing. And here lies the importance of soil, its quality and the immediate environment as a core element in the issues related to sustainability and society.

The need to protect the environment will necessarily involve the knowledge and the use of natural soil functions to ensure higher quality soils, better produce quality, water and air quality and quality of life etc., through proper management and rational use of soils. Since soil itself is a component of the ecosystem comprising of soil, plant, animal, human and climate continuum, an impact on any constituents is likely to have some effects on the other parts of system. The environment of each of these components is intimately linked to one another so much so that a change in one environment might adversely or benignly affect other constituents. It is similar to throwing refuse into neighbor's premises to keep one's own environment clean leading to ultimately to the peril of all. Goswami and Rattan (1992) defined soil health, "as being a state of dynamic equilibrium between organisms and its environment in which all the metabolic activities of the former proceed optimally without any hindrance, stress or impedance from the latter".

The importance of soils to human well-being and cultural enrichment is glorified in the ancient Indian scriptures dating back to the dawn of civilization. The "Prithvi Sutra" in *Artharva Veda* narrates hymn as a prayer of goddess Prithvi, *"O mother, with yours oceans, rivers, and other bodies of water, you give us land to grow grains, on which our survival depends"*, and *"May you, our motherland, on whom are born five races of mankind, be nourished by the cloud, loved by the rain"* (Dwivedi and Tiwari 1987). Mahatma Gandhi reiterated this concept even more directly by stating *"Earth has plenty for every man's needs but not for every man's greed"* (Gandhi and Gandhi 1994).

2.1 Physical and Chemical Characteristics of Soil

Soil and land, though related, are two different entities. Land is two dimensional entity representing geographical area and landscape, while soil is a three-dimensional body with length breadth and depth and is hidden below the land surface. It is largely hidden from the outside world until it is lost and goes out of the site. It is recognized by digging a pit and exposing its profile. ‘Soil’ a dynamic phase of the earth’s outer crust serves as a delicate interface between atmosphere, hydrosphere, lithosphere, and biosphere where various biogeochemical processes occur that aid in its development to support key ecosystem functions such as plant productivity, biotic activity and water quality. The soil fractionates the hydrologic cycle in the terrestrial area into runoff, percolation and evapotranspirational components. It is a source as well as a sink to absorb, desorb, fix or release mineral elements and gases and grow plants and decompose crop residues incorporated into it. It is a living factor where millions of tiny organisms are ceaselessly working day and night, transforming the organic matter and participating in carbon and nitrogen cycles. In our habitat planet of Earth, life could not exist without soil and no soil without life. It is no wonder that the astronauts landing on the moon searched for evidence of soil, life and water there and brought with them soil samples to make detailed studies for an evidence of life. Soil is a limited and non-renewable resource but a pivot for agriculture, food security, nutritional security, environmental safety and quality of life. Since soil is pivot of natural resources, its management has to be fully integrated with eco-friendly techniques.

2.1.1 Physical Properties of Soils

The physical properties of a soil play an important role in determining its suitability for crop production. The characteristics like the supporting power and bearing capacity, tillage practices, moisture storage capacity and its availability to plants, drainage, and ease of penetration by roots, aeration, retention of plant nutrients are all intimately connected with the physical properties of the soil. Soil as medium of plant growth should also be physically fertile. The soil which supports plants is a variable mixture of solids (mineral and organic matter), liquid (water) and gases (air); and therefore is called three phase system. In a representative silt loam soil, about 50% of the total volume is occupied by solids. At the optimum moisture for plant growth, 25% of the volume occupied by water and 25% by air. Soil particle of larger in size like sand are generally visible through unaided eye (naked eye), whereas particles of colloidal in nature like clay can be seen only through electron microscope. The liquid phase, consisting of soil water also contains dissolved salts and thus it is called soil solution. The gaseous phase consists of soil air of varying composition of oxygen and carbondioxide different from that of atmospheric air. The important soil physical attributes are (a) bulk density, (b) texture, (c) structure,

(d) water holding and transmission characteristics, (e) soil color, (f) soil aeration and (g) soil temperature, which are briefly described below:

(a) Soil Bulk Density The bulk density of soil is defined as its oven dry weight per unit volume which is normally expressed as g cm^{-3} or Mg m^{-3}. Typically, a silty loam soil has bulk density around 1.33 g cm^{-3}. Soil organic matter, texture, density of mineral matter and compaction influence bulk density of soil. Soils rich in organic matter content have lower bulk density and with increase in soil profile depth, the bulk density increases gradually. Compact soils like clayey soil have higher bulk density than the light textured of sandy soil. Bulk density of a soil influences several important soil properties related to rhizosphere quality and consequently, to plant growth. Its value of more than 1.6 g cm^{-3} may restrict proper root growth in the soil.

(b) Soil Texture Quantitatively, soil texture refers to the relative proportion/distribution of different particles size fractions, which are specifically referred to as coarse sand (2.0–0.2 mm), fine sand (0.2–0.02 mm), silt (0.02–0.002 mm) and clay (less than 0.002 mm). The soil texture is considered a permanent natural attribute of the soil and therefore most often is used to specify its physical composition, having a bearing on soil behavior. Sand particles can hold little water, and thus soils dominated by sand are prone to drought. Clay is the fraction having negative charge and greater surface area per unit mass and influence the soil behavior most. The water gets absorbed by clay particles and hydrated resulted in swelling and shrinkage of soil upon wetting and drying of soil. Normal agricultural practices do not generally change soil texture. However, frequent inundation of land by flood water and application of large quantities of inorganic amendment materials (e.g. coal ash) may alter soil texture to a large extent.

(c) Soil Structure Physically a soil is a blend of inorganic particles, decaying organic materials, air and water. The inorganic particles of various sizes (sand, silt or clay fraction) generally cluster together to form complex and irregular patterns of secondary particles which are called aggregates or peds. The term 'soil structure' refers to the arrangements of these primary and secondary particles into a certain structural pattern. Soil structure greatly influences many soil physical processes such as water retention and movement, porosity and aeration, transport of heat, etc. The various soil management practices such as tillage, cultivation, application of fertilizer and manures, amendments (liming, gypsum) and irrigation, bring about changes in soil structure that influence other soil properties, thereby affecting root growth, water & nutrient uptake, crop growth and yield. Organic matter plays important role in the formation of stable soil aggregates of granular and crumb type soil structure. The organic root exudates and microbial decomposition products of organic materials play very important role in stable aggregates formation.

(d) Water Holding and Transmission Characteristics In soil, water gets absorbed on the surface of the soil particles as well as in the spaces between adjoining particles (intervening voids) as pore water. Distribution of particle size fractions

(texture) and their aggregation (structure) determine air-filled and water filled pore spaces in soil which ultimately regulate the exchange of gases between atmosphere and lithosphere. The major soil physical properties like soil consistency, compaction, swelling and shrinkage properties determine soil water retention and its transmission. The water is held in three forms depending on the tenacity with which it is absorbed: hygroscopic water, capillary water and gravitational or free water. The first two molecular layers of water on soil particles which have high suction are termed as 'Hygroscopic water'. Water held in capillary pores with suction ranging from 0.1 to 31 bar is known as '*Capillary water*'. With further increase in water content, when the macropores also get filled with water, it is more and more loosely held. Under the action of gravitational force, this loosely held water is liable to move downwards and hence it is known as '*Gravity water or free water*'. The drainage or deep percolation loss of water following irrigation results from the downward movement of the gravity water.

(e) Soil Color Soil color is often a ready clue to its condition and some important properties. For example, dark-colored surface soils absorb more solar radiation than light colored soils and so get warm up faster. Color is often a diagnostic feature in major soil classification and interpretation. Soil color is included in the description of a soil profile. It is also indicative of other factors such as presence of excessive salts (white), erosion etc. For determination of soil color, standard system using '*Munsell*' colour chart is used.

(f) Soil Aeration A typical mineral soil, on volume basis, contains half solid and half pore space. The solid part is consisted of mineral material and organic matter, space part is filled with water and air interchangeably. Soil aeration refers to the exchange of oxygen and carbon dioxide between the soil pore spaces and atmospheric air. The process controls the deficiency of oxygen consumed during respiration of plant roots and soil microorganisms and prevents toxicity of carbon dioxide evolved during respiration in the soil air. The term '*air capacity*' is used to describe aeration status of soil. The air capacity of soil refers to the volume of pore space filled with air when the soil is under a tension of 50 millibar. This value is also called as '*non-capillary porosity*'. It corresponds to a pore size of 0.06 mm or larger in diameter. The composition of soil is more or less similar to that of the atmospheric air except the content of carbon dioxide. In the upper layer of a soil having stable structure and ample macropores, the composition of air is taken as: nitrogen and inert gases = 79.1%, oxygen = 20.6% and carbon dioxide = 0.25%.

2.1.2 Soil Chemical Property

Soil chemical environment is regulated by number of soil chemical properties, such as, soil pH, electrical conductivity, clay mineralogy and cation exchange capacity, soil organic matters content, total and available plant nutrient status which are briefly described below.

(a) Soil pH The concentration of hydrogen and hydroxyl ions of a solution determines acidity or alkalinity. Soil reaction is denoted by pH of a soil water suspension and is an important indicator of soil health. Soil pH is the negative logarithm of the activity of H^+ in solution. Soil having pH value of <5.5 is considered acid soil. Soil pH is determined by active and potential acidity. While active acidity refers to H^+ ion concentration in soil solution, potential acidity indicates H^+ ions adsorbed on soil colloids. The former can easily be determined by measuring soil solution pH; whereas, the determination of the potential acidity additionally requires exchange of H^+ by other cation species prior to the pH measurement. The soil pH values may vary widely from pH 3 (e.g. in acid sulphate and podzolic soils) to 10 (e.g. in sodic soil). The presence of weak acids (HCO_3^-) and strong bases (Na^+ or K^+) in soil solution of alkali soils may results in high soil pH as high as pH 10.5. The soil solution H^+ ion concentration has a marked influence on nutrient release dynamics through its effect on weathering and mineralization process. It also has significant influence on soil microbial diversity composition and plant growth. Most of the crop plants grow well under near neutral soil pH value of 6.0–7.2. At low pH, Fe, Mn, and Al are highly soluble and attain toxic levels whereas at high pH, these ions become deficient. Nitrification is slow below pH 5.5. Under acid conditions, the pH-dependents NH_4^+ fixation between lattices of expanding layer silicates decreases with decreasing pH. The availability of phosphorus, primarily $H_2PO_4^-$ and HPO_4^{2-} ions, is highly pH dependent. Its availability in many soils is maximum when the pH is neutral or slightly acidic and it declines as the soil becomes strongly acidic or strongly alkaline. In general, the availability of micronutrients present in cationic forms such as Fe, Mn, Cu and Zn, increases with increase in soil acidity, whereas the availability of those present as anions, namely molybdenum and boron decreases. Therefore, the leguminous plant growth and nodulation are mostly affected by Mo deficiency rather than Al toxicity in acid soil. It has been estimated that 40, 48 and 12% of acid soil in India falls under mildly acidic (pH 5.5–6.5), moderately acidic (pH 5.0–5.5) and strongly acidic (pH <5.0) soil, respectively. Neutralization of soil acidity is normally accomplished through application of agricultural lime materials as well as compost material. Carbonates, oxides and hydroxides of calcium and magnesium are referred to as agricultural lime.

(b) Electrical Conductivity It is the measure of soluble salt content in the soil. Salt affected soils are characterized by an excess of neutral soluble salts mostly of chlorides and sulphates of calcium, magnesium and sodium. Based on the nature of soluble salts, such soils termed as *'halomorphic soils'* are generally classified into *saline and alkali* soils. Saline soils are mainly distributed in arid and semi arid regions characterized by the presence of white patches of salt crust on the surface. Growth of most crop plants is adversely affected by the presence of excessive soluble salts. Due to high evaporation and upward capillary movement of soil moisture in arid and semi-arid region, salts get accumulated on the surface soil. In dry period saline soils often show a white efflorescence of salts on their surface. For this reason they are also sometimes known as '*white alkali soils*'. When

appreciable quantities of Na^+ are present on exchange sites, Na^+ can come into solution and hence Na_2CO_3 and $NaHCO_3$ can be formed. These high soil pH conditions and presence of relatively lower salt concentrations lead to the deflocculation of clay and organic matter particles. The soil structure thus becomes water unstable. The soils are black due to the dispersed humic particles and are sometimes known as '*black alkali soils*'. The electrical conductivity is expressed as millimhos/cm at 25 °C or deci-Siemens/metre (dSm^{-1}) at 25 °C. As per the US Salinity Laboratory Staff (1954), saline soils have pH of saturation paste less than 8.5, Exchangeable Sodium Percentage (ESP) less than 15 and EC more than 4 dSm^{-1} at 25 °C. The alkali soils are characterized by pH more than 8.5, ESP more than 15 and EC lower than 4 dSm^{-1} at 25 °C. Management of saline soil is basically done by the removal of salts from saline soils through the process of leaching with water and drainage. Reclamation of alkali soils requires neutralization of alkalinity and replacement of most of the sodium ions from the soil exchange complex by the more favorable calcium ions, through the application of gypsum (calcium sulphate).

(c) Clay Mineralogy and Cation Exchange Capacity Nutrient status and behavior in soils depend on a large number of soil properties, of which types of clay mineral and total quantity of clay are most important. Soil clays consist of two main kinds of particles: (i) silicate lattice minerals (kaolinite group, montmorillonite group and illite group) and (ii) hydrous oxides of Fe and Al. The kaolinitic group of clay minerals is built up of flat units each consists of one silica sheet and one alumina sheet tightly bound to each other by equally shared oxygen and atoms. This kind of structure is characteristic of a 1:1 clay mineral. The montmorillonite group of minerals consists of two silica sheets with one alumina sheet held between them by mutually shared O-atoms. This is typical of a 2:1 clay mineral structure. In this case the units are rather weakly held to each other and easily expandable. Water and cations can thus readily be absorbed between them in these 'inner surface'. The particles are much smaller than the kaolin clays. The illite group resembles montmorillonite. It is a 2:1 clay mineral but the units are bound to each other by K^+ ions. They are therefore less expandable and their properties lie between montmorillonite and kaolinite. The hydrous oxides as their name suggests are oxides associated with water molecules. The oxides are mainly those of Fe and Al with general formulae $Fe_2O_3.H_2O$ and $Al_2O_3.H_2O$. Hydrous oxides have much less marked colloidal properties than lattice clays. They do, however, carry negative charges. The soils in which they particularly predominate are present in the tropics and semi-tropics. The presence of hydrous oxides in these soils gives rise to a high absorption capacity. The cation exchange capacity of soil is determined by the nature of clay minerals present in soil (kaolinite group: 3–15 cmol(+) kg^{-1}, montmorillonite group: 80–200 cmol(+) kg^{-1} and illite group: 20–50 cmol(+) kg^{-1}). Soils high in clay content have tendency to get dispersed with consequent increase in total surface area and micropore spaces. As a result, soil retains more water (in the capillary pores) as the clay content increases. Thus, soils rich in clay content have high cation exchange capacity and water holding capacity than soils low in clay.

(d) Soil Organic Matter Organic materials are intrinsic and essential components of all soils. It is the key component in soil that supports all life and makes soil a dynamic living system. The immense importance of soil organic matter (SOM) accrues from the facts that it (i) provides food source for soil microorganisms and soil fauna, (ii) act as store house for nitrogen, phosphorus and sulphur supply to higher plants (iii) improves various chemical and physical properties of the soil. SOM contains group of carbon containing compounds originated from living beings, of which 10–15% is non humus substance and rest 85–90% is humic substances. Soil organic matter also plays immense role in reducing toxicity of various organic and inorganic pollutants that enters the agricultural land through anthropogenic activities.

(e) Total and Available Plant Nutrient Status Soil fertility refers to the intrinsic capacity of a soil to supply all essential plant nutrients in adequate and suitable proportion for their optimal growth. Plant nutrient exist in soil both in organic and inorganic forms which constitute reserve pool for the use by microorganism and plant growth. Plant nutrients are exclusively absorbed by roots in their inorganic forms. On the contrary and unlike green plant, other living beings like man, animal and microorganisms need additional food stuff directly from organic source. An essential plant nutrient may be defined as a chemical element which is absolutely necessary for supporting normal growth, metabolism and completion of life cycle of plants. Additionally, specific function(s) of these essential elements in plants cannot be substituted by other elements. In other words, the deficiency symptom caused by any of the essential elements cannot be corrected by supplying other element. Most of the essential elements take part in plant metabolism and forms integral part of structural components of cell, enzymes and energy related compounds. Among the essential nutrients, C, H, O are derived from air and water and therefore, these are termed as non-mineral essential nutrients. All other nutrient elements are derived from soil and therefore they are called mineral essential elements. These mineral essential elements are classified as primary elements (N, P, K), secondary elements (Ca, Mg, S) and trace elements or micronutrients (Zn, Cu, Fe, Mn, Cl, B, Mo, Ni) in accordance with their amount of uptake. The essential micronutrients and other beneficial elements required for growth and metabolism of plants, animals and human beings are given in Table 2.1.

Management of the fertility of Indian soil demands it's built up and sustenance at a high level to produce adequate food for the ever increasing population. Fertilizers, therefore, constitute one of the critical farm inputs and play a significant role for achieving high crop productivity. Soil test based fertilizer recommendations ensure

Table 2.1 Essential trace element minerals for plants, animals and human beings

Species	Micronutrients	Beneficial elements
Plants	Zn, Cu, Fe, Mn, Cl, B, Mo, Ni	Si, Na, V, Sr, Co
Human beings	Fe, Mn, Zn, Cu, Se, Cr, Mo, I, F	As, B, Br, Cd, Pb, Ni, Li, Si, Zn V Co
Animals	Fe, Zn, Cu, Mn, Se, I, Mo, Co	Cr, Si, Pb, Ni, V, As, B, Sn

balanced use of fertilizers and increase yields and profits. Balanced fertilizers encompass, besides major nutrients, secondary and micronutrients whose deficiencies appear in different soils in various agro-eco-regions due to intensive cultivation. Soil testing laboratories in India generally follow identical critical limits for sufficiency or deficiency of plants nutrients, irrespective of the mineralogy or texture of soil or even crop species. However, the ideal soil testing should not only take into account of amount of available nutrient reservoirs, but also the environmental conditions of the rhizosphere which influence loss or transformation of nutrients to less available forms. Integrated plant nutrient supply management envisages the use of inorganic chemical fertilizers in combination with organic manures; legumes based cropping system, biofertilizers and other locally available nutrient sources for sustaining soil fertility and crop productivity. The combined applications of organic manure and inorganic chemical fertilizers generally produces higher crop yields than when each is applied alone. This increase in crop productivity may be due to the combined effect of nutrient supply, synergism and improvement in soil physical and biological properties. Farmyard manure (FYM) constitutes an important component for integrated nutrient management for maintaining soil fertility and yield stability.

2.2 Soil Organic Matter and Its Role in Soil Productivity

Organic materials are intrinsic and essential components of all soils. This key component in soil supports all terrestrial life forms directly or indirectly and makes soil a dynamic living system. The immense importance of soil organic matter (SOM) can be perceived from the following roles played in the ecosystem:

(a) It provides food for soil micro- flora and macro- fauna and helps in maintaining the biological activity within soil.
(b) It is the storehouse of nutrients for their supply to higher plant. Plant available inorganic nitrogen is normally quite low in soil and most of the plant requirement is met from its organic source through mineralization process (conversion of organic to inorganic nutrient form). Further, major fraction of fertilizer derived inorganic nutrients is immobilized by soil organic matter, as a result of which their loss from the system is minimized.
(c) It improves cation exchange capacity, soil buffering capacity and enhances capability to trap and exchange nutrient cations like potassium, calcium, magnesium and micronutrients like zinc, manganese, copper, iron etc.
(d) SOM helps in releasing nutrients slowly from minerals through weathering process in synchrony with their uptake rate by plant.
(e) Plant growth and development are enhanced through physiological and nutritional effects of some organic substances that are directly utilized by plants.

(f) Another important influence of SOM is improvement in soil physical properties like soil structure, porosity, water holding capacity, infiltration, soil temperature, etc., which facilitate optimum root growth and nutrient uptake.
(g) Soil organic matter also influences a variety of pedogenic processes like weathering of minerals, and their transportation leading to the soil formation.

2.2.1 Nature and Amount of Soil Organic Matter

The term SOM usually refers to those organic materials which are a part of the soil matrix. This includes: (a) the litter layer which consists of dead plant residues at the uppermost layer of the soil, (b) partly decomposed plants residues formed by the action of soil microorganisms and fauna on the litter, (c) biological organic molecules which are components of plants or animal tissues, e.g. proteins, carbohydrates, lignins, lipids, peptides, amino acids, organic acids, alcohols, etc., which are collectively called as non-humic substances, and (d) humic substances which are the stable end products of decomposition of plant and animal residues. Depending on the climatic condition, soil organic matter content in soil varies widely. In general, cool and temperate regions contain relatively more organic matter (5–10%) in the surface soil due to lower mineralization, whereas, cultivated surface soils in the tropical region contains only less than 1% SOM owing to high rate of decomposition under prevalent conditions of high temperature and moisture.

Climate (temperature and rainfall) plays one of the most important roles in determining the extent of SOM accumulation. It has been observed that SOM contents decrease to 2–3 times for each 10 °C increase in mean temperature. Apparently, the decomposition loss of SOM (to CO_2) increases more rapidly with temperature than the rate of its formation. Therefore, in warmer climates, SOM levels are generally lower than in cooler regions. In general, the activity of soil microorganisms and decomposition of SOM increases with rise in temperature from 0 to 35 °C; but above this temperature, the decomposition gets suppressed. The decomposition of OM is also influenced by rainfall and therefore, by soil moisture levels. Intensity of decomposition increases with moisture and reaches maximum at 60–80% of the maximum water holding capacity.

The second important factor influencing OM levels in soil is the nature and amount of vegetation. Vegetation provides the basic input in the form of leaf litter, branches, roots. The effect of vegetation on OM accumulation is best seen in the tropics. In the tropical evergreen forests, where litter input is very high, the soils are very rich in OM; soils under grass cover in the same region have much lower OM levels since litter input to the soil is relatively less; cultivated areas with limited vegetative cover have very low levels of OM.

Other factors, which influence OM accumulation, are soil minerals and soil texture. OM is easily lost from light textured sandy soils whereas loamy soils tend to accumulate OM. This can best be seen along river banks in areas otherwise covered by sandy soils.

Topography also influences the OM levels through soil drainage and moisture levels. Soils in depressions and valleys may accumulate more OM than those at the slopes. If all other factors remain constant, OM accumulation reach equilibrium levels with time. Further increase will not occur after a certain period of time, which is normally a few years. However, the stable fraction (humus) can persist in the soil for as long as 250–1900 years. This stable fraction takes much longer to accumulate and has an average '*turn-over period*' of about 150 years.

2.2.2 *Humic Substances*

Soil organic matter contains both humic and non humic substances which play an important role in nutrient release dynamics influencing plant growth. Humic fraction is relatively stable in soil than non-humic substances and constitutes largest component of SOM (85–90%). This fraction plays a major role in maintaining soil fertility and crop productivity. In general, humic substances are formed by the decomposition of dead plant debris and animal residues by soil microorganisms and fauna. They are colloid-sized, polymeric substance having dark colors (black and brown-black to yellow). The three soil fractions (on the basis of solubility in acid and alkali) of humic substances are (i) fulvic acid (FA) (ii) humic acid (HA), and (iii) humin. Fulvic acid is the most mobile fraction being soluble in water, acids and alkalis and is yellow to brownish yellow in color. Humic acid is soluble in alkali and insoluble in water and acids. It has a dark brown to black color. The most insoluble fraction of humus is humin, which is insoluble both in acid and in alkali. This fraction is strongly bound to the soil mineral matrix and is difficult to extract.

2.3 Organisms in Soil and Their Activities

Land is a dynamic living ecosystem that forms a habitat for millions of living organisms. Organisms present in soil are classified into two main groups: (i) soil flora, belonging to plant kingdom, and (ii) soil fauna, the animal forms. These are further divided into two sub-groups: macroorganisms, i.e. those organisms which are big enough to be seen by an unaided eye and microorganisms which are so small that these can be seen only after magnification using a microscope.

Soil organisms can also be classified on the basis of requirements of molecular oxygen, temperature and mode of nutrition, etc. Organisms that need O_2 for respiration and cannot survive without it are called *'obligate aerobes'* and those which are aerobic but also adapt to grow under anaerobic environment using oxidized substances like NO_3, SO_4, CO_2 etc. as terminal electron acceptor in place of O_2 during respiration are classified as '*facultative anaerobes*'. An example of such organisms of great agricultural substance in soil is the denitrifying bacteria. Certain group of microorganisms like *Clostridia, Actinomyces, Bacteroides,*

Fusobacterium and Methanobacteria thrives well in the absence of oxygen are termed as '*obligate anaerobes*'.

The organisms have specific temperature requirements for their growth. Any change from the optimum temperature will not kill the organisms but will reduce their growth rate. *Psychrophiles* have optimum temperature for growth below 10 °C, while the group of organisms with optimum temperature between 20–35 °C is termed as *mesophiles*. This group is highly dominant and numerous in most of cultivated soils of India. A temperature higher than 45 °C favours development of *thermophiles*, commonly encountered in the compost pits.

Based on the mode of nutrition, soil organisms are classified into two groups- '*heterotrophs and autotrophs*'. '*Heterotrophs*' derive energy by oxidation of organic compounds while *autotrophs* derive their carbon from CO_2 for cell synthesis. The group of autotrophs is further sub divided into *chemoautotrophs* which get energy from the oxidation of simple reduced inorganic compounds like iron,sulphur etc. and *photoautotrophs*, which derive energy from sunlight by the process of photosynthesis.

2.3.1 Macroorganisms in Soil

The macroorganisms in soils include Acari, Collembola, Enchytracidae, Isoptera, Isopoda, Amphipoda, Diplopoda, Earthworm, Coleoptera, Mollusca, etc. Population or biomass estimation of these soil animals is highly difficult because they are not uniformly distributed in the soil and are highly mobile. Even though fewer in numbers than microorganisms, these organisms are very useful in soils as (i) they help in the decomposition of organic residues by mixing, churning or fragmentation as they eat on plant material, (ii) they form burrows and tunnels which increase soil aeration, drainage and turn in large amount of surface soil, (iii) enrich soil with organic material (as their food, ingested soil in the guts of earthworms gets converted into worm casts called '*mull humus*'), and (iv) Some of these organisms like mites, termites and acrinemites feed on soil microorganisms including plant pathogens.

Roles of Earthworms The total biomass of earthworms in soil ranges from 110 to 1100 kg ha^{-1} (furrow slice). Their population decreases on cultivation due to reduction in organic matter, with tillage which causes mechanical injuries and upon pesticide application which is toxic in nature. Earthworm cast is a rich source of nutrients, particularly N, P, Ca etc., and contains more bacteria and organic matter. Earthworms do intimate mixing of organic matter with mineral matter which increases the stabilization of clay bound carbon, depending on soil type. Earthworm-worked soils generally have high porosity, increased water holding capacity, higher water infiltration rate, more water stable aggregates and increased availability of plant nutrients. These organisms may also affect microbial population as they ingest microbes along with soil and organic matter.

Roles of Termites Termites also influence soil properties and processes through four types of activities: (i) physical modification of soil profiles by constructing mounds, sheeting and foraging runways, (ii) changes in soil texture emanating from movement of clay fractions from subsoil for construction they make (iii) changes in the nature and distribution of organic matter and plant nutrient through litter brought into nests which is digested by termites and decomposed by the microbes *in situ* (the termite-worked soil has higher CEC and exchangeable bases than surrounding soil), and (iv) the changes in soil drainage and moisture regimes upon constructing subterranean galleries.

2.3.2 Microorganisms in Soil

Bacteria The number of bacteria is highly variable depending upon soil type, nature of crop cover and climatic conditions. Generally, soils with low organic matter and sandy texture have very low population. The different bacterial genera commonly occurring in diverse soils are: *Pseudomonas, Arthrobacter, Clostridium, Bacillus, Achromobacter, Micrococcus and Agrobacterium.* The presence of enterobacteria (not strictly soil bacteria) in soils, is of great ecological significance, as it is indicative of fecal contamination. Their presence is a potential health hazard, particularly in growing crops whose edible plant parts come in direct contact with soil. Thus enteric pathogens may get into food chain and cause human diseases.

Due to their large number and rapid rate of multiplication, bacteria play a very significant role in carrying out various biochemical reactions controlling availability of plant nutrients. The processes of N_2 fixation, phosphate solubilization, organic matter decomposition and synthesis of humus, nitrification and denitrification, protein decomposition and ammonification, etc., lead to transformation of various macro- and micro-nutrients in soil and help in plant nutrition. Certain free-living soil bacteria can reduce the atmospheric nitrogen to ammonical form, thus enriching soil with plant available nitrogen. These bacteria belong to the genus *Azotobacter* and *Azotomonas*, and grow abundantly in soil which are relatively high in organic matter. The genus *Beijerinckia* and *Derxia* are unique; they can fix nitrogen in soils which are highly acidic (pH 4.0). *Clostridium* fixes nitrogen under anaerobic conditions and can be of economic significance in flooded soils. Besides these free-living bacteria, there is another group belonging to the genus *Rhizobium* which fixes nitrogen in symbiosis with the legumes forming tiny out-growths on roots called legume root nodules, the sites of nitrogen fixation. *Azospirillum*, a spiral bacterium fixes nitrogen in association with plants in rhizosphere after entering into roots without forming nodule and the process is called '*associative nitrogen fixation*'. The soil bacteria responsible for transformation of nitrate to oxide of nitrogen and gaseous nitrogen resulting in loss of fertilizer nitrogen are termed 'denitrifying bacteria'. The microbiological oxidation of ammonium to nitrate is one of the most important transformations taking place in soil which helps plant nutrition since nitrogen is predominantly taken up as NO_3 ion. This transformation is mediated by a group of bacteria called '*nitrifying bacteria*'.

Actinomycetes They are more common in dry soils and in undistributed pastures and grasslands. Like bacteria, they are more common in neutral to slightly alkaline soils. They are aerobic organisms and therefore their number is less in lowlands. As they can withstand drought conditions very well, they occur more frequently in soils undergoing dry spells. Actinomycetes are nutritionally heterotrophic and are highly adaptive to degrade wide range of organic substances, particularly those which are difficult to be decomposed by other soil organisms.

Fungi Soil fungi can now grow in wide range of soil pH but their population is more under acidic conditions because of severe competition with bacteria at neutral pH. A majority of fungi are aerobic and prefer to grow at optimum soil moisture. The contribution of these organisms in biochemical transformation under excessive moisture is negligible. Fungi are heterotrophs and therefore derive nutrition from either living plant tissues as a 'parasite', causing plant diseases or dead tissues as a 'saprophyte'. Saprophytic fungi perform a very important function in the decomposition of organic matter, particularly plant residues.

Some fungi form a symbiotic association with roots of higher plants facilitating uptake of plant nutrients, particularly of those which are less mobile. This association is known as 'mycorrhizal association'. The beneficial effect of these fungi on nutrient uptake has been attributed to three factors: (a) increased absorption of available nutrients from soil as the fungus changes root morphology; (b) increasing the nutrient availability by solubilizing insoluble nutrients like P which thus become available to plant, and (c) increasing the nutrient uptake due to faster intracellular nutrient mobility.

Algae Soil algae are chlorophyll containing organisms. They are autotrophic, and therefore their development is not restricted by organic carbon supply. They are abundant in habitats exposed to light and have sufficient moisture. Blue green algae have been reported to fix about 20 kg N per ha. Submerged soil condition in the rice fields provides an ideal environment for the growth of algae as the soil seldom dry out. A nitrogen fixing algae, '*Anabaena azollae*', forms a symbiotic association with a fresh water fern, Azollae and fixes nitrogen.

2.4 Soil Forming Processes

Soil is defined as outermost layer of earth's crust consisting of complex mixture of unconsolidated mineral and organic matter produced by the combined action of water, wind and organic matter decay. Soils are formed from hard rocks, loose and unconsolidated transported inorganic material and accumulated organic residues. Even the loose mineral materials from which soils are formed are originated by the

Table 2.2 Some important pedogenic processes

Structure development	Grouping of individual particles (clay, silt and sand) together with humus and free sesquioxides into aggregate or peds of fairly distinctive size and shape
Humification	Transformation of raw organic matter into humus
Translocation of lime	Removal of lime from the upper part of profile and its partial or total accumulation in the lower part. The process leads to the formation of a *kankar* or calcic horizon
Leaching	Removal of a constituent from soil solution e.g. soluble salts
Salinization	Accumulation of soluble salts in soil
Clay migration or lessivage	Removal of clay, particularly of fine clay in suspension from the upper layer of soil profile and its accumulation in the lower part
Braunification/rubification/ ferruginisation	Release of iron from primary minerals and their dispersal as coatings on soil particles or as complexes with organic matter/ clay or as discrete aggregates to impart a brown to red color to the soil
Laterization	Removal of silica from soil and accumulation of sesquioxides (goethite, gibbsite, etc.) with or without the formation of iron stone and concretions
Podzolization	Removal of iron and aluminum, often as complexes with humus, from the upper part and its deposition at some depth
Regur formation	Formation of intensively dark color complex of smectitic clay and humus. This dominant process occurs in black cotton soils
Gleization	The reduction of iron under anaerobic (waterlogged) conditions with production of bluish to greenish grey color with or without mottles or ferro-manganese concretion

Buol et al. (1997)

weathering of rock masses to stones, gravels, sands, silts, clays and soluble salts. The unconsolidated material overlaying the rocks is known as regolith. The profile of a soil with well marked horizons called A, B and C, tells the history of its formation and bears the imprint of many physical, chemical and biological processes which have led it to the present form. The common soil forming processes are given in Table 2.2. Several of these processes operate simultaneously though some may start acting sequentially. An example of the latter is translocation of clay within a profile occurs after leaching of lime or soluble salts. However, in the long run, one or two of these processes dominate and lead to the development of a soil with a distinctive profile dictated by factors of soil formation.

2.5 Types of Soil and Their Distribution

2.5.1 Soil Order

USDA Soil Taxonomy, the most accepted system of soil classification worldwide was developed by United States Department of Agriculture and the National

Cooperative Soil Survey. This system of soil classification is based on soil properties along the profiles (Table 2.3) and grouped under several levels namely: Order, Suborder, Great Group, Subgroup, Family, and Series (Table 2.4).

Table 2.3 Brief descriptions of the major soil orders in India according to soil taxonomy

Soil orders	Description
Alfisols	Soils are enriched with Fe and Al oxide minerals but lacks in $CaCO_3$ in the upper layer. Soils have diagnostic argillic horizon (rich with migrated clay), rich in exchangeable cations (Ca^{2+}, Mg^{2+}, K^+, Na^+) contributing to base saturation of greater than 35%. Downward migration of clay is generally less during pedogenic process as compared to Ultisol.
Ultisols	Soils are typically characterized with a humus rich surface layer having lower base saturation (<35%) and mineral nutrients over a clayey B horizon. They are highly weathered and acidic soil, mostly distributed in the humid areas of the tropical and subtropical regions. The soil is enriched with kaolin type of clay mineral and metal oxide (mainly of Fe & Al) imparting red color to it. Ultisols are more extensively leached than Alfisols.
Oxisols	Oxisols are highly weathered soils having oxic subsurface diagnostic horizon with very little variation in texture along soil depth. Oxisol are formed under tropical and subtropical climate along the equatorial region. Clay-size particles dominated by hydrous oxides of iron and aluminum. Kaolinite dominates the clay minerals. Oxisol differs with Alfisol and Ultisol by not having argillic horizon in the lower depth.
Vertisols	Soils are characterized by high clay content (>30%) which is dominated by 2:1 type of clay minerals (smectite). High cation exchange capacity, low permeability, slickensides, high base saturation (Ca dominated) and medium to low organic matter content (0.5–3%) are the characteristic features of Vertisol. This soil type develops wide and deep cracks during dry season.
Mollisols	This soil type is typically identified by humus rich (imparting dark color) surface horizon developed under native grassland or forest region in the subhumid to semiarid region. Such soil is rich in base cations (>50% base saturation with dominance in exchangeable calcium and magnesium) and evolved from calcareous rich colluvial deposits. This soil type has higher humus content than Alfisol, lacks swell-shrink characteristics (as in Vertisol) and possesses higher CEC than Ultisol.
Inceptisols	Inceptisols are formed under varying climatic condition except under aridic condition. These soils are mostly formed from colluvial alluvial and loess materials and lacks diagnostic argillic pedoturbation features. These soils are relatively young and lack characteristics surface and subsurface horizons. However soil profile indicates accumulation of humus and clays which are poorly developed to indicate horizons.
Entisols	Entisols are recently formed soils from the underlying sandy parent material with little or no evidence of pedogenic horizon development in the profile. Such soils are mostly formed in the floodplain area having significant erosion-deposition activity. Entisols differ from Inceptisols by a minor subsurface build-up with transported clay.
Aridisols	Soils of arid region are formed from wide variety of parent material under extremely dry climate. Such soils are low in humus and rich in sand sized particle. Significant accumulation of migrated clay or gypsum or cemented clays are often observed in the lower layers.

Adapted from Sehgal (2002)

Table 2.4 Taxonomical order of Indian soils and their distribution

Soil order	Area (million ha)	(%)
Entisols	78.7	23.94
Inceptisols	129.6	39.43
Vertisols	26.6	8.09
Aridisols	13.3	4.05
Mollisols	1.6	0.49
Ultisols	8.4	2.55
Alfisols	42.2	12.84
Others[a]	28.2	8.58

Bhattacharyya et al. (2013)

[a]Includes glaciers, sand dunes, mangrove swamps, salt waste, water bodies, rock land and rock outcrops

From soil pollution and remediation point of view, the issues 'how a particular soil is vulnerable, resilient and resistant to pollutants/disturbances' and 'how it responses to specific management interventions' are mainly influenced by soil texture and organic carbon content of the soil. For example though coarse textured soil (Entisols or Inceptisols) is easily disturbed by polluting activities; management or remediation actions can easily restore it to some extent. Fine textured soil (Vertisols) can resist soil pollutants some extent due to its buffering capacity. However after disturbance, it is either very difficult or sometimes not possible to restore soil functions through management or remediation. Higher organic matter in a soil type (Mollisols) causes higher microbial activity, CEC and metal chelation capacity and all these lead to increase in resilience and resistance of soil to pollution.

2.5.2 *Major Soil Types of India*

The Indian soils have also been classified by the committee constituted by Indian Council of Agricultural Research (ICAR) under major types as Alluvial Soils, Black Soils, Red Soils, Laterite Soils, Mountain Soils, and Desert Soils. Physico-chemical properties, mineralogical composition, physiography, geological formations, climate were the basis for this type of classification. The extent of area and distribution of these soil types are presented in Table 2.5.

2.5.2.1 Alluvial Soil

Alluvial soil is the most important (from food production point of view) and the largest soil group of India covering about 40% of the total land area of the country. This soil type supports nearly half of Indian population and contributes to major share of agricultural wealth. The alluvial soils have been formed from the deposition materials brought by river, glaciers and sea waves; and thus the parent material

Table 2.5 Different soil types of India along with their corresponding USDA nomenclature system

Major soils (traditional name)	Extent		Distribution in states	Soil orders US soil taxonomy
	000 ha	%		
Alluvial	100,006	30.4	Andaman and Nicobar Islands, Andhra Pradesh, Arunachal Pradesh, Assam, Bihar, Delhi, Goa, Gujarat, Haryana, Himachal Pradesh, Jammu and Kashmir, Karnataka, Kerala, Madhya Pradesh, Maharashtra, Manipur, Meghalaya, Mizoram, Nagaland, Odisha, Punjab, Puducherry, Tamil Nadu, Tripura, Uttar Pradesh, West Bengal	Alfisols, Entisols, Inceptisols, Aridisols
Coastal alluvial	10,049	3.1	Andaman and Nicobar Islands, Andhra Pradesh, Gujarat, Karnataka, Kerala, Lakshadweep, Odisha, Puducherry, Tamil Nadu, West Bengal	Inceptisols, Aridisols, Entisols
Red	87,989	26.8	Andaman and Nicobar Islands, Arunachal Pradesh, Andhra Pradesh, Assam, Delhi, Gujarat, Goa, Himachal Pradesh, Karnataka, Kerala, Madhya Pradesh, Manipur, Meghalaya, Nagaland, Mizoram, Puducherry, Rajasthan, Tripura, Tamil Nadu, Uttar Pradesh,	Alfisols, Entisols, Inceptisols, Ultisols, Mollisols, Aridisols
Laterites	18,049	5.5	Andhra Pradesh, Karnataka, Kerala, Maharashtra, Odisha, Puducherry, Tamil Nadu, West Bengal	Alfisols, Inceptisols, Ultisols
Brown forest	540	0.2	Karnataka, Maharashtra	Mollisols,Inceptisols
Hill	2262	0.7	Manipur, Nagaland, Odisha, Tripura, West Bengal	Inceptisols, Entisols
Terai	326	0.1	Sikkim, Uttar Pradesh	Mollisols, Entisols
Mountain meadow	60	–	Jammu and Kashmir	Mollisols
Sub-montane	104	–	Jammu and Kashmir	Alfisols
Black	54,682	16.6	Andhra Pradesh, Bihar, Gujarat, Madhya Pradesh, Maharashtra, Odisha, Puducherry, Rajasthan, Tamil Nadu, Uttar Pradesh	Inceptisols, Vertisols, Entisols, Mollisols, Aridisols
Desert	26,283	8.0	Gujarat, Haryana, Punjab, Rajasthan	Aridisols, Entisols, Inceptisols
Others[a]	28,305	8.6	–	–
Total	328,700	100	–	–

Bhattacharyya et al. (2013)

[a]Includes glaciers, sand dunes, mangrove swamps, salt waste, water bodies, rock land and rock outcrops in different states

of this soil has always been transported from elsewhere. In India, it is distributed widely in northern plains (starting from Punjab in the west to Assam and West Bengal in the east) and river valleys (Deltas of Mahanadi, Godavari, Krishna and Kaveri rivers in the Peninsular India). Alluvial soils are also found in coastal and northern parts of Gujarat, Great Northern plain and lower valleys of Narmada and Tapti. Alluvial soil is considered superior to other soil types due to the optimum combination of good chemical and physical conditions for proper root growth as well as having high fertility. This soil is composed of fine particles of sand, silt and clay with dominance of silt fraction along with moderate organic matter content. Geologically alluvial soil is classified into old alluvium termed as 'bangar' soil and new alluvium known as '*khaddar*' soil. Most of the new alluvial soils are found in the flood plains and deltas area along the lower part of the course of rivers and may even have nearly ten thousand years old deposits. Old alluvial soils are found on the higher side of the river valley and termed locally as '*bhangar*'. The areas having *khadar* soil are flooded almost every year and remain young with sandy/sandy loam in texture and light in colour. *Bangar* soil is silty clay/clay loam in texture with darker in color and may contain alkaline effloresces. Such old alluvial soils are relatively less fertile than new alluvial khaddar soil. Alluvial soils are generally deficient in organic carbon and nitrogen; however the soil is fertile due to presence of adequate phosphorus, potassium and calcium. Illite is predominant mineral in the clay fraction. These are deep loamy in texture having moderated cation exchange capacity, moderate to high water holding capacity and with proper irrigation it yields bumper produce of rice, wheat, sugar cane, maize, cotton, oilseeds, jute, tobacco etc. The Indo-Gangetic plains of north India are the most fertile lands and are largely irrigated, contributing 65% of the total food basket.

2.5.2.2 Black Soil

In contrast to alluvial soil, black soil has developed in-site through intense weathering processes and is locally termed as 'regur' soil (derived from Telugu word 'reguda'). Cotton is the major crops grown in black soil and hence alternatively termed as 'black cotton soil'. Parent material of such soil was formed due to solidification of lava during volcanic activity. The black soil is distributed mostly in Deccan trap covering large areas of Maharashtra, Madhya Pradesh, Chhattisgarh and Gujarat. It is also found in some parts of Krishna and Godavari river valleys, covering parts of Andhra Pradesh, Karnataka and some parts of Tamil Nadu. Black cotton soil is high in clay content and has high water holding capacity. Generally, black soils are rich in potash and calcium, but poor in phosphate content. Due to its high clay content, such soil develops deep and wide cracks during hot dry season; but becomes sticky during rainy season imparting poor physical condition in the context of crop cultivation. Therefore workable soil moisture range for tillage operation is very narrow and the soil is required to be tilled immediately after the first or the pre-monsoon shower. Black soil found near the valleys or lowlands is deep, dark and highly fertile; whereas, such soil located upland is low in fertility.

Due to its high water holding capacity and rich in nutrients, black soil is best suited for producing wheat, cotton, sugar cane, rice, millets, linseed, tobacco and oilseeds.

2.5.2.3 Red Soil

The red soil occupies about 10% of the total geographical area in India and is spread across almost whole of the Tamil Nadu, parts of Karnataka, southeast Maharashtra, eastern parts of Andhra Pradesh, Madhya Pradesh, Orissa and Jharkhand. It also extends to large parts of south Bihar, western districts of West Bengal, the eastern parts of Rajasthan; few districts of Uttar Pradesh; and parts of Nagaland, Assam, Mizoram, Manipur, Tripura and Meghalaya. Red soil has been formed from the weathering of iron rich acid granites (igneous) parent material and gneisses (metamorphic) rocks in a warm, moist climate under deciduous or mixed forests. Though soil texture of red soils is mostly loamy in texture, it may vary widely from sandy to clayey in nature. Such soil may have thin organic surface layer. It may be rich in potash, but poor in organic carbon, nitrogen, phosphorus and calcium. As result such soil has low fertility and is difficult to till. The red soils respond well to fertilizers and irrigation and give excellent yields of wheat, rice, pulses, cotton, millets, tobacco, oil seeds, potatoes and fruits.

2.5.2.4 Laterite Soil

The word 'laterite' is derived from a Latin word '*later*' which means 'brick' due to the tendency of the soil to harden upon exposure. The laterite soil is formed from the parent material rich in iron and aluminum under high temperature and rainfall condition of climate. It is also found in the high altitude areas of plateau regions and has been widely spread on the summits of the Eastern Ghats, Western Ghats, Rajmahal Hills, Satpuras, Vindhyas and Malwa plateau. It is abundantly found in southern Maharashtra, and hilly parts of Orissa, Meghalaya and Assam, Karnataka, West Bengal, Kerala, Andhra Pradesh and Bihar. Laterite soils are poor in soil fertility due to prominent leaching process that takes place in intense rainfall areas. During leaching process, calcium and silica is lost from soil profile leaving behind high iron and aluminum content. Hence, laterite soils are poor in lime (calcium and magnesium), organic carbon and nitrogen, while rich in iron, aluminum and potash.

Red and laterite soils (Alfisols) have good amount of sesquioxides, low in organic matter and have a large proportion of kaolinite and non-expanding illitic clay fractions. Laterite and associated soils are less water retentive and are thus more drought-prone. The physical constraints to crop production in red and lateritic soils are hardening of soil, low water-holding capacity, reduced soil volume due to concretions, occurrence of hard plinthite or petroplinthite, drought related stress, low cation exchange capacity, low organic matter, high acidity, iron and aluminium toxicity, high phosphorus fixation and severe erosion. Laterite soil is composed of less clay and more gravel which leads to poor soil fertility. But under proper manuring and irrigation, the laterite soil is productive for crops like cashew nuts, ragi, rice, tea, coffee, rubber, coconut, arecanut, etc. In addition, laterite soil is highly suitable for making bricks.

2.5.2.5 Mountain Soil

The mountain soil is normally found on the hill slopes covered with forests and such soil accounts for around 8.7% of the total geographical area of India. It is distributed in valley basins of Himalayan region, in the Western and Eastern Ghats and in some parts of the Peninsular India. The mountain soil typically contains high humus due to organic matter deposition from forest cover. However, it is poor in potash, phosphorus and calcium. Mountain soil is highly heterogenous and its characteristics vary from place to place due to changes in parent material, climate and vegetation. The mountain soil is normally acidic, grey in colour and its textural class varies widely from loamy/silty on valley sides to coarse-grained in the upper slopes. Such soils are highly suitable for plantation crops like tea, coffee, spices and tropical fruits in Karnataka, Tamil Nadu and Kerala. In the Himalayan region, crops like wheat, maize, barley and temperate fruits can be produced under proper fertilizer management.

2.5.2.6 Desert Soil

Arid and semi-arid regions having less than 50 cm of annual rainfall have predominantly sandy textured desert soils. The desert soil is mostly found in Rajasthan and the adjoining areas of Haryana, Punjab and the Rann of Kachchh in Gujarat. Desert soils are mostly developed in-situ where sands originate from mechanical disintegration of parent rocks under hot-dry climate. However, this may also develop from wind-laden loess materials. The soil in the desert area has nearly 90–95% sand particle and with very less amount of clay content (5–10%). Due to high evapotranspiration and less rainfall, some desert soils are saline in nature. Such soils are generally poor in organic matter and nitrogen. Phosphate content may be as high as that in alluvial soil. Due to water shortage in arid regions and also poor water holding capacity of desert soil, additional irrigation facility is required to produce variety of crops like wheat, millets, barley, maize, pulses, cotton etc.

2.5.3 Major Soil Types Under Different Agro-Ecological Zones (AEZ) of the Country

On the basis of climate characteristics, length of crop growing period and soil type, total geographical area of India has been classified into 20 agro-ecological zones (AEZ) for devising effective land use plan and farming strategy for various regions (Table 2.6 and Fig. 2.1) (Sehgal 2002).

Table 2.6 Taxonomical order of Indian soils and their distribution

AEZ No.	Location	Climate type	Soil type
1.	Western Himalayas	Cold arid ecoregion	Shallow, loamy skeletal soils
2.	Western Plain, Kachchh and part of Kathiawar Peninsula	Hot arid ecoregion	Desert and saline soils
3.	Deccan plateau	Hot arid eco-subregion	Red and black soils
4.	Northern Plain (and Central Highlands) including Aravallis	Hot semi-arid ecoregion	Alluvium derived soils
5.	Central (Malwa) Highlands, Gujarat plains and Kathiawar Peninsula	Hot semi-arid ecoregion	Medium and deep clayey black soils
6.	Deccan Plateau	Hot semi-arid ecoregion	Shallow and medium (dominant) black soils
7.	Deccan Plateau (Telangana) and Eastern Ghats	Hot semi-arid ecoregion	Red and black soils
8.	Eastern Ghats and Tamil Nadu Uplands and Deccan (Karnataka) Plateau	Hot semi-arid ecoregion	Red loamy soils
9.	Northern Plain	Hot sub-humid (dry) eco-region	Alluvium derived soils
10.	Central Highlands (Malwa and Bundelkhand)	Hot sub-humid (dry) ecoregion	Red and black soils
11.	Chattisgarh Mahanadi Basin	Hot sub-humid (dry) ecoregion	Red and yellow soils
12.	Eastern Plateau (Chhotanagpur) and Eastern Ghats	Hot sub-humid ecoregion	Red and lateritic soils
13.	Eastern Plain	Hot sub-humid (moist) ecoregion	Alluvium derived soils
14.	Western Himalayas	Warm subhumid (to humid with inclusion of perhumid) ecoregion	Brown forest and Podzolic soils
15.	Assam and Bengal Plain	Hot sub-humid (moist) to humid (inclusion of perhumid) ecoregion	Alluvium derived soils
16.	Eastern Himalayas	Warm perhumid ecoregion	Brown and red hill soils
17.	North-eastern Hills (Purvachal)	Warm perhumid ecoregion	Red and lateritic soils
18.	Eastern Coastal Plain	Hot subhumid to semiarid ecoregion	Coastal alluvium derived soils
19.	Western Ghats and Coastal Plain	Hot humid-perhumid ecoregion	Red, lateritic and alluvium derived soils
20.	Islands of Andaman-Nicobar and Lakshadweep	Hot humid to perhumid island ecoregion	Red loamy and sandy soils

Adapted from Sehgal (2002)

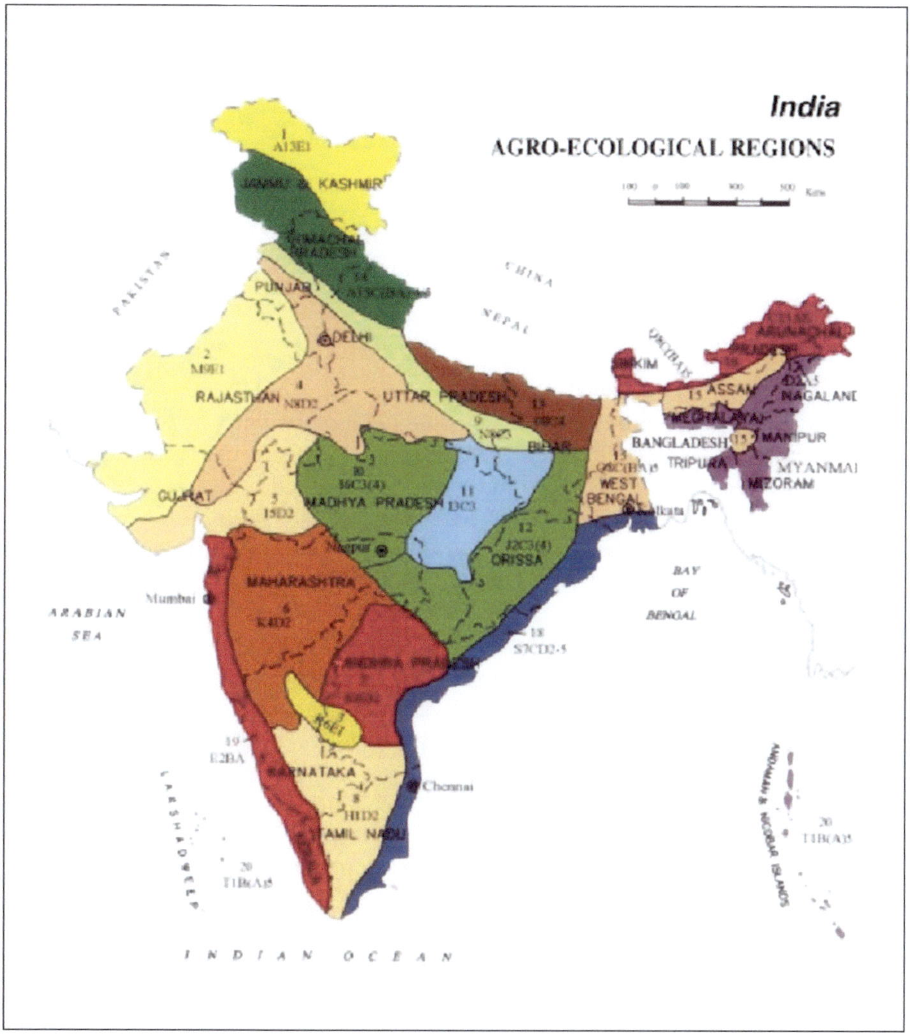

Fig. 2.1 Agro-Ecological Regions of India (Adapted from Sehgal 2002)

2.6 Relationship Between Soil and Vegetation and Concept of Top Soil

Soil, which is considered as common terrestrial substrate serves as a medium for plant growth. Therefore, soil properties like texture, structure, water retention, depth and nutrient status have major influence on species diversity and vegetation establishment of any site. On the other side, vegetation and its species diversity plays significant role in different pedogenic processes leading to soil formation.

Fundamental soil equation (Jenny 1958) states nature of soil in a given location is dependent of several factors:

$$Soil = \int (climate, parent\ material, relief, organisms, time)$$

Similarly, Major (1951) proposed equation for plant community in which same environmental factors responsible for Jenny's above soil formation equation are also responsible for the vegetation that is produced :

$$Vegetation = \int (climate, parent\ material, relief, organisms, time)$$

Above equations also indicate that specific plant community and soil type in an area are mutually associated for their development. This further indicates that all strata on the landscape with the same type of soil should have same vegetation community, the same plant association, and the same habitat type. However several studies conducted in this area either supported such association or failed to establish any relationship between these two entities (soils and vegetation) (Hironaka et al. 1990). Studies of Koptsik et al. (2003) revealed a significant correlation between the species diversity of plant community and soil properties. Soil factors acidity, concentrations of exchangeable calcium, potassium, and magnesium, influenced variations of species diversity indices for vascular plants and bryophytes in forest ecosystem. Likewise strong relationship was observed between plant diversity index and soil functional diversity (Rodríguez-Loinaz et al. 2008). On the contrary, inconsistent results were observed while predicting tree site index using selected soil properties (Copeland 1958; Monserud et al. 1990).

Several authors (Daubenmire et al. 1968; Daubenmire 1979; Neiman 1988; Jensen et al. 1990; Tisdale and Bramble-Brodahl 1983) have attempted to correlate predominant vegetation community and soils. The results have been mostly inconclusive (Hironaka et al. 1990). These failures indicate that either relationship between soil and vegetation does not exist or some underlying fundamental causes have been overlooked. Johnson and Simon (1987) observed influence of ash and loess (aeolian deposit) on species diversity of grasses and shrubs. For example, bluebunch wheatgrass, scabland and xeric shrubland communities existed on soils having little influence from loess deposits; whereas fescue-prairie junegrass and mesic shrubland communities occur on soils that are considerably influenced by loess. Likewise, forest species diversity was considerably influenced by ash (volcanic origin) content of the soil. However, researchers were unable to extrapolate such findings (correlations between soils and vegetation) outside their area of study. Similarly, a study from central western New South Wales, USA showed poor associations between soil landscapes and vegetation types; where specific type of vegetation was associated with several soil landscapes or soil types and specific soil type or landscape was normally linked with several vegetation types (Rankin et al. 2007). In a study at Himalayan region of China, high tree diversity, height and vegetative cover were linked to good soil quality and favorable topographic positions with lower solar incident radiation, runoff and soil erosion potential (Xu et al. 2008). Plant diversity was mainly correlated with soil water content, and clay content. In a flood plain area of Argentina, nature and type of plant communities varied with drainage, pH and exchangeable sodium percentage in the soil of the area and such vegetative diversity was related to hydromorphic, hydrohalomorphic or halomorphic characteristics of soil (Debelis et al. 2005). Changes in soil chemical environment due to entry of pollutants can have tremendous impact on species diversity and density of natural vegetation. Only tolerant species and varieties proliferate, while the population of sensitive species goes down.

Information generated thus indicates that correlation between selected soil properties and vegetation had been successfully established by several researchers in different parts of the world. However, these conclusions were found only locally significant and could not be applicable elsewhere. Because of huge variation that persists in classification units of soil and vegetation, universal relationship or correlation has not been found to exist between soils and habitat type. Further it is relatively difficult to confirm the correctness of above assumption pertaining to vegetation due to absence of undisturbed vegetation across various geographical and climatic regions.

References

Bhattacharyya T, Pal DK, Mandal C et al (2013) Soils of India: historical perspective, classification and recent advances. Curr Sci 104:1308–1323

Buol SW, Hole FD, McCracken RW (1997) Soil genesis and classification, 4 th edition. Iowa State University Press, Ames

Copeland OL Jr (1958) Soil-site index studies of western white pine in the northern Rocky Mountain region. Soil Sci Soc Am Proc 22:268–269

Daubenmire R (1979) Steppe vegetation of Washington. Tech. Bull. 62. Washington State University, Pullman, Washington Agricultural Experiment Station, p 131

Daubenmire R, Daubenmire D, Jean B (1968) Forest vegetation of eastern Washington and North Idaho. Tech. Bull. 60. Washington State University, Pullman, Washington Agricultural Experiment Station, p 104

Debelis SP, Bozzo AA, Barrios MB et al (2005) The relationship between soil characteristics and vegetation as a function of landform position in an area of the flooding pampa. Span J Agric Res 3(2):232–242

Dwivedi OP, Tiwari BN (1987) Environmental crisis and Hindu religion (verses from Artharva Veda, 12.1). Geetanjali, New Delhi, p 202

Gandhi M, Gandhi MK (1994) An autobiography: the story of my experiments with truth. South Asia Books, Columbia, p 754

Goswami NN, Rattan RK (1992) Soil health – key to sustained agricultural productivity. Fert News 37:53–60

Hironaka M, Fosberg MA, Neiman KE (1990) The relationship between soils and vegetation. Symposium on Management and Productivity of Western-Montane Forest Soils, Boise, ID, April 10–12

Jenny H (1958) Role of the plant factor in the pedogenic functions. Ecol 39:5–16

Jensen ME, Simonson GH, Dosskey M (1990) Correlation between soils and sagebrush-dominated plant communities of northeastern Nevada. Soil Sci Soc Am J 54:902–910

Johnson CG, Simon SA (1987) Plant associations of the Wallowa-Snake Province, Wallowa-Whitman National Forest. R6-ECOL-TP-255a-86. US Department of Agriculture, Forest Service, Pacific Northwest Region, Portland, pp 400

Koptsik SV, Koptsik GN, Livantsova SY et al (2003) Analysis of the relationship between soil and vegetation in forest biogeocenoses by the principal component method. Russ J Ecol 34:34–42

Major J (1951) A functional, factorial approach to plant ecology. Ecol 32:392–412

Monserud RA, Moody U, Breuer DW (1990) A soil-site study for inland Douglas-fir. Can J For Res 20:686–695

Neiman KE (1988) Soil characteristics as an aid to identifying forest habitat types in northern Idaho. Res. Pap. INT-390. US Department of Agriculture, Forest Service, Intermountain Research Station, Ogden, p 16

Rankin MO, Semple WS, Murphy BW et al (2007) Is there a close association between soils and vegetation? a case study from central western new South Wales. Cunninghamia 10:99–214

Rodríguez-Loinaz G, Onaindia M, Amezaga I et al (2008) Relationship between vegetation diversity and soil functional diversity in native mixed-oak forests. Soil Biol Biochem 40:49–60

Sehgal J (2002) Soil classification. Fundamental of Soil Science, ISSS, New Delhi, pp p19–p53

Tisdale EW, Bramble-Brodahl M (1983) Relationships of site characteristics to vegetation in canyon grasslands of west Central Idaho and adjacent areas. J Range Manag 36:775–778

US Salinity Laboratory Staff (1954) Diagnosis and improvement of saline and alkali soils. USDA handbook 60. U.S. Government Printing Office, Washington, DC

Xu XL, Ma KM, Fu BJ et al (2008) Relationships between vegetation and soil and topography in a dry warm river valley, SW China. Catena 75:138–145

Chapter 3
Impacts of Soil Pollution and Their Assessment

Abstract Impact assessment of pollution is important step towards initiating appropriate remedial measures. However impact of soil pollution may not necessarily be uniform across the agro-ecological regions due to complex relationship among soil types, climate, crop types, nature and level of contamination as well as soil and crop management methodologies adopted. Moreover land resources are habitat of numerous organisms having varying degree of sensitivity towards contaminants. Hence, choosing right approach for assessing impact of pollution becomes important for satisfying various stakeholders affected by soil pollution. This chapter covers different types of impact of pollution on soil quality, economic yield of crop, food quality, biodiversity, fertilizer nutrient use efficiency, economy of the country etc. Pollution has already made adverse impacts on environment including soil in different countries. Widely cited instances of different types of pollution across the world and their impacts on soil and crop has also been discussed in this chapter.

Keywords Impact • Soil • Pollution • Fertility • Biological activity • Environmental tragedies

Global environmental concerns and related movements have widely been related with the famous publication of "Silent spring" authored by Rachel Carson in the year 1962, wherein the author thoroughly described the widely used pesticide dichloro-diphenyl-trichloroethane (DDT) entering into the food chain of human and animal, causing genetic damage and cancer (Carson 1962). This issue evolved with changes in environmental aspects (activities, products or services that can interact with the environment) in broader sense, and the need for ecological balance and clean environment has been increasingly felt.

Soil is an important component of the biosphere where all the three components lithosphere, hydrosphere and atmosphere interface. Hence disturbing any component of the biosphere affects the other components severely because of the interlinking of different ecological functions operating within. A range of chemical substances is liberated into the environment by the activities of human at global level and can be seen ubiquitously including in soil system. Anthropogenic activities leading to soil pollution have severe impacts on the quality of living organisms ranging from impairment of functional quality to complete elimination of a species

J.K. Saha et al., *Soil Pollution - An Emerging Threat to Agriculture*, Environmental Chemistry for a Sustainable World 10, DOI 10.1007/978-981-10-4274-4_3

in most severe cases. As there has been wide variation in the diversity and functional quality of living organisms across the geographical dimensions due to the influence of climate and essential natural resources, identification of the impact of soil pollution is many a times complex. This chapter briefly describes on different ways to identify such impacts. Although several impacts can be perceived within short span of time, some of these require longer period for expression due to their slow rate of change. Now the era of modern technologies and significant research in different branches of science have helped us to find out environmental changes of even few weeks. Soil pollution has wide ranges of impacts on agroecosystems which can be identified and quantified in following different ways.

3.1 Loss of Crop Productivity and Crop Diversity

Some of the direct impacts of soil pollution can be seen in agricultural production system like loss/decline in soil productivity and loss in crop diversity. Pollution of soil and water drastically has estimated to reduce the crop yields by about 15–25% over the years and total cropped area has decreased significantly. Within 20 years, the total cropped area has fallen about 10.5% in Vellore district (Dakiky et al. 2002; Nath et al. 2005). In this tract, use/release of textile effluent in the agricultural field has ruined the traditional cropping sequence with loss of crop productivity and diversity. The toxicity of heavy metals mainly chromium (Cr) in irrigation water and soil has replaced wheat crop to a large extent with other hardy horticultural shrubs nearby Kanpur tannery on the bank of Ganga River.

In surrounding agricultural land of Ratlam industrial area, use of polluted groundwater has adversely affected farming through changes in cropping pattern, reduction in crop yield, as well as reduced longevity of irrigation infrastructure (Saha and Sharma 2006). Area under vegetable and pulse cultivation reduced considerably and the same under fallow has increased during rabi season as many farmers prefer to keep their land fallow instead of using polluted groundwater as irrigation. The yields of the traditional crops such as soybean, gram wheat, methi and garlic are less in polluted area (Table 3.1). Onion grown using polluted groundwater, has a very short keeping quality and starts rotting within 10–15 days after harvest; thus inflicting heavy economic losses to the farmers. The iron pipes of the tubewell rusted frequently incurring losses on account of repair and maintenance of irrigation infrastructure.

Similar situation also prevailed in other part of Indian states *i.e.* Vellore (TN), Howrah (WB). Loss of crop productivity in contaminated agricultural fields can be assessed through following methods:

Table 3.1 Estimation of loss due to reduction of yield of crops

	Average yield (Mg ha^{-1})		
Crop	Unpolluted area	Polluted area	Loss (US$ per ha)
Wheat	3.6	2.9	101
Gram	12.5	9[a]	108
Soybean	15	11.5	93
Methi	18	12	200
Garlic	60	45	667

Saha and Sharma (2006)
[a]When irrigation is given at later stages

3.1.1 *Farmer's Interaction/Survey*

In the villages, local agriculture officers measure the crop productivity loss by recording views of key persons (old people or head of village local body) due to having the vast knowledge of natural calamities as well as soil and crop history of particular area. Though this type of estimation is not exact, this can be a preliminary method for further investigation. During crop productivity assessment by this method, distinction should be made between biological and economic yield and correction must be made for threshing and moisture content.

3.1.2 *Crop Cutting Survey*

This is a popular method of crop loss measurement. The crop yield estimation in many country is mostly carried out on the basis of sample survey approach. The estimates of yield rates are obtained on the basis of scientifically designed Crop Cutting Experiments (CCE) conducted in which, a particular shape and size of plots are mentioned (Table 3.2). Same size and shape of plot cannot be taken for all crops. As plot size increases, the estimated yield and standard deviation of yield decreases and yield estimation attains a stable value when the plot size is significantly large. Minimum recommended shape and size varies widely from 1/160 of an acre for paddy to 1/10 of an acre for cotton. Limited evidences, however show that farmers' estimates may no more be biased than crop cutting; but require fewer resources and supervision (Poate 1988). Crop cutting survey data from polluted area should be compared with those from nearby unpolluted area in order to quantify the pollution impact.

Selecting CCE Field for Crop Loss Measurements

- The area of the selected field should be more than the area of CCE plot, so that at least a one CCE plot must be accommodated in the selected field.

Table 3.2 Parameters of crop cutting experiments

Name of crop	Shape	Length (m)	Breadth (m)	Diagonal (m)
Paddy, wheat, jowar, bajra, ragi, maize, groundnut, tobacco, sugarcane, okra, green gram, chillies, mesta, horse gram, black gram, bengal gram, sunflower	Square	5	5	7.07
Red gram, sesamum, caster, cotton	Square	10	10	14.14

Panse (1947)

- If the field is sown with mixed crop. The experimental crop must constitute at least 10% of its cropped area.
- In case experimental crop is damaged completely, yield must be recorded as zero.

3.1.3 Use of Satellite Imagery

Nowadays imageries from remote-sensing satellites are being used for assessing the crop growth condition and crop productivity loss. However, this method is suitable when vast area is affected with soil pollution. Important advantage of this method is that extent of area affected can be estimated. For more reliability of this method, this technique should be supported with ground truth. Remote sensing based methodologies have distinct advantages, e.g. in providing rapid objective assessments without an investigator bias, longitudinal assessments (reporting changes over time at the same location), providing assessments on the hostile terrains, and rapid assessments of the extent of affected areas (Srivastava 2015). Environmental impact of geothermal pollution activities in the Mt. Amiata area was studied using on-site spectral analyses of various ecological components like vegetation and lichens (Manzo et al. 2013). Spectral indices have been found sensitive to environmental pollution and their responses are statically correlated to chemical and ecophysiological analyses within a notable distance. Remote sensing technique has been used to generate pollution potentiality map of Bhatinda district of Punjab by relating groundwater quality (water quality index) with land use/land cover (land cover index) (Srivastava et al. 2012). Researchers concluded from their study indicated that most of the study area exhibited poor groundwater quality due to very high pollution.

3.2 Decline in Nutrient Use Efficiency

Nutrient use efficiency is defined as quantum of food/fodder production per unit of nutrient applied externally. Lower nutrient use efficiency indicates increase in fertilizer requirement for same level of productivity; thereby increase in cost of production. Different soil contaminants interact with soil components, microorganisms and root cells in several ways at multiple levels to affect nutrient transformation processes in soils and their absorption by roots. Heavy metals have been found to affect uptake of essential nutrients by plants which may consequently reduce use of applied fertilizers (Aslam et al. 2014). Increase in soil salinity in polluted soils affect absorption of water and hence, affect nutrient uptake from soil. Several contaminants are harmful to microorganisms (like Cd, Cr, Cu) and thereby affect several microbially mediated nutrient transformation processes like nitrogen fixation, nutrient mineralization etc. (Bondarenko et al. 2010; Arora et al. 2010; Bianucci et al. 2011). Alkalinity development (as observed in soils near Ratlam and Nagda industrial area) enhance volatilization loss of ammoniacal form of N from soil and thereby reduces N use efficiency.

3.2.1 Measurement in Decline in Nutrient Use Efficiency (NUE)

Use efficiency may be defined as yield per unit input. In agricultural nutrient management, this is usually related to the input of fertilizer. Improvement in NUE is an essential pre-requisite for enhancing crop production in underdeveloped countries having largely resource poor farmers. The efficient utilization of applied fertilizers can be calculated with the help of partial factor productivity (PFP) and agronomic efficiency (AE), which are computed using following formula:

$$PFP\ (kg\ yield\ per\ kg\ of\ nutrient\ applied) = \frac{Y_f}{N_a}$$

$$AE(kg\ yield\ per\ kg\ nutrient\ applied) = \frac{Y_f - Y_c}{N_a}$$

where 'Y_f' stands for yield from a N-fertilized plot, 'Y_c' stands for yield in control plot, and 'N_a' stands for amount of N applied in kg/ha. Comparing PFP and AE from polluted area with those from unpolluted area may indicate impact of soil pollution on agriculture.

3.3 Quality of Food: Vegetable, Fruit and Grain

Environmental pollution impacts can also be perceived as symptoms on crop maturity through assessment of quality of vegetable and fruit like shelf life, size, taste and aroma etc. Polluted soil may produce vegetables, having lower storage life, foul smell and gives unpleasant aroma during cooking. Significant increases in lipid peroxidation and H_2O_2 were observed in turnip and lettuce due to irrigation by industrial wastewater, which indicated oxidative damage to the vegetables (Hassanein et al. 2013). The significant increases in putrescine in shoots and roots of these vegetables grown with industrial wastewater indicated decreased shelf life of vegetables grown on polluted land. As transfer of heavy metals from soil to edible parts of vegetable is higher as compared to grain crops, contaminated soil or effluent (for irrigation purpose) contaminates vegetables with several heavy metals. The effluent containing various pollutants may reduce the fruit size, distort colour and may result in early drop of flowers (Fig. 3.1), poor quality aroma and marketability of fruits. Fruits grown with polluted irrigation water may emit foul smell when stored for few days. Keeping quality of onion has been found severely reduced when the crop is grown with polluted groundwater near Ratlam industrial area (Saha and Sharma 2006).

In the adjoining area of mega cities, farmer's use untreated sewage water as a source of irrigation water. It provides farmer's security in terms of ensured nutrient enriched water supply (Ghosh et al. 2012) and also provides the society with a reliable and inexpensive system for wastewater treatment and disposal (Feigin et al. 1991). However, untreated wastewater also contains considerable pathogens

Fig. 3.1 Early drop of immature fruits (*yellow colour*) due to use of sewage water coming from Vapi city of Gujarat

responsible for several human diseases related to gastroenteritis, skin, and helminthes; and therefore sewage farming poses severe risk to farmers and their family. Also consumption of vegetable crops grown in peri-urban area with untreated sewage water has been reported to enhance intake of heavy metals as well as to spread diseases (particularly when eaten uncooked) (Khurana et al. 2004, Gupta et al. 2010).

3.3.1 *Produce Quality Assessment Method*

Samples can be taken from the field following standard randomization method and can be analyzed for different quality parameters like organic acids, mineral, fat, protein, carbohydrates and water contents. Size of fruits can be measured by length using scale and by diameter using vernier-calliper. Most of the farmers short the inferior quality of fruits and vegetables on the basis of experience with respect to different size. In some crops, grading machines are used for categorizing in various sizes. The change in fruit taste can be measured by total soluble solids, vitamins and sugar content. In cereal crops, impact of environmental pollution may be assessed through size of grain, colour, test weight, duration of crop maturity, protein content, spoilage time, moisture percent, taste and mineral content. Guidelines for acceptable quality of food material in India can be found in The Prevention of Food Adulteration Act & Rules (as on 1.10.2004) (DBT 2016). This can be used as identification tool for deterioration of quality of agricultural produce.

3.4 Contamination of Food with Toxic Compounds Like Heavy Metals

Due to mammalian toxicity of several heavy metals, their contamination in food materials is an important concern at national as well as at international level and are monitored by government agencies. Heavy metal concentrations in plants grown in contaminated and with wastewater-irrigated land have been found significantly higher than in plants grown in the reference soil (Sharma et al. 2006; Sachan et al. 2007; Chary et al. 2008).

3.4.1 *Identification of Contamination of Food with Heavy Metals*

Contamination of agricultural produce can be confirmed through solubilization of samples with acids and subsequent analysis for total contents using Inductively Coupled Plasma-Optical Emission Spectrophotometer (ICP-OES) or other sensitive instruments. For the purpose of confirming contamination, an upper limit of background concentration (C_{ul}) of any heavy metal should be determined using following formula. Any sample having concentration of metal more than this limit is said to be contaminated.

$$C_{ul} = \mu + 3\sigma$$

where, mean (μ) and standard deviation (σ) are with respect to heavy metal concentrations in large number (>30) of food samples collected from uncontaminated area.

3.4.2 Contamination Beyond the Permissible Limits of Government Quality Control Regulations

Some countries have formulated quality control regulation in respect of maximum permissible heavy metal contents in certain food products/food additives for the purpose of protecting human population. In India, heavy metal limits in certain beverages/food additives/food products are mainly controlled by Prevention of Food Adulteration Act No. 37 of 1954 (Table 3.3). Codex Alimentarious Commission (CAC 2011) has also prescribed International Food Standards (hereafter will be referred as 'FAO/WHO CODEX limit') on some limited heavy metals contents in food items. However, limited number of toxic elements (As, Cd, Pb, Hg, Cu, Zn and Sn) are covered under this guideline for food crops grown on land. Maximum safe limits in different agricultural produce (edible) have been prescribed only for As, Cd and Pb based on maximum safe intake rate of heavy metals by human. Guidelines for other toxic heavy metals (Hg, Cu, Zn and Sn) were issued as Provisional Maximum Tolerable Daily Intake (PMTDI), Provisional Tolerable Weekly Intake (PTWI), Provisional Tolerable Monthly Intake (PTMI) etc.

The daily intake of heavy metals through the consumption of a food item is calculated according to the equation:

Table 3.3 Maximum permissible concentration limits of heavy metals as specified in the Indian Prevention of Food Adulteration (PFA) Act No. 37 of 1954

Heavy metals	Maximum permissible concentration (mg kg^{-1})
Arsenic	1.1
Cadmium	1.5
Copper	30
Mercury	1.0
Methyl mercury	0.25
Lead	2.5
Zinc	50

Awashthi (2000)

$$\text{Daily intake of metal}\ (\mu g\ day^{-1}) = [\text{Daily consumption of the food item}\ (g\ day^{-1})] \times [\text{heavy metal concentration in the food item}\ (\mu g\ g^{-1})].$$

Daily vegetable consumption can be obtained by conducting a survey in the study area. Such computed daily intake of heavy metals through the consumption of all food items and drinking water can be compared to the PMTDI values to assess the impact of soil pollution on human population in the affected area.

Instances of heavy metal contamination in food crop beyond the permissible limits have been reported from different parts of India. The vegetable samples from five broad sampling areas in Delhi (Faridabad, Okhla, the Yamuna flood plain, Najafgarh and Alipur) contained unsafe levels of Pb and Zn. In these areas waste water was being used for vegetable cultivation (DFID 2003). About 72% samples (of total 222 samples collected) of palak contained Pb concentrations that exceeded the Indian Prevention of Food Adulteration Act (PFA) permissible limit of 2.5 mg kg^{-1}. About 24% of the samples exceeded the PFA permissible limit by more than two-fold (*i.e.* had concentrations of >5.0 mg kg^{-1}). If the more stringent FAO/WHO CODEX limit of 0.3 mg/kg is used in Table 3.4, then 100% of the palak samples exceeded safe limits. About 21% of 609 palak samples analysed showed that Zn concentrations exceeded Indian PFA limits (50 mg kg^{-1}). About 3% had concentrations at twice the PFA limit. Though cadmium in all the 260 samples of palak had concentrations within the PFA limits (1.5 mg kg^{-1}), the content in 70% samples exceeded the much more stringent FAO/WHO CODEX limit (0.2 mg kg^{-1}). In several vegetable samples at market sites of Varanasi, concentrations of Cu in cauliflower, and of Zn and Cd in both spinach and cauliflower had exceeded the PFA standard (Sharma et al. 2009). In contrast, Pb in vegetables tested from both field and market sites was below the PFA limit, but was considerably higher than the current FAO/WHO CODEX standards. Heavy metals accumulation in vegetables were higher at market sites than those at the crop production sites. The contributions of these vegetables to dietary intake of Cu, Zn, Cd and Pb were 13%, 1%, 47% and 9% of provisional tolerable daily intake, respectively.

3.5 Rejection of Export Consignment

Soil pollution can also exert significant adverse impact on country's export of food items. Food export consignments may be rejected by exporting countries due to presence of pesticides residues, heavy metals etc. In India, the reasons for rejection varies from presence of pesticide residues in grapes beyond permissible levels under the EU food laws to the presence of non-permissible artificial colour, antibiotics, heavy metals etc. (Table 3.5). Several importing countries have implemented stringent quality control measure in respect of heavy metal contents, which raises concerns for agricultural products exporting countries.

Table 3.4 Maximum permissible concentration limits of heavy metals in agricultural food as specified in the FAO/WHO International Food Standards

Contaminants	Food	Maximum level (μg/g)	Provisional maximum tolerable daily intake (μg/kg bw/day)
As	NI[a]		3[b]
Cd	Fruiting and bulb vegetables	0.05	0.8
	Leafy vegetables, wheat	0.20	
	Potato, pulses, root and tuber vegetables, stalk and stem vegetables, cereal grains,	0.1	
Pb	Fruits, fruiting and bulb vegetables, root and tuber vegetables	0.1	<1.9
	Leafy vegetables,	0.3	
	Legume vegetables, pulses, cereal grains	0.2	
Hg	NI		0.57 (inorganic Hg)
Methylmercury	NI		0.00022
Cu	NI		50–500
Zn	NI		300–1000
Sn	NI		2000

CAC (2011)
[a]*NI* Not indicated for agricultural produce
[b]Benchmark Dose Lower Limit (μg/kg bw/day) for a 0.5% increased incidence of lung cancer

Table 3.5 Export consignments from India rejected since 2003

Importing Countries	Consignments	Products	Reasons
US	50	Pickles, preservatives, ayurvedic supplements	Heavy metals, labelling, microorganisms
EU	30	Honey	Antibiotics, nitrofurans, heavy metals
		Grapes	Pesticides
		Egg power	Anitibiotics
		Chilli powder	Sudan dye
		Soft drinks	Color and flavor
		Marine products	Antibiotics, nitrofurans
		Dairy products	Lead, antibiotics

Source: The Times of India, 15 May, 2014

3.6 Soil Hardness/Compaction

Friability of soil is an important soil quality parameter influencing root growth and consequently plant growth and crop yield. On the contrary, soil compaction is the process causing densification to a soil as air is displaced from the pores between the soil grains. Several industrial activities adversely affect soil friability and result in compaction or hardness. For example, dusts (containing mostly mineral matters) emitted from thermal power plants, heavy traffics and other vehicular activities, mining activities gets deposited on nearby agricultural land and imparts soil hardness due to decrease in soil organic matter. Most of the industrial activities deteriorate irrigation water quality in terms of increasing Na salt concentration through discharge of liquid effluent. Use of such high Na containing irrigation water for crop production enhances salinity, raises pH and high ESP on the exchangeable sites. Alkaline soil condition destroys soil structure through dispersion of clay minerals and thereby enhances soil hardness. Soil organic matter content is a major soil quality parameter influencing soil structure and friability. Lower soil organic carbon is also one of the prime factors for soil hardness. Polluted soils also affect growth of vegetation and root biomass which results in lower C sequestration in soil profile (and consequently lower content of soil organic matter).

Excessive soil hardness impedes root proliferation and consequently, decreases the plant's ability to take up nutrients and water. This also results in poor infiltration (hence, more soil erosion, less storage of water in soil profile, and lower groundwater recharge). From the standpoint of crop production, the adverse effect of soil hardness on water flow and storage may be more serious than the direct effect of soil compaction on root growth. In dry years/seasons, this can lead to stunted, drought stressed plants due to decreased root growth resulting in yield reductions. Soil hardness in wet years decreases soil aeration resulting in increased denitrification (loss of oxide of nitrogen to the atmosphere). Soil hardness considerably increases draft power requirement of tillage implements. Hence, its measurement is important during investigation on the impact of soil pollution.

3.6.1 Measurement of Soil Hardness

A soil penetrometer is an instrument used for estimating soil hardness or compaction. This shaft with cone is inserted into the soil, where amount of pressure (kg cm^{-2}) required to cause penetration is taken as measure of soil hardness. The penetrometer is designed to mimic a plant root. It consists of a 30-degree circular stainless steel cone with a driving shaft and a pressure gauge. The tip is slightly wider than the driving shaft to limit friction of the shaft with the soil. The driving shaft is usually graduated every 3 inches to allow the determination of depth of penetration. Root penetration decreases linearly with penetration resistance, until almost no root penetrate into soil with a penetration resistance of 300 psi. The measurements should be taken when the whole profile is at field capacity (approximately 24 h after a soaking rain).

3.7 Decrease in Available Soil Moisture Content

Available soil moisture content is an important factor influencing agricultural productivity and fertilizer use efficiency. Depletion in available soil moisture content increases the frequency of irrigation for agricultural crop production. Early wilting is one of the visible symptoms in relation to lower available soil moisture content. Entry of various pollutants in the agroecosystem causes degradation of soil structure which affects available moisture retention in the soil profile. As absorption of soil moisture by plant roots also depends on osmotic potential, increased soil salinity (hence increased osmotic potential) reduces availability of soil moisture. Entry of salt loaded effluent and wastes of industrial origin therefore, reduces available soil moisture content in soil. Such impact is more severe in arid and semi-arid climatic region and on heavy textured soils.

Several soil water/moisture determination methods are available viz., thermogravimetric method, electrical resistance method, nuclear techniques, time domain reflectometry, microwave remote sensing, soil psychrometer, tensiometer etc. Soil moisture content can be measured in cheaper way with the help of Tensiometer under under field condition and by gravimetric method in laboratory as described below.

3.7.1 Tensiometer

A tensiometer consists of a vacuum gauge connected by a tube to a porous ceramic tip. The tube is filled with water and is normally transparent (made of plexiglass) so that water level within it can easily be seen. The ceramic tip is permeable and is saturated by the water in the tube. When inserted at the desired depth of soil profile, water from the tube permeates to surrounding area through saturated ceramic tip which is in contact with the unsaturated soils in the root zone. As water moves out from the tube, a partial vacuum pressure is created which is measured by the gauge. Vacuum pressure created inside the tensiometer can also be measured with the help of mercury manometer. However, this type of tensiometer is discouraged keeping in view of extreme toxicity of the mercury to environment. This measurement is not a direct measurement of soil water content. Rather, it is a measurement of soil water tension (also called soil moisture tension). Soil moisture tension increases when there is less water in the soil. As a result, the tensiometer gauge reads high for dry soils and low for wet soils.

3.7.2 Gravimetric Method

The soil moisture content may be expressed as the ratio of the mass of water present to the dry weight of the soil sample, or by volume basis as ratio of volume of water to the total volume of the soil sample. To determine any of these ratios for a particular soil sample, the mass of soil moisture must be determined by drying the soil to constant weight and measuring the soil sample mass after and before drying. The water mass (or weight) is the difference between the weights of the wet and oven dry samples. The criterion for a dry soil sample is the soil sample that has been dried to constant weight in oven at temperature between 100 and 110 °C (105 °C is typical).

$$Soil\ moisture(\%) = \frac{(wt.\ of\ wet\ soil) - (wt.\ of\ dry\ soil)}{(wt.\ of\ dry\ soil)} \times 100$$

3.8 Waterlogging due to Poor Infiltration

Infiltration refers to the entry of water into the soil. The rate at which water enters is referred to as the rate of infiltration. Problem of water-logging occurs if the irrigation water does not enter the soil rapidly enough during a normal irrigation cycle to replenish the soil with water needed by the crop before the next irrigation. Poor infiltration rate reduces the quantity of water stored into the lower depth of soil profile for later use by the crop. Hard pan due to continuous use of poor quality water in Bikaner region of Rajasthan (India) was a well known event in 2003–2004, which drew wide attention by media (Kar et al. 2009). Shallow water table enhanced chances of contamination in ground water, but poor infiltration rate of soil made it water logging or swampy lands. Hard pan results from silica cementation and clay deposition. It may be due to leaching following episodic sheet flooding. Low salinity water (less than 0.5 dS m^{-1} and especially below 0.2 dS m^{-1}) tends to leach surface soils free of soluble minerals and salts, especially calcium, reducing their strong stabilizing influence on soil aggregates and soil structure. At lower salts and calcium content, the fine soil particles (clay) disperse, which fill many of the smaller pore spaces, sealing the surface and greatly reducing the infiltration rate. Such soil crusting often results in poor crop emergence, in addition to a reduction in water storage in the soil profile causing water stress between irrigations. The reduced infiltration rate, if happen to be due to poor quality of applied water, is generally a problem within the upper few centimetres of soil but occasionally may occur at lower depths.

With the help of infiltrometer, the rate of water infiltration into the soil is measured. Two factors, salinity and SAR, must be considered together for a proper evaluation of the ultimate effect on water infiltration rate. Indirectly, comparison of

SAR and ESP measured in soil surface of polluted soil and unpolluted soil gives indication of poor infiltration rate.

3.9 Impacts on Soil Fertility Parameters

Soil fertility refers to the ability of the soil to supply essential plant nutrients and water in adequate amounts and proportions required for plant growth and reproduction in the absence of toxic substances which may inhibit plant growth. A fertile soil has good physical and chemical environment for growth of plant roots as well as for beneficial soil microflora and fauna responsible for carrying out nutrient transformation leading to optimum supply of plant nutrients. Hence, self reliance on food production for a country depends a lot on the availability and maintenance of fertile land. Fertility of a soil is identified with its physical, chemical and biological characteristics in a agro-ecosystem. Some of these soil characteristics are easy to measure and therefore, are included in the fertility evaluation programme of the soil testing laboratories in the many countries for the purpose of recommending fertilizers to the crops. Entry of pollutants in agroecosystem affects different soil properties resulting in loss of soil fertility and hence crop productivity. Different soil physical and chemical parameters are described below which can be used for evaluating the impact of soil pollution.

3.9.1 Soil pH

The soil pH is a measure of the acidity or alkalinity in its chemical environment and is defined as the negative logarithm (base 10) of the activity of hydronium ions (H^+ or, more precisely, H_3O^+) in a solution. In general, pH below 7 is acidic and above 7 is alkaline. Different industrial activities have tremendous effect on soil pH. Acid precipitation nearby industrial area (due to atmospheric emission of SO_2 from smelter industries) and its consequences on soil acidification is of worldwide concern in the developed countries of temperate region. Continuous use/entry of alkaline industrial wastes like fly ash, steel/basic slag and cement kiln dust may alter pH of particularly acid soil. Likewise, continuous use of acidic industrial effluent (like that from tannery industry, electroplating industry, distillery industry, acid mine effluent) contaminated water for irrigation may remove essential basic cations and lower soil pH. Most of the essential plant nutrients in soil are optimally available in pH range of 6–7.5. Acidification may aggravate heavy metal toxicity to plant, while depleting availability of phosphorus and molybdenum. On the other hand, alkalization of soil may lead to deficiency of most of the plant nutrients and encourage volatilization of N from fertilizer sources. Soil reaction (pH) can be measured in soil water suspension (1:2 or 1:2.5) and with the help of a pH meter.

Table 3.6 Soil pH

Category	pH range
Ultra acid	<3.5
Extreme acid	3.5–4.4
Very strong acid	4.5–5.0
Strong acid	5.1–5.5
Moderate acid	5.6–6.0
Slight acid	6.1–6.5
Neutral	6.6–7.3
Slightly alkaline	7.4–7.8
Moderately alkaline	7.9–8.4
Strongly alkaline	8.5–9.0
Very strongly alkaline	>9.0

Soil Survey Division Staff (1993)

The United States Department of Agriculture (USDA) Natural Resources Conservation Service, classifies soil pH ranges as described in Table 3.6.

3.9.2 Electrical Conductivity (EC)

Soil electrical conductivity is an indirect measurement of the concentration of salt in soil solution (salinity) and is an important indicator of soil health. Electrical conductivity is defined as the ability of soil suspension to conduct (transmit) an electrical current and is commonly expressed in units of milli-siemens per meter ($mS\ m^{-1}$). High soil salinity (as indicated by high EC) affects crop selection, crop yields, plant nutrient availability, and activity of soil microorganisms that influence key soil processes including the emission of greenhouse gases such as nitrogen oxides, methane, and carbon dioxide as described in Chap. 4. High salts in soils occur naturally in arid and semiarid climates due to their net upward movement in absence of adequate rainfall. Salt levels can also increase as a result of fertilization, poor quality irrigation water, and land management. Although EC does not provide a direct measurement of specific ions or salt compounds, it has generally been correlated with concentrations of NO_3^-, K^+, Na^+, Ca^{+2}, Mg^{+2},Cl^-, SO_4^{-2}, and NH_4^+. Contribution of some other contaminant ions in EC measurement is also possible in severely polluted soils. For certain non-saline soils, determining EC can be a convenient and economical way to estimate the amount of nitrogen (N) available for plant growth. In soil, EC is measured with the help of conductivity meter in saturation paste or suspension (1:2 or 1:2.5 soil: water ratio).

3.9.3 Sodium Adsorption Ratio (SAR)

It is a measure of the suitability of water for use in agricultural irrigation, as determined by the concentrations of solids dissolved in the water. It is also a measure of the sodicity of soil, as determined from analysis of water extracted from the soil.

The formula for calculating sodium adsorption ratio is:

$$\text{S.A.R.} = \frac{Na^+}{\sqrt{\frac{1}{2}\left(Ca^{2+} + Mg^{2+}\right)}}$$

where, concentrations of Na, Ca, and Mg are in milliequivalents/liter.

In general, the higher the sodium adsorption ratio, the less suitable the water is for irrigation. Irrigation using water with high SAR value may require soil amendments to prevent long-term damage to the soil. If irrigation water with a high SAR is applied to a soil for years, the sodium in the water can displace the calcium and magnesium in the soil. This will cause a decrease in the ability of the soil to form stable aggregates resulting in a loss of soil structure and tilth. This will also lead to a decrease in infiltration and permeability of the soil to water leading to problems related to crop production.

3.9.4 Bulk Density (BD)

Soil density is expressed in two well accepted concepts as particle density and bulk density. In the metric system, particle density can be expressed in terms of grams per cubic centimeter (g cm^{-3}). Bulk Density is the oven dry weight of a unit volume of soil inclusive of pore spaces is called bulk density. The bulk density of a soil is always smaller than its particle density. The bulk density of sandy soil is about 1.6 g cm^{-3}, whereas that of organic matter is about 0.5 g cm^{-3}. BD is measured with pycnometer. Bulk density normally decreases, as mineral soils become finer in texture (Table 3.7). The bulk density varies indirectly with the total pore space present in the soil and gives a good estimate of the porosity of the soil. Bulk density is of greater importance than particle density in understanding the physical behavior of the soil. Generally soils with low bulk densities have favorable physical conditions. Some of the factors affecting the BD *are* pore space, soil organic matter and texture.

Table 3.7 Soil BD of various textural classes

Textural class	Bulk density (g cm^{-3})	Pore space (%)
Sandy soil	1.6	40
Loam	1.4	47
Silt loam	1.3	50
Clay	1.1	58

Source: http://agridr.in/tnauEAgri/eagri50/SSAC121/lec09.pdf

3.9.5 Available Major and Micro Nutrients

Available plant nutrient contents are important defining parameters of the soil fertility; these affect the crop productivity of a particular area. Contents of major nutrients (N, P, K, Ca, Mg, S) and micronutrients (Cu, Zn, Fe, Mn, Ni, Co, Mo, B, Cl) in soil vary from region to region and their requirement for crop are also affected by crop and variety, management and environmental factors. For their quantification, soil testing with the help of standard procedures is the prime way. Procedures and corresponding limits of different fertility parameters are mentioned in Tables 3.8 and 3.9.

3.9.6 Soil Organic Matter

Soil organic matter (SOM) consists of plant and animal residues at various stages of decomposition, cells and tissues of soil organisms, and substances synthesized by soil organisms. SOM exerts numerous positive effects on soil physical and chemical properties, as well as the soil's capacity to provide regulatory ecosystem services. Quantity of SOM is regarded as being critical parameter for different soil ecosystem functions and resilience capacity. Soil organic matter can be measured in mainly 4 different ways.

(i) Weight loss due to removal of organic matter by ignition at high temperature: A known weight of sample in a ceramic/porcellin crucible is heated to a temperature between 350 and 440 °C overnight (Nelson and Sommers 1996), which is subsequently cooled in a desiccator and weighed. Organic matter content (expressed as %) is calculated as loss in weight during ignition divided by the initial sample weight times 100. All weights should be corrected for moisture content prior to organic matter determination. Ignition temperature should be maintained below 440 °C to avoid any error due to destruction of inorganic carbonates present in the sample.

(ii) Weight loss due to removal of organic matter, either by oxidation with H_2O_2: The method involves the addition of concentrated H_2O_2 (30% or 50%) to a pre-weighed soil sample. H_2O_2 is incrementally added to the sample until frothing ceases. The samples may be heated to 90 °C during the process to enhance speed of oxidation. After completion of oxidation process, the sample is dried at 105 °C, cooled in a desiccator, and weighed. Organic matter

Table 3.8 Standard method of soil fertility analysis

Characteristics	Method/references
pH (soil:water 1:2.5)	Jackson (1973)
EC (soil:water 1:2.5) (dS m^{-1})	Jackson (1973)
Mechanical analysis	Bouyoucos hydrometer (1962)
Organic carbon (%)	Walkley and Black (1934)
Available phosphorus (kg ha^{-1})	Olsen et al. (1954)
Available nitrogen (kg ha^{-1})	Subbiah and Asija (1956)
Available potassium (kg ha^{-1})	Jackson (1973)
DTPA extractable micronutrient (mg kg^{-1})	Lindsay and Norvell (1978)

Table 3.9 Rating chart for soil test values of nutrients

	Rating		
Nutrient	Low	Medium	High
Organic carbon (%)	<0.5	0.50	>0.75
Alkaline $KMnO_4$-N (kg ha^{-1})	<280	280–560	>560
Olsen's P(kg ha^{-1})	<10	10–25	>25
AmmoniumAccetate –K (kg ha^{-1})	<120	120–280	>280
Sulphur (mg kg^{-1})	<10	10–20	>20
Zinc (mg kg^{-1})	<0.6	0.6–1.2	>1.2
Iron (mg kg^{-1})	<4.5	4.5–9	>9
Manganese (mg kg^{-1})	<5	5–10	>10
Copper (mg kg^{-1})	<0.2	0.2–0.4	>0.4
Boron (mg kg^{-1})	<0.4	0.4–0.8	>0.8
Molybdenum (mg kg^{-1})	<0.02	0.02–0.04	>0.04

Kundu et al. (2012)

content is calculated as loss in weight during ignition divided by the initial sample weight times 100%. The major limitation of this method is that the oxidation of the organic matter may not be complete and the extent of oxidation varies markedly with soil type.

(iii) Oxidation of organic matter by chromic acid, either by heat of dilution or applying external heat followed by either determination of the quantity of chromic acid consumed for oxidation through titration method (Walkley and Black 1934) or by photometric determination of Cr^{+3} produced due to oxidation of organic matter (Datta et al. 1962). This method though rapid, suffers from several limitations like incomplete oxidation of organic matter, use of toxic heavy metals (Cr) and subsequent disposal of residues, interference with other ions.

(iv) Oxidation by chromic acid (after carbonate removal) and measurement of CO_2 evolved: Soil samples are pre-treated with dilute acid (HCl or H_3PO_4) to remove inorganic carbonates and subsequently, organic matter in these samples are oxidized by chromic acid as above. Carbon dioxide evolved during the process are measured using in a Van Slyke-Neil apparatus by measuring the change in pressure with a gauge (Nelson and Sommers 1996).

(v) Dry combustion and measurement of CO_2 evolved (after removal of carbonates): Generally a automated instrument (TOC analyzer or CHN analyzer) is used for this type of measurement. Soil samples are pre-treated with dilute acid (HCl or H_3PO_4) to remove inorganic carbonates. Subsequently, these are combusted at elevated temperatures (between 900 and 1500 °C) in a furnace in the presence of catalysts and pure oxygen stream (Nelson and Sommers 1996). Evolved CO_2 gas is measured using either thermal conductivity or infrared detectors.

3.9.7 *Available and Total Heavy Metals Contents*

Collection of soil samples is an important step prior to analysis for heavy metals content. Sampling area, frequency, time of collection, depth of sampling depend on the objectives of the assessment and therefore, needs to be given equal or more importance than analytical methods.

3.9.7.1 Soil

Plant Available Heavy Metals There are several special considerations for the assessment of hazards due to heavy metal contamination (McLaughlin et al. 2000), including the examination of effects on different components of the environment like humans, animals, plants, and microorganisms. Plants essentially take up heavy metals from the soil via soil solutions (Schindler and Sposito 1991; McBride 1991). It has been generally accepted that phytotoxicity of a metal is related to its free ion activity in the solution to which the plant root is exposed (Lund 1990; Parker and Pedler 1997) and also to soluble metal-organic complexes and metal-inorganic ligand complexes (Smolders and McLaughlin 1996; Huang et al. 1997; Jones 1998). Recent studies have also shown that less aggressive reagents that extract predominantly soluble and exchangeable fractions, most notably, dilute salt solutions such as 0.01 M $CaCl_2$, are frequently better at predicting plant availability of excess heavy metals in soils than the traditional, more aggressive tests involving the use of reagents such as hydrochloric acid, diethylene triaminepentaacetic acid, and ethylene diamine tetraacetic acid (Houba et al. 2000; McBride et al. 2003, 2009; Menzies et al. 2007; Meers et al. 2007). After a comprehensive review of literature, McLaughlin et al. (2000) concluded that soil tests based on the determination of the most easily extractable metal pools (like the neutral dilute salt-extractable fraction) appear to reflect the current toxicity better than total metal concentrations or the amounts of metals removed by strong extractants. To estimate plant-available fractions of metals, air-dried soil was extracted with 0.01 M $CaCl_2$ solution in the ratio of 1:2.5 after shaking for 4 h.

Metals in extracting solution can be determined using Atomic Absorption Spectrophotometer equipped with graphite furnace (GF-AAS) or Inductively Coupled Plasma-Optical Emission Spectrometer (ICP-OES).

Total Heavy Metal Contents Heavy metals can be extracted from the oven-dried soil samples by using a wet digestion method involving HF, HNO_3, H_2SO_4, and $HClO_4$ as described by (Baker and Amacher 1982). Metals were detected in extracted samples by using inductively coupled plasma-optical emission spectroscopy.

3.9.7.2 Plant

Plant parts to be analyzed for heavy metals content depend upon the purpose of impact assessment and it may be edible parts (in case of food chain contamination assessment) or leaves and shoots (as in case of fodder for animal). As translocation of metals within different plant parts is an important phenomenon, their distribution may vary with plant parts. Therefore, leaf samples should be collected from a larger number of plants from an uniform positions in a representative area (like youngest full grown leaf). After collection, plant samples are washed thoroughly in deionized water, and dried in the oven at 70 °C. Thoroughly pulverized biomass samples are digested in di-acid (HNO_3:$HClO_4$ at 10:1 ratio) for the analysis of heavy metals.

3.9.8 Soil Enzyme Activity

Soil enzymes, mainly of microbial origin, are known to play a substantial role in maintaining soil health and its environment. Their activity releases nutrients into the soil by means of organic matter degradation. In the context of soil quality assessment, their activities can be used as measures of microbial activity, soil fertility, and toxic effects of pollutants. Activities of soil enzymes are assessed using specific chemical compound, known as substrate and are quantified by measuring the concentration of the product of the substrate specific biochemical reaction. Simple laboratory biochemical assays are available for a large number of soil enzyme activities, viz., dehydrogenase, urease, amidases, phosphatases, arylsulphatase, cellulases, glucosidases, and phenol oxidases (Dick et al. 1996; Tabatabai 1982). Role of these soil enzymes in soil process of agricultural importance is presented in Table 3.10.

Table 3.10 Soil enzymes as indicators of soil health

Soil enzyme	Enzyme reaction	Indicator of microbial activity
Dehydrogenase	Electron transport system	C-cycling
β-glucosidase	Cellobiose hydrolysis	C-cycling
Cellulase	Cellobiose hydrolysis	C-cycling
Phenol oxidase	lignin hydrolysis	C-cycling
Urease	Urea hydrolysis	N-cycling
Amidase	N-mineralization	N-cycling
Phosphatase	Release of PO_4^-	P-cycling
Arylsulphatase	Release of SO_4^-	S-cycling

Das and Verma (2011)

3.9.9 Nitrogen Fixation Capacity

Nitrogen fixation from its gaseous form in the atmosphere to its compounds which plants and other organisms can use (either NH_4^+ or NO_3^-) is mediated either by bacteria in symbiotic relationships with vascular plants, or by symbiosis between cyanobacteria and fungi (lichens) or plants, or by free living heterotrophic or autotrophic bacteria that are typically associated with soil. Ecological constraints that may limit the success of N-fixing plants include the availability of soil nutrients other than N (especially P or Mo), the existence of poor edaphic conditions such as high acidity, alkalinity or aridity, removal of N-fixing species by preferential grazing and removal of woody dicots (including woody legumes) by fire (Fredan 2011).

Species potentially able to fix N_2 (target species), or not (reference species) are sampled in each plot. The target species are legumes with a known nodulation capacity or without information on nodulation. Samples are collected from all the target species found in each site and also from a reference species. After collection, completely expanded healthy green leaves were selected, forming a composite sample of 200 g of leaves per plant. The leaf material was oven dried at 60 °C for 72 h and then ground to fine powder. Total N was analyzed by micro Kjeldhal methodology and ^{15}N abundance by mass spectrometry and expressed in "delta" notation, which is the deviation per thousand (%) of ^{15}N abundance of the sample in relation to the standard.

$$\delta = \left(\frac{R_{sample}}{R_{standard}} - 1\right) \times 1000$$

where, R_{sample} and $R_{standard}$ are the ratio ^{15}N:^{14}N of the sample and the standard (atmospheric N_2), respectively (Dawson et al. 2002).

The enzyme responsible for N_2-fixation (i.e. nitrogenage) also reduce acetylene (C_2H_2) to ethylene (C_2H_4) (Dilworth 1966). This process has also been used to have quantitative estimation of N_2-fixation. Although for quantitative determination of N_2-fixation ^{15}N techniques should be used, the acetylene reduction assay is still being used widely because it provides a highly sensitive and inexpensive way to quantify the relative nitrogenase enzyme activity in N_2 fixing samples of soil. In this method, soil samples are incubated in a closed chamber with small quantity of acetylene and after incubation period ethylene produced is measured using gas chromatography.

3.9.10 *Microbial Diversity Measurement*

Microbial diversity can be defined as the range of different kinds of unicellular organisms, bacteria, archaea, protists and fungi in soil. Various microbes thrive throughout the biosphere, defining the limits of life and creating conditions conducive for the survival and evolution of other living beings. The different kinds of microbes are distinguished by their differing characteristics of cellular metabolism, physiology, and morphology, by their various ecological distributions and activities, and by their distinct genomic structure, expression, and evolution. The diversity of microbes presently living on earth is known to be high and is thought to be enormous, but the true extent of microbial diversity is largely unknown. New molecular tools are now permitting the diversity of microbes to be explored rapidly and their evolutionary relationships and history to be defined.

Bacteria, fungi and actinomycetes are estimated by the serial dilution and plating technique as described by Rolf and Bakken (1987). Microbe specific media are used for culture of microorganisms which are plated on the glass petri dish. Small volume of water extract from soil (generally 0.1 mL) after 10^1 to 10^5 dilutions is poured on the surface of petri dish with the help of a micropipette and is evenly spread with the help of a sterile spreader and incubated at 30 ± 2 °C. Colony count was carried out daily up to 10 days. Total media specific microorganism are estimated from the number of colonies formed on the media plate after multiplying by dilution factor and results are reported as colony forming units (CFUs). Total bacterial population counts in diluted extracts can also be done through direct observation of cells in a specialized counting chamber slide viewed under phase contrast microscopy.

One gram of soil has been reported to contain up to ten billion microorganisms and thousands of different species. Major limitation of above technique is that most of the microorganisms (more than 99%) are unculturable and therefore true picture of microbial biodiversity is not obtained. In recent years, soil microbial diversity is being assessed in other ways using instrumental techniques. Using metagenomic tools, genetic materials recovered directly from environmental samples. This broad field is also referred to as environmental genomics, ecogenomics or community genomics. Polymerase chain reaction (PCR), a biomedical technology in molecular biology is used to amplify a single copy or a few copies of a piece of DNA across several orders of magnitude, generating thousands to millions of copies of a particular DNA sequence. With the help of various microbial tools and techniques, large number of species of microorganisms can be measured.

3.9.11 Soil Quality Assessment

Soil is a dynamic, non-renewable resource. If neglected or misused, soil ecosystem will cease to function properly and will result in adverse consequences on the cycling of nutrients and organic matter. Traditionally, the quality of soil has been mainly associated with its productivity, whereas more recently, the definition has been expanded to include the capacity of a soil to function within ecosystem environmental quality, and promote plant and animal health.

Under agroecosystem, major emphasis of soil quality assessment is to evaluate its fitness to support crop growth without resulting in soil degradation or otherwise harming the environment. At the beginning of soil-quality assessment, there is a need to have answer for two main questions: (i) how does the soil function; and (ii) what procedures are appropriate for making the evaluation. Subsequently, a range of parameter values that indicate a soil is functioning at full potential can be computed using landscape characteristics, knowledge of pedogenesis, and a more complete understanding of the dynamic processes occurring within a soil. Necessary steps for soil quality estimation can be mentioned as below (Table 3.11):

While selecting soil quality indicators, their desired attributes required to considered are like

- should represent desired soil function
- should be sensitive to changes in land use or management
- should be easily measured and reproducible
- should be reliable
- should be applicable under field conditions

Soil quality depends on a large number of chemical, physical, biological and biochemical properties, and its characterization requires the selection of the properties most sensitive to change in management practices, called indicators. Since it is impractical to measure every ecosystem or soil property, many researchers have proposed a minimum data set, which is the smallest set of soil properties or

Table 3.11 Sequential framework to evaluate soil quality

Steps	Sequential framework	Question implied by the framework
1	Purpose	What will the soil be used for?
2	Functions	What specific role is being asked of the soil?
3	Processes	What key soil processes support each function?
4	Properties/attributes	What are the critical soil properties and their critical or threshold levels?
5	Indicators, surrogates,or pedotransfer function	When the attribute is difficult to measure or not available, which indirect or related property or properties can be used in its place?
6	Methodology standardization	What methods are available to measure the attribute? Technical rules and protocols for soil sampling, handling, storage, analysis and interpretation of data.

indicators needed to measure or characterize soil quality. Identifying key soil properties or attributes that are sensitive to change in soil functions establish a minimum data set. Different kinds of land uses may require increased capacities of certain soil functions and, thus, prefer certain indicators over others for assessment. For example, a wetland soil has a different role than a agricultural soil for the soil function, partitioning of water. Therefore, infiltration would not be a useful indicator of wetland soil quality, but would be very useful for assessing the quality of most agricultural soils. Soil organic matter (SOM) is one of the more useful indicators of soil quality, because it interacts with other soil components; affecting water retention, aggregate formation, bulk density, pH, buffer capacity, cation exchange properties, mineralization, sorption of pesticides and other agrochemicals, colors (facilitate warming), infiltration, aeration and activity of soil organism.

3.10 Computation of Soil Quality Index from the Perspective of Capacity of Soil to Produce and Sustain Agricultural Productivity

The minimum data set of indicators are identified that best represent soil function and, therefore, only those indicators which are thought to contribute to the function of interest (management goals) are included for indexing soil quality. After identifying indicators that best represents soil functions, weight-factors are assigned to each attributes on the basis their importance and relative contribution to the soil function. Sum of all weight-factors are normalized to 100%. Once weightage is assigned to the indicators, each attributes are characterized into different classes with specified marks, called score (maximum mark for most desired class). Analytical values for each indicators under minimum data set are transformed into scrore depending on the class in which it belongs.The soil quality index (SQI) is calculated by using the following equation,

$$SQI = \sum W_i M_i$$

where,

W_i = Weight of the indicators
M_i = the score of the indicators classes

The SQI is calculated by two steps: (i) multiplying weight of indicator and score value of the indicator and (ii) summing up the multiplied values of all the indicators to derive the SQI value for a particular soil.

3.11 Major Instances of Soil Contamination in the World and Their Impact on Environment

Ever since human started understanding the natural processes operating in the environment and began manipulating environmental functioning through different inventions for enhancing both their own chance of survival as well as comfort, alteration and degradation of environment began, whose pace increased continuously as civilization progressed with time. However, extent of environmental degradation became severe as process & magnitude of industrialization took revolutionary turn since late eighteenth century. As researches on environmental consequences always lagged behind the degrading processes due to anthropogenic activities, regulations to control industrial activities were formulated and implemented much later and as a result, severe consequences on environment happened during this transition period and cases of severe environmental damages were reported from different parts of the world. Some of the widely reported instances of land/soil pollution are described briefly below:

3.11.1 The Love Canal Tragedy

Love Canal is one of the most widely discussed environmental tragedies in American history. At the beginning of the twentieth century, William T. Love dug a canal for the purpose of generating and supplying hydropower to meet the energy requirement of industrial and residential areas in his community on the edge of Niagara Falls in New York. However, Love couldn't complete his project due to certain economic and technical reasons. During the 1920s, the canal he dug was turned into a landfill operated by a chemical company where the premises had been filled with waste products resulting from the manufacturing of chemicals as well as with municipal wastes. Subsequently after few decades, several houses and a elementary school were built on the site. Over the years, the underground containers filled with approximately 21,000 tons of chemical waste were corroded. In 1977, a record rainfall brought about a tragic consequence. The waste began to leach into people's homes, backyards, and playgrounds. In American history, this tragedy is termed as one of the vast horrifying environmental disasters. The analysis of soil and sediment samples collected from Love Canal area indicated the existence of more than 200 various types of organic compounds. Among the pollutants, dioxin was found most toxic and its highest concentration (300 ppb) was found in the storm sewer in the adjoining area. Love Canal had been officially associated with high rates of birth defects, miscarriages, and other severe illness resulting from land contamination (Phillips et al. 2007). Following a federal emergency declaration in May 1980, several remedial measures were taken viz., landfill containment with leachate collection, treatment, and disposal; excavation and interim storage of the sewer and creek sediments; thermal treatment of sewers and creeks sediments.

These remedial measures eliminated the significant contamination exposure pathways, making the site safe for nearby residents and the environment.

3.11.2 Chernobyl Nuclear Disaster

World ill-famous Chernobyl disaster, known as the Chernobyl accident was a catastrophic nuclear accident and happened on 26th April 1986 at the nuclear power plant in Ukraine. Both acute radiation effects in the initial periods (radiation death of plants and animals, loss of reproduction, etc.) and long term effects (change of biodiversity, cytogenetic anomalies, etc.) have been observed in the affected areas. The disaster happened in the fourth plant of Chernobyl located in the city of Pripyat nearby Dnieper river, when sudden power surge followed by failure of the emergency shutdown system caused a strong explosion in the reactor vessel. More than 70% of area in Belarus, Russia and Ukraine were contaminated with the radioactive materials due to their aerial spread (IAEA 2006). The toxicity of disaster was huge, in which 237 plant workers were highly exposed, 134 were diagnosed with acute radiation syndrome and more than 28 died in this massive calamity. The zone was demarcated as a Chernobyl exclusion zone and more than 116,000 neighbouring residents were shifted to a safe place to prevent exposure to harmful radiation. Environmental samples were periodically examined to see the toxic effect of this accident on soil – plant –human system. The concentrations of radioactivity in air, agricultural plants, animals and soil were more in the initial months of the incidence, which might be due to surface deposition of contaminants. This early period of the accident was more toxic due to fast decay of ^{131}I. Such toxicity was also observed in the milking animals, when the radioactive iodine was speedily absorbed by milk. This impaired the normal functioning of thyroid gland in contaminated milk consuming people in the belt of Belarus, Russia and Ukraine. Apart from ^{131}I, radioisotopes of caesium (^{134}Cs and ^{137}Cs) also created the human health problems in the affected areas. The collected samples contained ^{137}Cs at concentration level of more than 100 Bq per kilogram, the limit requiring national remedial action.

The radionuclides were absorbed by the plant roots and transferred to various edible parts and therefore, soil and food contamination were regularly monitored in the affected area. The half life of cesium (^{134}Cs) is 2.1 years and in the mid 1990s the level of ^{137}Cs in soil is more toxic and harmfully affected the agricultural production system (IAEA 2006). The radioactive caesium content in foodstuffs was influenced not only by deposition levels but also by types of ecosystem, soil as well as management practices. The persistent problems of residual radioactivity in the affected areas occurred in extensive agricultural systems on soils with a high organic content and in animals grazing on unimproved pastures that are not ploughed or fertilized. The contaminated environment affected the primary production system, mostly soil related production activities. The long term intake of ^{137}Cs even at smaller concentration, contaminated milk and animal products, and food crops sufficiently contributed the harmful level of radioactive in human internal dose. The regular monitoring of affected area showed the decreasing concentration of ^{137}Cs in vegetables as well as animal foodstuffs. The trace levels of ^{137}Cs in the human body persisted for a longer time and affected the healthy function of organs. The other radioactive materials with longer half life like Strontium (^{90}Sr), plutonium isotopes and Americium (^{241}Am) levels were found at insignificant levels in human organs. In the Chernobyl exclusion zone, agencies suggested that the environmental remediation part is still required. The most effective countermeasures in the early phase were exclusion of contaminated pasture grasses from animals' diets and the rejection of milk.

3.11.3 Cadmium Contamination of Soils in China

The toxicity of Cd is considered most hazardous among the heavy metals as it has several adverse impacts on soil microorganisms, plants, animal and human health. A number of incidences of Cd contamination have been reported across the world. A highly discussed case of Cd contamination happened in 2006 at the Xianghe Chemical Factory, which is situated nearby the Liuyang River (Guangwei 2014). In the adjoining Shuangqiao village, people were mostly dependent on this industry for livelihood. The industry initially produced zinc sulphate in powdered and pellet form to be used as the animal food supplements. Afterward, it started producing indium, a highly priced metal (more than gold) in international market. However in the process, it also generated Cd as a byproduct. In the year 2009, adverse effects of Cd contamination was observed initially among Xianghe Chemical Factory workers and subsequently, the number of victims of the tragedy increased enormously in Shuangqiao village. The research agencies monitored the source and sink of Cd contamination and found that most of the soil and agricultural food crops were highly contaminated with this metal. More than 500 villagers were worst affected with Cd toxicity and several of them died during the course of the investigation. The run-off from the industries carrying sludge during rainy season contaminated 659 acres area nearby the plant. Because of this unfortunate tragedy, Shuangqiao village is infamously known as Cd village across the globe.

3.11.4 Thermal Power Plant Disaster

Thermal power plants are one of the backbone of industrial growth as well as upliftment of society. However, accidents in such seemingly innocuous industry can create havoc in the land of nearby areas. In this line, one of the major disaster happened in the morning of December 22, 2008 at Roane County, Tennessee, USA, due to rupture of ash dyke and as a result, 4.13 million m^3 of coal fly ash slurry spilled out and covered 84 acre area nearby the thermal power plant. The coal fired power plant was situated across the Clinch River in the

Kingston city. In order to prevent aerial release, the fly ash generated as a byproduct of coal combustion was being transported in the ash dyke pond as slurry which was stored in dredge cells in wet form. After rupture of ash dyke during the accident, the slurry got released in the Emory River and spread in the nearby fields. The quantum of release was about 4 million m^3 and slurry covered the land surface by up to 1.8 m high as measured by the survey of contaminated area. This disaster created massive damage to residential buildings, water bodies, agricultural fields and worst damage to roads, ruptured telephone and gas lines, railway track, trees and power lines. The disaster killed fish population in a large scale. An assessment study was carried out immediately after the tragedy, which indicated that surface release of coal ash with high levels of toxic elements (75 mg As kg^{-1}; 150 μg Hg kg^{-1}) and radioactivity ($^{226}Ra + ^{228}Ra =$ 8 pCi g^{-1}) to the environment had potential to generate atmospheric pollution through suspended fine particles posing a health risk to local communities, to contaminate of surface waters in low lying areas through leaching of contaminants, to poison fish with As and methyl mercury (Ruhl et al. 2009). However, an 18 months investigation after the accident indicated that contaminant levels in the downstream river water were relatively low (below the EPA drinking water and ecological thresholds); although elevated levels of arsenic (up to 2000 μg L^{-1}) were found in stagnated surface water bodies and in pore water extracted from the river sediments downstream from the spill (Ruhl et al. 2010). Most of the fly ash was removed from the deposition area by August 2010 through extensive dredging; however, about 10% of ash was left as river sediments to avoid disturbing legacy contamination in the downside with hazardous contaminants. Through a study, a concern was raised on the migratory potential of toxic trace metals from aquatic system of the affected site to the terrestrial system through food chain (Beck et al. 2013).

3.11.5 Seveso Disaster (Chemical Manufacturing Plant)

This disaster happened on July 10, 1976 in a small chemical plant nearby Milan, Italy. The industry was manufacturing 2,4,5-trichlorophenol from 1,2,4,5-tetrachlorobenzene with the help of nucleophilic aromatic substitution reaction with sodium hydroxide. An explosion in the industry released a toxic cloud into the atmosphere containing about 6 tonnes of TCDD (2,3,7,8 -tetrachlorodibenzo-p-dioxin), which contaminated a densely populated area of about 18 km^2, lying downwind from the site (Mocarelli et al. 1991; Consonni et al. 2008).

For effective remediation plan, entire affected area was categorized into three groups i.e. A, B and R, on the basis of concentrations of TCDD in surface soil. The class A area with the 736 resident population was having >50 μg m^{-2} TCDD concentration in soil. Class B area had 4700 residents and the concentration of TCDD was in the range between 5 and 50 μg m^{-2} in upper soil layer. The class R area with resident's population of 31,800 was having lower concentration of TCDD (<5 μg m^{-2}). In that tragic disaster, about 3300 animals mostly poultry and rearing rabbits died. The toxicity of this chemical was widely found among inhabitants in the area and more than 15 children were hospitalized due to skin inflammation and irritation (De Marchi et al. 1996). Examination of about 1600 peoples of the affected area indicated that 447 people suffered skin lesions or chloracne (acne-like eruption of blackheads, cysts, and pustules). The government agencies issued the awareness guidelines regarding the level of pollution and associated health problems. The local peoples were advised not to touch and eat the vegetables and fruits grown in this contaminated area. To prevent the spread of toxicity due to TCDD into the food chain, a larger number of animals (more than 80,000) were slaughtered in the year 1978. During 50 days after accident, class A area was evacuated completely and fenced. A study conducted after 14 year of the disaster showed that thousands of peoples in the area was suffering from toxicity mainly in the form of chloracne. In the adjoining area of the accident, the toxic effects of TCDD like neuropathy and liver enzyme induction, birth defects etc. were observed among the population. The chemical contamination also enhanced the incidences of mortality due to cardiovascular and respiratory diseases and the psycho-social consequences also dominated in these areas. Apart from this, occurrence of diabetes in the human population was found enhanced. The mortality due to gastrointestinal cancer and cancer in lymphatic and hematopoietic tissue was also increased.

3.11.6 Organic Pollutants Contamination from PCB Manufacturing Industry at Brescia (Italy)

The case is related to food and human exposure of a large contamination of agricultural soil caused by industrial PCB mixtures.The source the pollutants was an industrial plant inside Brescia city at the North-West of Italy, where about 31,000 ton of polychlorinated biphenyls (PCBs) were produced during 1930–1983. During monitoring programs for construction of a municipal waste incinerator near the industrial plant, high PCB levels (3000–6000 folds higher than the allowable limit of 0.001 mg/kg) were observed in the soils of an agricultural area where several small farms were located (Brievik et al. 2002). The contamination process was not caused by a single major release of products; but probably due to continuous leakage during operational phase of industry. Although production of PCB was terminated in 1984, the contamination remained undetected for many years. The vehicle of the propagation of the contamination into the environment was the dispersion with irrigation water of contaminated sediments of a water stream passing through the plant. The farmers had been consuming food, such as milk and meat, produced in their own farms for a long time. Total levels of PCDD/Fs varied from 8 to 592 pgTE(WHO)/g for soil samples and when the dioxin-like PCBs (dl-PCBs) are included, the levels varied from 14.6 to 1033.7 pgTE(WHO)/g (Turrio-Baldassarri et al. 2007). In several cases, the legal limit was exceeded by more than one order of magnitude, with the highest contamination in some agricultural areas and in the surrounding zones. In the forage samples, total levels of PCDD/Fs varied from 0.29 to 2.04 pgTE(WHO)/g and, when dl-PCBs are included, this range increased from 2.04 to 4.75 pgTE(WHO)/g. Vapourization from contaminated land followed by condensation on leaf surface was ascribed as probable way PCB contamination of the forage. The main component of the contamination source is probably a heavy PCB mixture, such as Aroclor 1262. The most severely contaminated area had some very peculiar features: four small farms were located in it, and they included a sort of self contained food chain. Cows and calves were grown and fed, partially or totally, on contaminated forage, while the farmers as well as their family members consumed most of the meat and milk produced in their farms. Due to a relatively closed system of food chain, families experienced a long-lasting exposure to these chemicals, and exhibited a mean blood level of PCB, PCDD and PCDF of 419 pgTE(WHO) g^{-1} lipid, roughly eight times higher than the population of uncontaminated area from the same region (Turrio-Baldassarri et al. 2008). The contamination levels in the peri-renal fat, liver and milk of the overall 28 contaminated bovines indicated that TEQ levels varied from 30 to 81 pgTE (WHO) g^{-1} in perirenal fat, from 107 to 138 pgTE(WHO) g^{-1} fat in liver and from 45 to 50 pgTE(WHO) g^{-1} fat in milk, which were roughly tenfold higher than the European limits. Non-ortho dioxin-like (dl)PCBs (predominantly PCDF) were by far the largest contributors to toxicity in all the biological samples from forages, bovines and consumers of contaminated food. The indicator PCB levels varied within the range 226–664 ng g^{-1} in peri-renal fat; 929–1822 ng g^{-1} fat in liver; 183–477 ng g^{-1} fat in milk and these levels were about 100 times higher than the regional background (Turrio-Baldassarri et al. 2009) indicating grave environmental consequences due the manufacturing industry.

3.11.7 Soil Contamination at Palmerton, Pennsylvania, USA due to Zinc Smelter

In the year 1889, a Zn smelter started in Pennsylvania, USA and continuously operated up to 1980. During its operation, plant emitted toxic metals Zn and Cd as oxides at high concentration (more than permissible limits of the toxic metals as per the industrial norms). The rate of emission of Zn, Cd, Pb were 3575, 47 and 95 Mg per year respectively for the more than 80 years of duration (USEPA 1987). Spread of these pollutants extended up to about 50 km away from the plant (Tan 1988) and forest soils at 100 km away from the plant recorded significantly higher concentrations of these metals (Anderson et al. 1980). This resulted in ecosystem collapse and prevented plant growth across a 3000-acre area. Due to lack of protective vegetative cover, erosion of contaminated soil impacted the nearby Lehigh River and Aquashicola Creek. Forage samples collected from farms nearby Palmerton area showed high concentration of Zn and Cd, whereas Pb concentration ranged up to 25 mg kg^{-1} (Chaney 1988). In the town and lawn area of Palmerton, very high level of contamination with Zn (10,000 mg kg^{-1}) and Cd (100 mg kg^{-1}) was observed in the soils. The concentration of heavy metals was dangerously high in the soils at close vicinity of the smelter; the Zn concentration ranged from 26,000 to 80,000 mg kg^{-1} and Cd reached to the level of 900–1500 mg kg^{-1} (Buchauer 1973, Jordan 1975). Higher concentration of Zn, Cd and Pb reduced the microbial activities in soil (Strojan 1978).

3.11.8 Soil Contamination at Noyelles-Godault, France due to Lead Smelter

A lead smelter company operated for more than a century in northern France and emitted considerable amount of heavy metals in surrounding areas (18.4 tons of Pb, 26.2 tons Zn, 823 kg Cd emitted in the year 2001). As a result, top 30 cm soils of area surrounding the smelter plant had been highly contaminated with heavy metals Cd, Pb and Zn (to a lesser degree with As, Hg, Sb and In) (Godin et al. 1985; Sterckeman et al. 2000, 2002). Even though smelter plants were shut down from operation in 2003, these left behind a large area of spoiled agricultural and urban soils as well as a serious socio-economic crisis. The epidemiological studies carried out during the industrial activity showed that Cd concentration in blood was higher among children (2–4 years old) within 4 km of a zinc smelter compared with children living more than 4 km away (Leroyer et al. 2000, 2001). The geometric mean of the blood-lead levels of children living on the polluted site was 39.5 μg l^{-1} and that of the non-exposed children was 30.6 μg l^{-1}. About 10–15% of children had blood Pb levels higher than the clinically identified critical level of 100 μg Pb L^{-1}. The soil particles of lawn and kitchen garden of the area had 47–68% human bioaccessible heavy metals. Due to such high bioaccessibility in the contaminated soil, Pb was the main concern for children and induced a relatively high health risk through soil particle ingestion (Pelfrêne et al. 2013). On the other hand, total soil Cd concentrations in most soils exceeded the health risk level derived through site-specific assessment criteria using the SNIFFER method, particularly from consumption of vegetables grown on it.

The environmental problems including adverse effect on human health induced by high contamination of the area had been reported even after closing-down of smelter plants, especially in the urban soils, where metal concentrations of agricultural soils were found very high (Douay et al. 2008a). Various studies have shown that vegetables grown in kitchen gardens close to smelter plant, both before and just after closing, frequently exceeded the European legislation limits for Cd and Pb concentrations (Douay et al. 2008a; Pruvot et al. 2006). The Cd and Pb contents in the agricultural soils of the contaminated area were respectively 20 and 15 times above their contents in those from unpolluted area. Similarly, contents of Cd and Pb in kitchen garden soils located near the smelter were respectively about 16 and 10 times above the reference garden soils (Pruvot et al. 2006). The highest concentrations in crops (agricultural crops, home grown vegetables and grass of lawns) were found in the polluted area around the smelter (up to 8–16 times above the reference areas according to the element). Seven years after the close-down of smelting activities, the agricultural and urban soils were strongly contaminated by Cd, Pb, and Zn; the effect being more on kitchen garden topsoils in the nearby urban area (Douay et al. 2013). A high proportion of the vegetables produced in the kitchen gardens did not conform to the European foodstuff legislation indicating persisting nature of the risks of environment and the population's health due to heavy metals pollution in soil. The contaminated dusts existing in the area immediately after closure of smelting operation were also found a potential source of contamination of vegetables in the surrounding area (Douay et al. 2008b).

3.11.9 Itai-Itai Disease in Japan due to Cadmium Pollution

Itai-itai disease refers to osteomalacia which is associated with softening of bones as a result of renal tubular dysfunction. The disease was first reported among residents in a number of cadmium polluted areas throughout Japan, more severely in the districts of Toyama near the Jintsu River, in Tsushima of Nagasaki region, in the Kakehashi river basin of Ishikawa region, and in the Ichi river basin of Hyogo region (Yoshida et al. 1999).

The disease was caused by the toxicity of Cd, mostly in Cd contaminated sites or mining area of gold, zinc, copper and silver. The mining area released huge amounts of Cd into the environment particularly into famous river Jintsu and its tributaries. The contaminated river water was used mainly for irrigation of rice fields and also for drinking, washing, fishing purpose etc. by the downstream populations. Due to continuous irrigation with Cd contaminated river water, high Cd was found in the rice grains and its consumption resulted Cd accumulation in the people. Because high cadmium levels had been found in rice grains, problem of cadmium toxicity was not only confined to the locality in which the cadmium was originally dispersed into the soil but was also reported from other areas adjoining to it (Friberg et al. 1971).

3.11.10 Red Mud Disaster in Hungary

Red mud is the fine fraction by-product of the Bayer process for refining bauxite for alumina production. A major industrial disaster occurred in Ajka (Hungary) in October 2010 when dam of the waste reservoir of the alumina refining plant collapsed releasing about 1 million m^3 of highly alkaline, metalliferous red mud sludge. The sludge flooded three downside villages. Ten people died, and about 150 people were injured in the accident mainly from chemical burn due to high pH (>12) of the waste material. The low-lying agricultural lands in the region of 10 km^2

area on the flood plains of the Torna Stream and Marcal River were impacted by the spill, alongside up to 120 km of river channel where the red mud was also deposited. Besides caustic burn related injury, no major health effect was observed aftermath of accident. Sludge flow into the nearby river severely destroyed aquatic life. After 2 years of accident, river water recorded moderately higher levels of As and Ni compared to those observed in the pre-spill period (Nagy et al. 2013). The contaminated land area was sealed to prevent cultivation. Long-term salinization of soils was identified as major risk affecting plant growth. Salinity build up in soil and release of oxyanion-forming metals and metalloids (e.g., Al, As, Cr, Mo, and V) in the soil–water environment were the potential impact of the accident due to red mud pollution (Mayes et al. 2016).

Extensive remedial measures were carried out in the area which included collection & removal of red mud and acid neutralization of leachate. These measures had resulted very rapid recovery toward pre-spill conditions of soils, stream sediments, waters and aquatic biota (fungi, invertebrates and fish) (Mayes et al. 2016).

References

Anderson AM, Johnson AH, Siccama TG (1980) Levels of lead copper, and zinc in the forest floor of the northeastern United States. J Environ Qual 9:293–296

Arora NK, Khare E, Singh S, Maheshwari DK (2010) Effect of Al and heavy metals on enzymes of nitrogen metabolism of fast and slow growing rhizobia under explanta conditions. World J Microbiol Biotechnol 26:811–816

Aslam U, Ahmad I, Hussain M et al (2014) Effect of heavy metal pollution on mineral absorption in sunflower (*Helianthus annuus* L.) hybrids. Acta Physiol Plant 36(1):101–108

Awashthi SK (2000) Prevention of food adulteration act no. 37 of 1954. Central and state rules as amended for 1999. Ashoka Law House, New Delhi

Baker DE, Amacher MC (1982) Nickel, copper, zinc and cadmium. In: Page AL, Miller RH, Keeney DR (eds) Methods of soil analysis, Part 2, Chemical and microbiological properties, Agronomy monograph no. 9. ASA-SSSA, Madison, pp 323–336

Beck ML, Hopkins WA, Jackson BP (2013) Spatial and temporal variation in the diet of tree swallows: implications for trace-element exposure after habitat remediation. Arch Environ Contam Toxicol 65:575–587

Bianucci E, Fabra A, Castro S (2011) Cadmium accumulation and tolerance in Bradyrhizobium spp. (peanut microsymbionts). Curr Microbiol 62:96–100

Bondarenko O, Rahman PKSM, Rahman TJ, Kahru A, Ivask A (2010) Effects of rhamnolipids from Pseudomonas aeruginosa DS10-129 on luminescent bacteria: toxicity and modulation of cadmium bioavailability. Microb Ecol 59:588–600

Bouyoucos GJ (1962) Hydrometer method improved for making particle size analysis of soils. Agron J 54:464–465

Brievik K, Sweetman A, Pacyna JM, Jones KC (2002) Towards a global historical emission inventory for selected PCB congeners- a mass balance approach 1. Global production and consumption. Sci Total Environ 290:181–198

Buchauer MJ (1973) Contamination of soil and vegetation near a zinc smelter by zinc, cadmium, copper and lead. Environ Sci Technol 7:131–135

CAC (2011) Working document for information and use in discussions related to contaminants and toxins in the GSCTFF (CF/5 INF/1), Codex Alimentarious Commission (Joint FAO/WHO Food Standards Programme Codex Committee On Contaminants In Foods) Fifth Session, The Hague, The Netherlands, 21–25 March 2011

Carson R (1962) Silent spring. Houghton Mifflin Company, New York

Chaney RL (1988) Effect of zinc smelter emissions on farms and garden of Palmerton, PA. Trace Subst Environ Health 22:263–280

Chary NS, Kamala CT, Raj DS (2008) Assessing risk of heavy metals from consuming food grown on sewage irrigated soils and food chain transfer. Ecotoxicol Environ Saf 69:513–524

Consonni D, Pesatori AC, Zocchetti C et al (2008) Mortality in a population exposed to dioxin after the Seveso, Italy, accident in 1976: 25 years of follow-up. Am J Epidemiol 167:847–858

Dakiky M, Khami A, Manassra A, Mereb M (2002) Selective adsorption of chromium(VI) in industrial wastewater using low-cost abundantly available adsorbents. Adv Environ Res 6:533–540

Das SK, Varma A (2011) Role of enzymes in maintaining soil health. In: Shukla G, Varma A (eds) Soil enzymology. Springer, Berlin/Heidelberg

Datta NP, Khera MS, Saini TR (1962) A rapid colorimetric procedure for the determination of organic carbon in soils. J Indian Soc Soil Sci 10:67–74

Dawson TE, Mambelli S, Plamboeck AH et al (2002) Stable isotopes in plant ecology. Annu Rev Ecol Evol Syst 33:507–559

DBT (2016) The prevention of food adulteration acts and rules (as on 01.10.2004). Indian Biosafety Rules & Regulations, Department of Biotechnology, Govt of India. http://dbtbiosafety.nic.in/act/PFA%20Acts%20and%20Rules.pdf. Accessed on 8 Oct 2016

De Marchi B, Funtowicz S, Ravetz J (1996) Seveso: a paradoxical classic disaster. In: Mitchell JK (ed) The long road to recovery: community responses to industrial disaster, United Nations University Press, Tokyo/New York/Paris. http://archive.unu.edu/unupress/unupbooks/uu21le/uu21le00.htm#Contents

DFID (2003) Heavy metal contamination of vegetables in Delhi, Executive summary of technical report. Department for International Development, London

Dick RP, Breakwell DP, Turco RF (1996) Soil enzyme activities and biodiversity measurements as integrative microbiological indicators. In: Doran JW, Jones AJ (eds) Methods of assessing soil quality. Soil Science Society of America, Madison, pp 247–271

Dilworth MJ (1966) Acetylene reduction by nitrogen fixing preparations from clostridium pasteurianum. Biochim Biophys Acta 127:285–294

Douay F, Roussel H, Pruvot C, Waterlot C (2008a) Impact of a smelter closedown on metal contents of wheat cultivated in the neighbourhood. Environ Sci Pollut Res 15:162–169

Douay F, Roussel H, Pruvot C, Loriette A, Fourrier H (2008b) Assessment of a remediation technique using the replacement of contaminated soils in kitchen gardens nearby a former lead smelter in northern France. Sci Total Environ 401:29–38

Douay F, Pelfrêne A, Planque J et al (2013) Assessment of potential health risk for inhabitants living near a former lead smelter. Part 1: Metal concentrations in soils, agricultural crops, and homegrown vegetables. Environ Monit Assess 185:3665–3680

Feigin A, Ravina I, Shalhevet J (1991) Irrigation with treated sewage effluent: management for environmental protection. Springer, Berlin, p 224

Fredan MAA (2011) Nitrogen fixing legumes in the plant communities. Am J Environ Sci 7:166–172

Friberg L, Piscator M, Nordberg GF, Kjellström T (1971) Cadmium in the environment. CRC Press, Cleveland, p 166

Ghosh AK, Bhatt MA, Agrawal HP (2012) Effect of long-term application of treated sewage water on heavy metal accumulation in vegetables grown in northern India. Environ Monit Assess 184:1025–1036

Godin P, Feinberg M, Ducauze C (1985) Modelling of soil contamination by airborne lead and cadmiumaround several emission sources. Environ Pollut 10:97–114

Guangwei H (2014) The soil pollution crisis in China: a cleanup presents daunting challenge. Yale Environment 360, Yale School of Forestry & Environmental Studies, Yale University. Available via http://e360.yale.edu/feature/the_soil_pollution_crisis_in_china_a_cleanup_presents_daunting_challenge/2786/. Accessed 15 Sept 2016

Gupta S, Satpati S, Nayek S, Garai D (2010) Effect of wastewater irrigation on vegetables in relation to bioaccumulation of heavy metals and biochemical changes. Environ Monit Assess 165:169–177

Hassanein RA, Hashem HA, El-Deep MH, Shouman A (2013) Soil contamination with heavy metals and its effect on growth, yield and physiological responses of vegetable crop plants (turnip and lettuce). J Stress Physiol Biochem 9:145–162

Houba VJG, Temminghoff EJM, Gaikhorst GA, van Vark W (2000) Soil analysis procedure using 0.01 M calcium chloride as extraction reagent. Commun Soil Sci Plant Anal 31:1299–1396

Huang JW, Chen J, Berti WR, Cunningham SD (1997) Phytoremediation of lead-contaminated soils: role of synthetic chelates in lead phytoextraction. Environ Sci Technol 31:800–805

IAEA (2006) Environmental consequences of the Chernobyl accident and their remediation: 20 years of experience, Report of the Chernobyl forum expert group 'environment'. International Atomic Energy Agency, Vienna

Jackson ML (1973) Soil chemical analysis. Prentice Hall, Englewood Cliffs

Jones DL (1998) Organic acids in the rhizosphere – a critical review. Plant Soil 205:25–44

Jordan MJ (1975) Effects of zinc smelter emissions and fire on chestnut-oak woodland. Ecology 56:78–91

Kar A, Galg BK, Singh MP, Kathju S (2009) Trends in arid zone research in India. Central Arid Zone Research Institute, Jodhpur, p 481

Khurana MPS, Singh J, Nayar VK (2004) Assessment of heavy metal contamination in soils and plants irrigated with sewage waters containing industrial effluents in district Amritsar, Punjab. Indian J Environ Ecoplann 8:221–280

Kundu S, Coumar MV, Saha JK et al (2012) Assessing soil health of vertisol of AESR 10.1 using selected physical, chemical and biological attributes of soils. J Indian Soc Soil Sci 60 (4):281–287

Leroyer A, Nisse C, Hemon D et al (2000) Environmental lead exposure in a population of children in northern France: factors affecting lead burden. Am J Ind Med 38:281–289

Leroyer A, Hemon D, Nisse C et al (2001) Determinants of cadmium burden levels in a population of children living in the vicinity of non-ferrous smelters. Environ Res 87:147–159

Lindsay WL, Norvell WA (1978) Development of a DTPA soil test for zinc, iron, manganese, and copper. Soil Sci Soc Am J 42:421–428

Lund W (1990) Speciation analysis – why and how? Fresenius J Anal Chem 337:557–564

Manzo C, Salvini R, Guastaldi E, Nicolardi V, Protano G (2013) Reflectance spectral analyses for the assessment of environmental pollution in the geothermal site of Mt. Amiata (Italy). Atmos Environ 79:650–665

Mayes WM, Burke IT, Gomes HI et al (2016) Advances in understanding environmental risks of red mud after the Ajka spill. Hung J Sustain Metall. doi:10.1007/s40831-016-0050-z

McBride MB (1991) Processes of heavy and transition metal sorption by soil minerals. In: Bolt GH et al (eds) Interactions at the soil colloid-soil solution interface. Kluwer Academic Publishers, Dordrecht, pp 149–174

McBride MB, Nibarger EA, Richards BK, Steenhuis T (2003) Trace metal accumulation by red clover grown on sewage sludge-amended soils and correlation to Mehlich 3 and calcium chloride extractable metals. Soil Sci 168:29–38

McBride MB, Pitiranggon M, Kim B (2009) A comparison of tests for extractable copper and zinc in metal-spiked and field-contaminated soil. Soil Sci 174:439–444

McLaughlin MJ, Zarcinas BA, Stevens DP, Cook N (2000) Soil testing for heavy metals. Commun Soil Sci Plant Anal 31:1661–1700

Meers E, Samson R, Tack FMG et al (2007) Phytoavailability assessment of heavy metals in soils by single extractions and accumulation in *Phaseolus vulgaris*. Environ Exp Bot 60:385–396

Menzies NW, Donn MJ, Kopittke PM (2007) Evaluation of extractants for estimation of the phytoavailable trace metals in soils. Environ Pollut 145:121–130

Mocarelli P, Marocchi A, Brambilla P et al (1991) Effects of dioxin exposure in humans at Seveso, Italy. In: Banbury report 35: biological basis for risk assessment of dioxin and related compounds. Cold Spring Harbor Laboratory Press, New York

Nagy AS, Szabó J, Vass I (2013) Trace metal and metalloid levels in surface water of Marcal River before and after the Ajka red mud spill, Hungary. Environ Sci Pollut Res Int 20:7603–7614

Nath K, Saini S, Sharma YK (2005) Chromium in tannery industry effluent and its effect on plant metabolism and growth. J Environ Biol 26:197–204

Nelson DW, Sommers LE (1996) Total carbon, organic carbon, and organic matter. In: Page AL et al (eds) Methods of soil analysis, part 2, 2nd edn. American Society of Agronomy Inc, Madison, pp 169–185

Olsen S, Cole C, Watanabe F, Dean L (1954) Estimation of available phosphorus in soils by extraction with sodium bicarbonate, USDA circular nr 939. US Government Print Office, Washington, DC

Panse VG (1947) Plot size in yield survey. Nature 15:820

Parker DR, Pedler JF (1997) Reevaluating the free-ion activity model of trace metal availability to higher plants. Plant Soil 196:223–238

Pelfrêne A, Douay F, Richard A, Roussel H, Girondelot B (2013) Assessment of potential health risk for inhabitants living near a former lead smelter. Part 2: Site-specific human health risk assessment of Cd and Pb contamination in kitchen gardens. Environ Monit Assess 185:2999–3012

Phillips AS, Yung TH, Bosela PA (2007) Love canal tragedy. J Perform Constr Facil 21:313–319

Poate D (1988) A review of methods for measuring crop production from smallholder producers. Exp Agric 24:1–14

Pruvot C, Douay F, Hervé F, Waterlot C (2006) Heavy metals in soil, crops and grass as a source of human exposure in the former mining areas. J Soils Sediments 6:215–220

Rolf AO, Bakken LR (1987) Viability of soil bacteria: optimization of plate-counting technique and comparison between total counts and plate counts within different size groups. Microb Ecol 13:59–74

Ruhl L, Vengosh A, Dwyer GS et al (2009) Survey of the potential environmental and health impacts in the immediate aftermath of the coal ash spill in Kingston, Tennessee. Environ Sci Technol 43:6326–6333

Ruhl L, Vengosh A, Dwyer GS, Hsu-kim H (2010) Supporting information: the environmental impacts of the coal ash spill in Kingston, Tennessee: an eighteen-month survey. Environ Sci Technol 44:9272–9278

Sachan S, Singh SK, Srivastava PC (2007) Buildup of heavy metals in soil water plant continuum as influenced by irrigation with contaminated effluent. J Environ Sci Eng 49:293–296

Saha JK, Sharma AK (2006) Impact of the use of polluted irrigation water on soil quality and crop productivity near Ratlam and Nagda industrial area. IISS Bulletin, Indian Institute of Soil Science, Bhopal, p 26

Schindler PW, Sposito G (1991) Surface complexation at (hydr)oxide surfaces. In: Bolt GH et al (eds) Interactions at the soil colloid-soil solution interface. Kluwer Academic Publishers, Dordrecht, pp 115–145

Sharma PD, Singh M, Ali M (2006) Recycling and utilization of urban and rural wastes for the welfare of the society. In: Proceeding Indian Society of Soil Science, ICSWEQ, p 387–401

Sharma RK, Agrawal M, Marshall FM (2009) Heavy metals in vegetables collected from production and market sites of a tropical urban area of India. Food Chem Toxicol 47:583–591

Smolders E, McLaughlin MJ (1996) Chloride increases cadmium uptake in Swiss chard in resin-buffered nutrient solution. Soil Sci Soc Am J 60:1443–1447

Soil Survey Division Staff (1993) Soil survey manual, Soil conservation service. U.S. Department of Agriculture handbook 18. Soil Survey Division Staff, Washington, DC

Srivastava MK (2015) Crop monitoring for improved food security. In: Proceedings of the expert meeting Vientiane, Lao People's Democratic Republic, 17 February 2014. Food and Agriculture Organization of the United Nations and Asian Development Bank, Bangkok

Srivastava PK, Gupta M, Mukherjee S (2012) Mapping spatial distribution of pollutants in groundwater of a tropical area of India using remote sensing and GIS. Appl Geomat 4:21–32

Sterckeman T, Douay F, Proix N, Fourrier H (2000) Vertical distribution of Cd, Pb and Zn in soils near smelters in the north of France. Environ Pollut 107:377–389

Sterckeman T, Douay F, Proix N, Fourrier H, Perdrix E (2002) Assessment of the contamination of cultivated soils by eighteen trace elements around smelters in the north of France. Water Air Soil Pollut 135:173–194

Strojan CL (1978) Forest leaf litter decomposition in vicinity of a zinc smelter. Oecologia 32:203–212

Subbiah BV, Asija GL (1956) A rapid procedure for the determination of available nitrogen in soils. Curr Sci 25:259–260

Tabatabai MA (1982) Soil enzymes. In: Page AL, Miller RH, Keeney DR (eds) Methods of soil analysis, Part 2, Chemical and microbiological properties, Monograph no, vol 9. ASA-SSSA, Madison, pp 903–947

Tan PM (1988) Palmerton zinc superfund site remediation strategy. Trace Sub Environ Health 22:296–305

Turrio-Baldassarri L, Alivernini S, Battistelli CL et al (2007) A study on PCB, PCDF/PCDF industrial contamination in an mixed urban/agricultural area significantly affecting the food chain and the human exposure Part I: Soil and feed. Chemosphere 67:1822–1830

Turrio-Baldassarri L, Abate V, Battistelli CL et al (2008) PCDF/PCDF and PCB in human serum of differently exposed population groups of an Italian city. Chemosphere 73:228–234

Turrio-Baldassarri L, Alivernini S, Carasi S et al (2009) PCB, PCDD and PCDF contamination of food of animal origin as the effect of soil pollution and the cause of human exposure in Brescia. Chemosphere 76(2):278–285

USEPA (1987) Palmerton zinc superfund site remedial investigation/feasibility study for the blue mountain project. USEPA, Philadelphia

Walkley A, Black IA (1934) An examination of Degtjareff method for determining soil organic matter and a proposed modification of the chromic acid titration method. Soil Sci 37:29–37

Yoshida F, Hata A, Tonegawa H (1999) Itai-Itai disease and the countermeasures against cadmium pollution by the Kamioka mine. Environ Econ Policy Stud 2:215–229

Chapter 4
Major Inorganic Pollutants Affecting Soil and Crop Quality

Abstract Inorganic pollutants are released into the environment due to activities of mining, industry, transportation and urban activities. Environmental risks associated with inorganic pollutants vary widely due to several complex interactions at both intracellular and extracellular levels. Salts of alkali metals degrade physical and chemical environment of rhizosphere soil affecting water and nutrient uptake as well as reduce rate of several nutrient cycling processes mediated by microbes with overall impact on nutrient use efficiency. Toxic heavy metals and metalloids interact quite strongly with soil constituents as compared to salts of alkali metals, rate of which however, depend on the element and their speciation. Although their mobility in soil is low, these elements disrupt biochemical processes in organisms even at low concentration affecting physiological activities. Toxicity to organisms and transformation reactions for inorganic pollutants with soil constituents and their consequences on transfer to organisms including plants are discussed in this chapter.

Keywords Environmental risk • Salts • Heavy metals • Plant • Microorganisms • Human health

The population growth forced rapid growth of industrial sector due to larger demand for consumer goods. Consequently it is also releasing huge amount of potentially toxic compounds into the ecosystem affecting quality of life and ecological functions. The pollutants being released in the environment can be categorized on the basis of sources, types and mode of toxicity. Although considerable knowledge has been generated, researchers are continuously engaged in finding out the toxic effects of various pollutants on human activities or organism's health and also on the environment as a whole. Among the inorganic pollutants, heavy metals are having the prime place and affected the system most. The most common listed toxic metal pollutants in soil are arsenic, cadmium, chromium, copper, zinc, mercury. These pollutants can enter into the soil system through poor agricultural management practices, improper industrial effluent discharge, landfill leakage, unauthorized dumping of household wastes, dry and wet deposition, volcanic eruption and industrial emissions.

J.K. Saha et al., *Soil Pollution - An Emerging Threat to Agriculture*, Environmental Chemistry for a Sustainable World 10, DOI 10.1007/978-981-10-4274-4_4

The impact of pollutants may be categorized into long- and short-term damage on environment, soil biota, plant and human system. The toxicity depends on the type of metals, concentration and receiving host and also varies from species to species.

4.1 Understanding Soil Contamination and Human Health

Introduction of toxic substances into the soil ultimately reach the organisms including human being in several ways: direct deposition or through irrigation water, manures and other soil amendments with considerable impact on ecosystem. Crops grown on contaminated soil lead to accumulation of toxic substances in food grain which indirectly enter the human physiological process to affect the human health severely (Fig. 4.1). Hence it is essential to focus on soil-plant-human continuum of toxic substances.

4.2 Risks Associated with Soil Pollution

Environmental risk assessment may be defined as the "process of estimating the potential impact of a chemical or physical agent on a specified ecological system under a specific set of conditions" (NHMRC 1992). In the context of human health, risk assessment has been defined as the "characterization of the potential adverse health effects of human exposures to environmental hazards" (NRC 1983). Soil pollution risk establishment is very necessary to avoid human health exposure to

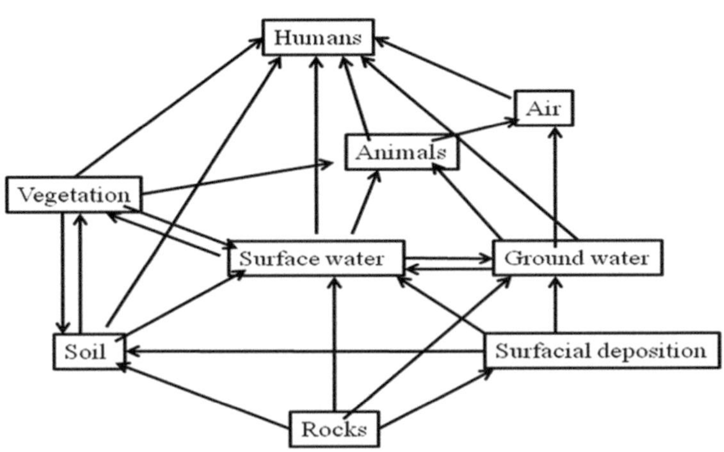

Fig. 4.1 Potential pathways for toxic elements to become concentrated in plants, soil, animals and humans (Adapted from Connor 1990)

many toxic substances in the soil. Although no specific established assessment method exists, the most common steps of risk assessment of contaminated site are the following (Markus and McBratney 2001):

(i) Identifying its specific purpose
(ii) Identifying critical receptors of the contaminants. Risk assessment should determine on whether the risk assessment is for the entire species of organisms or just the most sensitive species
(iii) Determining the level of acceptable risk
(iv) Exposure assessment to identify the pathways of exposure to individuals at risk and wherever possible, to quantify their individual contribution
(v) Toxicity assessment to determine the nature of adverse effects related to exposure and dose – response relationships for these effects
(vi) Risk characterization taking into account of all assumptions and uncertainties to produce a final estimate of the risk

Risk assessment in the polluted sites can be studied through single pollutant effect or combined effects. Ecotoxicological effect and biological availabilities are discussed in detail below:

4.2.1 Risks Related to One Soil Pollutant

Majority of the researches in this regards have been carried out in and around areas with high pollution activity and the survey information have been presented on single contaminant basis in the form of statistical parameters and spatial maps showing contaminant distribution. However, protocols for estimating the environmental or health risk of a particular contaminant are not uniform and well-defined. The USEPA adopted a modeling approach to quantify human exposure, which involve (i) a comprehensive analysis on the pathways of pollutant transfer to selected target organisms viz. humans, plants, animals and soil biota, (ii) identification of dose – response relationships for the receptors followed by an assessment of the likely harmful effects that metals may have on the target (iii) determination of the concentration of contaminant that would protect the most highly exposed individual from adverse effects (Ryan and Chaney 1997). Approach of environmental risk assessment of the Netherlands is based on the soil concentration at which half of the species in an ecosystem are potentially at risk (i.e., the Hazardous Concentration for 50% of species) (Theelen 1997).

Ordinarily, possible influence of nonchemical stressors on chemical toxicity is not considered during risk assessment and toxicity is assumed as an intrinsic property of the contaminant, where influence of receptor properties is reflected mainly through species-specific slopes of the dose-response curve.

4.2.2 Risks Associated with Multiple Pollutants in Soil

Consideration of multiple-stressor scenarios has been recognized as one of the steps to improve the method of assessment of toxicants and thereby reduce the uncertainty due to extrapolations inherent to ecological risk assessment procedure (Hope 2006). Current approaches to assess the risk of pollution to ecosystems are developed for single contaminant and determine its impact primarily as a function of properties of the contaminant. However, the cumulative impact of multiple contaminants and natural stressors may differ markedly from the impact of the single contaminant and can result in nonlinear effects with ecological surprises (Segner et al. 2014). So during the study of soil pollution, the combined effect of various pollutants/stressors should also be considered. Interactions of natural antagonists and pollutants have been shown to result in considerable impacts on ecologically relevant parameters. Screening for non-additive effects among many stress factors in simplified experimental designs by using the model of independent action has been suggested as a valuable strategy for an environmentally more realistic risk assessment of chemicals (Coors and De Meester 2008). Several researchers however, highlighted the problems in assessing multiple stressor effects as responses of biological receptors to stressors vary with biological scale, physiological and ecological traits, and receptor interactions and network organization (Munns 2006; Segner et al. 2014).

4.3 Inorganic Pollutants

4.3.1 Sodium, Salinity and Sodicity

Industries, particularly those associated with chlor-alkali, textiles, glass, rubber production, animal hide processing and leather tanning, metal processing, pharmaceuticals, oil and gas drilling, pigment manufacture, ceramic manufacture, soap and detergent production are the major consumer of salts (mainly NaCl) produced in the world today. Due to their high mobility in the soil, salt ions present in the industrial effluents when released into the environment, percolate through the soil profile and contaminate the groundwater. Such salt-loaded groundwater and surface water when used for irrigation to crops may lead to development of salinity/sodicity in soils of agricultural land.

At higher concentration in soil, Na causes damage to normal plant growth and such damage is also associated with high salinity and chloride. Thus, injury due to salinity and excess Na occur together (Bergmann 1992). The growth inhibition associated with high concentration of Na and salinity is often not attributed to specific toxic effect, but to the reduced water availability (physiological water stress). Moreover, when Na accounts for over 10–15% of the cation exchange capacity of clayey and loamy soil or over 20% in sandy soils, situation (known as sodicity) results in destruction of soil structure and consequently affects root growth due to compaction, poor permeability and aeration (Maliwal and Somani 2010). As soil pollution due to high salts affects plant growth (consequently, crop-cover and root anchorage), erodibility of land increases; and hence, accelerates soil erosion.

A review by Silva and Fay (2012) indicated detrimental effects of salinity on soil microbial community, as well as on their activities and genetic diversity. Considerable increase in N_2O production under anaerobic condition in both unfertilized and N-fertilized soil has been observed due to increase in salinity (Table 4.1) (Adviento-Borbe et al. 2006). Therefore, irrigation with salt loaded industrial/ municipal effluent for submerged rice cultivation may aggravate environmental pollution due to reduced N use efficiency and induced GHG emission. Activities of soil enzymes, responsible for mineralization of nutrients were found severely affected (Pathak and Rao 1998; Tripathi et al. 2007), thereby imparting additional dimension to the plant stress in the salt affected soil.

4.3.1.1 Risk Assessment Due to Excess Salts

Risk on crop growth due to salinity build-up in soil can easily be assessed through indirect measurement of electrical conductivity (EC) of soil solution, where electric current is directly proportional to ionic concentration. The quantity (ppm) of 'total dissolved solids (TDS)' in the form of soluble salts is approximately equal to "EC $\times$ 640", where EC is represented as mS cm^{-1}. Impact of salinity has been found related with soil texture and therefore, different salinity classes have been recommended for different textural classes for identifying levels of salinity (Table 4.2).

Remote sensing technique can identify salinity development in a shorter span of time on large area. Multi-temporal optical and microwave remote sensing can be successfully employed in detecting temporal changes of salt-related surface features. Airborne geophysics and ground-based electromagnetic induction meters, combined with ground data, have also shown potential for mapping depth of salinity occurrence (Metternicht and Zinck 2003).

4.3.1.2 Risk Assessment Due to Sodicity

Adverse effect on soil quality due to excess Na is generally assessed by its concentration in relation to the other competitive cations in either soil solution or on exchange sites. Relative concentration of Na on exchange sites (CEC) is termed as 'exchangeable sodium percentage (ESP)' and is expressed as:

$$ESP = \frac{Exchangeable\ Na}{CEC} \times 100$$

Table 4.1 Impact of soil salinity on microbial processes and N_2O emission

Process	$EC_{1:1}$ (dS m^{-1})	Decrease (−)/Increase (+)
Respiration	0.7 to 2.8	−17 to −47%
Decomposition	0.7 to 2.9	−2 to −25%
Nitrification	0.7 to 2.9	−10 to −37%
Denitrification	1.0 to 1.8	+32 to +88%
Anaerobic N_2O gas production (high nitrate)	0.02 to 2.8	+1500 to +31,500%
Anaerobic N_2O gas production (low nitrate)	0.5 to 2.0	+ 200 to +90,000%

Smith and Doran (1996) and Adviento-Borbe et al. (2006)

Table 4.2 Degree of salinity and their relationship with soil texture

	Level of salinity ($EC_{1:1}$, dS m^{-1})				
Texture	Non-saline	Slightly saline	Moderately saline	Strongly saline	Very saline
Coarse to loamy sand	≤1.1	1.2–2.4	2.5–4.4	4.5–8.9	≥9.0
Loamy fine sand to loam	≤1.2	1.3–2.4	2.5–4.7	4.8–9.4	≥9.5
Silt loam to clay loam	≤1.3	1.4–2.5	2.6–5.0	5.1–10.0	≥10.1
Silty clay loam to clay	≤1.4	1.5–2.8	2.9–5.7	5.8–11.4	≥11.5

Smith and Doran (1996)

where, both 'Exchangeable Na' and CEC are expressed as milliequivalent of Na^+ or H^+ per 100 g of soil.

Since the determination of CEC is time consuming and expensive, a practical and satisfactory correlation between the Sodium Adsorption Ratio (SAR) and ESP has been established by several researchers. The SAR is determined as:

$$SAR = \frac{[Na^+]}{\sqrt{\frac{([Ca^{2+}]+[Mg^{2+}])}{2}}}$$

Where $[Na^+]$, $[Ca^{2+}]$ and $[Mg^{2+}]$ are the concentrations (expressed as millimoles per liter) of the respective cations in soil solution. Several researchers have established linear regression model to predict ESP based on experimentally determined SAR values, for example:

$$ESP = \frac{100 \times (-0.0126 + 0.01475 \times SAR)}{1 + (-0.0126 + 0.01475 \times SAR)} \quad \text{...... Oster and Sposito (1980)}$$

$$ESP = 1.95 \times SAR + 1.8 \quad \text{...... Rengasamy and Churchman (1999)}$$

Considerable build-up of salinity and sodicity was observed on agricultural lands of several villages nearby an industrial area due to irrigation of crops with polluted

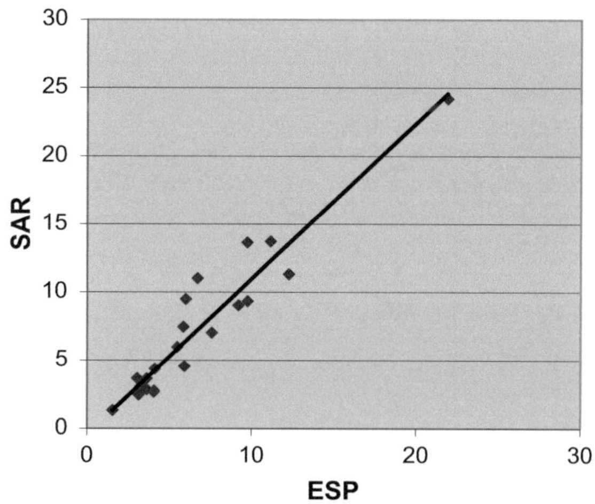

Fig. 4.2 Relationship between exchangeable sodium percentage (ESP) and sodium absorption ratio (SAR) of soil solution in polluted groundwater irrigated area

groundwater (pH 7.70; EC 2.84 mS cm^{-1}) (Saha and Sharma 2006). While SAR values were generally more in 15–30 cm layer, the salinity level was more in upper 15 cm layer. In that area with predominantly clayey soil, ESP was related to SAR of the soil solution (Fig. 4.2) according to the regression equation:

$$\text{ESP} = -0.2977 + 1.1621^{*}\text{SAR}\ (\text{r} = 0.96^{**})$$

Sodium increases dispersibility of soil clays and affects soil physical properties by breaking down aggregated structure, reducing permeability & porosity as well as putting the land more vulnerable to soil erosion. Such impact is more in soils predominated with montmorillonite clays than in soils with illite-vermiculite clays, and much more than in soils with kaolinite-sesquioxide clays, which are the least affected (Ayers and Westcot 1976). The threshold SAR and EC values for sandy soil were 16 and 4.0 dS m^{-1} respectively; while such threshold values for clays were 8 and 1.5 dS m^{-1} (Hanson et al. 1999). In a polluted area in India affected with sodicity, first few spells of rain increased pH of soil considerably (on an average by about 0.67 unit) due to replacement of Cl^- with HCO_3^- in the soil solution (Saha and Sharma 2006), which was further corroborated in the leaching experiment under laboratory condition (Fig. 4.3). However, change in pH due to leaching of Na-salts during monsoon also depends on the reserve pool of counter cations Ca and Mg as well as on the extent of sodicity in the soil (Saha 2005). While monsoon rainfall (971 mm) caused decrease in pH of soil previously irrigated with polluted groundwater, it resulted increase in pH of the soils previously irrigated with industrial effluent (Table 4.3).

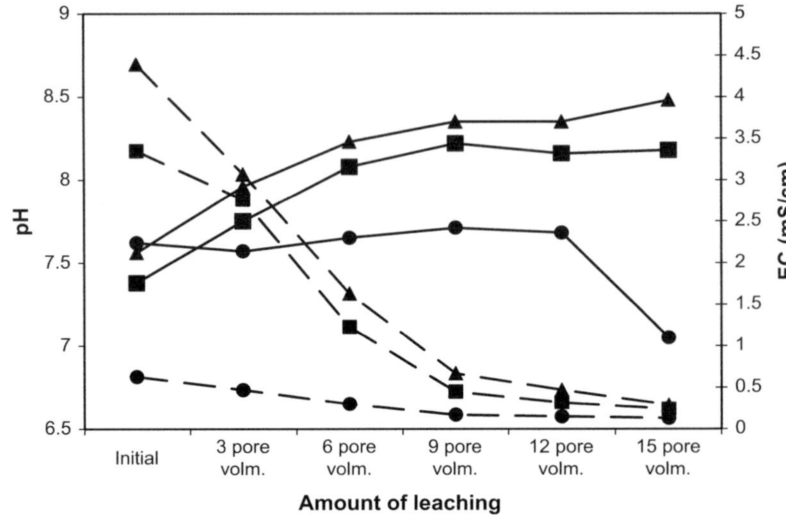

Fig. 4.3 Changes in pH and EC due to leaching of salts (*Solid lines* represent pH and *broken lines* represent EC values; '●' represents soil from unpolluted area, and '■' and '▲' represent soils from polluted area.)

Table 4.3 Changes in salinity and sodicity parameters of polluted soils during monsoon rain

	Soils previously irrigated with textile effluent		Soils previously irrigated with polluted groundwater	
	Pre-monsoon	Post-monsoon	Pre-monsoon	Post-monsoon
pH	7.90	8.11	8.37	7.94
EC (mS/cm)	4.44	0.52	0.84	0.64
SAR	24.03	1.71	6.43	1.91
ESP	31.18	2.28	6.96	2.20

4.3.2 *Heavy Metals and Metalloids*

Heavy metals are naturally present in the deeper part of the earth from where they are being mined out and released into the biosphere in the form of waste either during the manufacturing process or after human use of the manufactured products (Table 4.4). There are several industries and sources which release heavy metals in the environment, like mines, smelters, thermal power plant, metallurgical industries, electronics, textiles, phosphatic fertilizers, municipal solid wastes (MSW), sewage and industrial sludge etc. Burning of fossil fuel at stationery sources (as in coal burning at power plants) contributed maximum emissions of Cr, Hg, Ni, V, Se; vehicular traffic contributed maximum for Pb; copper production contributed maximum for As, Cd and Cu; and zinc production contributed maximum for Zn. After being emitted to the atmosphere, trace metals are subjected to transport within air masses and a large part of this gets deposited on the land mass not only

Table 4.4 Sources of hazardous elements

Metal	Industry
Chromium (Cr)	Mining, industrial coolants, chromium salts manufacturing, leather tanning
Lead (Pb)	Lead acid batteries, paints, e-waste, smelting operations, coal- based thermal power plants, ceramics, bangle industry
Mercury (Hg)	Chlor-alkali plants, thermal power plants, fluorescent lamps, hospital waste (damaged thermometers, barometers, sphygmomanometers), electrical appliances
Arsenic (As)	Geogenic/natural processes, smelting operations, thermal power plants, fuel burning
Copper (Cu)	Mining, electroplating, smelting operations,vanadium (Va) spent catalyst, sulphuric acid plant
Nickel (Ni)	Smelting operations, thermal power plants, battery industry
Cadmium (Cd)	Zinc smelting, waste batteries, e-waste, paint sludge, incinerations and fuel Combustion
Zinc (Zn)	Smelting, electroplating

Adapted from CPCB (2010)

around the source but also in area far away, even crossing boundaries of the country/ continent (Pacyna and Pacyna 2001). Significant part of the metal loaded effluents, generated particularly from small scale industries in developing and under-developed countries are released untreated into land and water bodies. In most of cases, metals are present in dilute and small quantities in polluted water bodies and may not cause any harm to plant growth immediately when used for irrigation. However, their immobility and consequent persistence imply that concentrations may become elevated in the long run to such an extent that they begin exhibiting toxic effect on plant, soil microorganisms and food chain. Compost is most commonly used organic amendment for improving soil productivity. However, such material prepared from MSW and sewage-sludge contains several harmful heavy metals in high concentrations (Sullivan and Miller 2001; Saha et al. 2010a). Also, some of the metals are impurities/constituents of extensively used agrochemicals like fertilizers (e.g. Cd in phosphatic fertilizer), pesticides (e.g., Zn, Cu, Sn, Hg, organic pollutants) etc. and contaminate the rhizosphere when these are used in intensive agriculture. Long-term exposure to heavy metals has been reported to affect human and animal health adversely (ATSDR 2005). Among the heavy metals, Ni, Co, Cr and Cu are relatively more toxic to plants and As, Cd, Pb and Hg are relatively more toxic to higher animals (McBride 1994).

The presence of heavy metals in soil affects the edaphological environment in several ways like contamination of food as well as a decrease in crop productivity (Rooney et al. 2007) and diminution of soil microbial activity (Chaudri et al. 1993; McGrath 1993; Akerblom et al. 2007). Heavy metals have a strong affinity for chelating substances, including the enzymes implicated in cellular metabolism and consequently by blocking these enzymes, they cause adverse physiological changes leading to even death of cell. These were found to affect soil microbial activity much earlier than their adverse effect on plant growth were expressed (Saha et al. 2013). In long-term experiments, adverse effects on soil microbial biomass, as well as on symbiotic N_2-fixation by blue-green algae and *Rhizobium* were observed due to heavy metals accumulation from sludge application (Brookes and McGrath 1984; Brookes et al. 1986; Martensson and Witter 1990). As soil microorganisms play important role related to plant growth like nutrient mineralization from organic matter, native P solubilization, N_2 fixation from atmosphere, indirect adverse effect of heavy metal accumulation on plant growth can be inferred. A comprehensive review indicated inhibitory effect of Cr on uptake of several essential nutrients by plants resulting in their deficiency and crop growth (Shanker et al. 2005).

Contamination of food with heavy metals is another serious issue in management of polluted soils. Transfer of metals to edible part of plant depends on their concentration in soil, interaction with soil constituents and ability of root to exclude or inactivate metals and their translocation from root to edible parts. Fine-texture, higher pH, organic matter and CEC favoured lower transfer coefficients for Cd, Cr and Hg in flowering Chinese cabbage, resulting from their high sorption capacities (Liu et al. 2007). Due to higher pH and clay content, Vertisol had higher capacity to absorb Cd and Pb and as a result, expressed toxicity to soybean crop at higher levels of contamination as compared to Inceptisol and Alfisol (Adhikari et al. 2012).

4.3.2.1 Behaviour of Some of the Potentially Toxic Heavy Metals, Metalloids and Selenium in Soil and Their Impact on Plants

Cadmium

Chemical properties of Cd and Zn are quite similar because both of these belong to same group (IIb) in the periodic table. Cadmium is thus able to mimic the behavior of the essential element for Zn in its uptake and metabolic functions. This element is universally acknowledged to be a dangerous heavy metal due to its toxicity in human, animal and plant even at low concentrations. This is partly due to the ease with which it is taken up by plants but also to the relative speed with which it is translocated to the above ground parts (Table 4.5). Relative mobility of Cd is higher than that of Pb or Zn.

Transfer coefficient is calculated as the ratio of the concentration of a metal in a plant to the total concentration of that metal in soil. A higher transfer coefficient indicates a greater mobility of metal from soil into plant. Transfer coefficient for Pb in uncontaminated soil was about 2.5–4.4 times higher compared to Pb treated soil which ranged from 0.06 to 0.11 (Saha et al. 2010b). In contrast, transfer coefficients for Cd in treated soils were 1.1–1.7 times higher compared to control, which

Table 4.5 Transfer coefficients (ratio of concentration in plant to soil) according to Sauerbeck (1986)

Transfer coefficients	1–10	0.1–10	0.1–1	0.01–0.1
Heavy metal	Cd, Ti, Zn, B	Mo, Se	Cu, Ni, V	Co, Cr, Hg, Pb, Sn, As, Be, F

indicates that applied Cd was either equally or more available to plant in comparison to native form. Very low values of transfer coefficient for Pb indicate that soil-plant barrier for its translocation is very high as compared to Cd. Cadmium, which has lower affinity for metal sorbing soil constituents (i.e. oxides, humified natural organic matter) has the greatest potential for transmission to food chain in levels that pose risk to consumers (Chaney and Ryan 1994; Chaney et al. 1999).

In nearly all the publications on the subject, the soil pH is listed as the major soil factor controlling both total and relative uptake of Cd. Cadmium is more mobile in acidic soils within the range of pH 4.5–5.5; whereas in neutral to alkaline soil, it is rather immobile. At pH < 6, its high mobility/availability is attributable to the fact that Cd^{+2} is adsorbed relatively weakly on major soil particles like organic matter, silicate clays and iron-aluminium oxides. At the similar level of soil cadmium, its availability is higher at low pH and lower contents of clay, organic matter and oxides. Above pH 7, Cd^{+2} is co-precipitate with $CaCO_3$ or precipitate as $CdCO_3$ with consequent decline in solubility/availability. For this reason, liming acid soils is recommended as ameliorant in Cd contaminated soils. However, as the pH is increased in the alkaline range, monovalent hydroxy ion species are likely to occur (e.g., $CdOH^+$), which could not easily occupy the sites on cationic exchanges complexes increasing its mobility. Geogenous cadmium and that accumulated in the soil over very long periods of time are absorbed by the plant more slowly than recent inputs.

Although content of Cd in natural soil is reported worldwide to vary in the range 0.06–1.1 $\mu g\ g^{-1}$, its concentration exceeding 0.5 $\mu g\ g^{-1}$ in soils is suspected to be contaminated from anthropogenic sources. Plant species vary greatly in their ability to translocate Cd at a given soil concentration, like: lettuce> spinach, tomato, carrot > rape, cabbage, radish> maize, mustard> sunflowers, bean, pea, cucumber> wheat, oat. The symptoms of cadmium toxicity in plants include inhibited growth with or without chlorotic blotches.

Chromium

Worldwide soil Cr in the unpolluted area range from 7 to 221; although those derived from serpentine may contain several per cent Cr. Chromium in soils exists in two oxidation states: in +3 oxidation state as the Cr^{+3} cation, and in +6 oxidation state as chromate anion ($CrO_4{}^{-2}$). Normal soil conditions generally favour Cr^{+3} form which is highly immobile and hence, unavailable to plants due to strong complexation/chemisorptions with soil organic matter, oxides, silicates even under acidic condition as well as due to formation of relatively insoluble compounds such as chromite $Fe\text{-}Cr_2O_4$. The concentrations of soluble Cr(III) in equilibrium with insoluble compounds are < 10^{-9} M (0.05 ppb) in water at pH 6 to less than 10^{-15} M at pH 8 (Richard and Bourg 1991). Chromates (Cr^{+6}) are soluble and highly toxic to organisms; and are only stable under alkaline condition in the soil. When hexavalent chromate ion enters in to the soil along with polluted irrigation water (e.g. tannery effluent), most or all of these are reduced to Cr^{+3} quite rapidly (within 1 or 2 days) especially under acidic condition and in presence organic matter. Due to such transformation of soluble Cr(VI) into insoluble Cr(III) compounds, plants do not accumulate Cr even when it is present in the soil in high concentrations and tolerate relatively high levels of Cr applied to the soil through wastewaters of sludge. Soil organic matter plays important role in minimizing risk associated with Cr contamination. Further, availability of Cr decreases rapidly with increasing pH value. Chromium applied to the soil surface is usually absorbed in the top 5–10 cm. At higher pH, a small fraction of Cr^{+3} may be oxidized to soluble hexavalent CrO_4^{-2} and such oxidation is promoted by Mn oxides. The oxidation of Cr(III) to Cr(VI) is limited by the concentration of water-soluble chromium, pH, initial available surface area, and the ionic strength (Apte et al. 2005). A large portion of chromium in soil will not be oxidized to Cr(VI) even in the presence of manganese oxides and favorable pH conditions, due to the unavailability of mobile Cr(III).

Cultivated crop plants can be placed in the following order of decreasing chromium (VI) uptake: cauliflower > beet root > radish > carrot > tomato > maize > lettuce > barley > moong bean. Chromium toxicity manifests itself in plants, as it greatly inhibits growth and photosynthesis, and develops chlorosis, small brownish-red or purplish leaves and necrotic lesions. Root growth is greatly inhibited by high chromium concentrations.

Lead

Worldwide mean total lead content of uncontaminated soils varies between 10 and 84 ppm, but more commonly below 50 ppm. Soils other than pure sandy soil have a high fixing capacity of lead which increases with increase in clay and humus content owing to strong affinity of Pb for minerals (−OH groups of silicates and hydroxides) and organics (−OH and – COOH). Organic matter content determines the immediate relative metal retention while iron oxides determine the redistribution of metals with time (Diagboya et al. 2015). About 60–80% of dissolved Pb exists as organic complex, while the proportion may reach to 80–100% at near-neutral pH (Sauvé et al. 1997). Free Pb^{+2} concentration in soil solution depends mainly on its total content and pH. Under oxidizing conditions, Pb^{+2} becomes less soluble as pH is increased. In alkaline soils, however, its mobility may increase by the formation of soluble Pb-organic and Pb-hydroxy complexes. Being a strong chalcophile, Pb solubility is extremely reduced under reducing condition due to its precipitation as sulphide.

Plants transfer little Pb^{+2} from soil to aboveground parts; enriching it at the root surface and immobilizing in the cell wall of the root. Therefore, very high concentrations of Pb in the soil are necessary to exhibit toxicity. Soil factors such as high cation exchange capacity, alkaline pH, high organic matter and P-content in the soil antagonize its uptake by plants. It implies that if wastes rich in phosphorus and organic matter (such as sewage waters and sludge) are applied to the soil, very little hazards due to Pb are expected. However at very high Pb concentration in soil under favourable chemical condition (low pH, light textured, less SOC and oxides), plant can accumulate this metal in the above ground parts. The main consequences of Pbtoxicity in plants are disruption of Ca metabolism, inactivation of various enzymes, reduced CO_2 assimilation and consequently suppression of photosynthesis.

Nickel
Soils throughout the world contain Ni within wide range from 4 to 55 mg kg^{-1}, depending upon type of parent material and soil forming processes. Ni behaves largely like Zn in the soil-plant system, but it forms stronger chelates with soil organic matter, thereby showing resemblance to Cu. Having similar electronic configuration and electronegativity, Ni^{+2} forms stable complexes with organic matter similar to Cu^{+2}. Therefore, this metal occurs mainly in organically bound forms in surface soil horizon, a part of which may exist as soluble chelates. High soil organic matter level in Ni contaminated soil may mobilize it through formation of soluble complex, particularly at higher pH. Fraction of soil Ni occluded in Fe and Mn oxides also seems to be available to plants. Generally, solubility of Ni is inversely related to soil pH and also to its sorption on Fe & Mn oxides which is generally pH dependent. The metal does not play essential role in plant metabolism; although evidence of its beneficial effects exists. Nickel is readily taken up by plants from soil and its absorption is positively correlated with its concentration in soil until certain Ni concentration in plant tissues is reached. Its content in the range of 50–100 mg kg^{-1} within plant (on dry weight basis) is indicative of its toxicity in plants. Restricted growth of plants and chlorosis in leaves are common symptom of its toxicity. Possibility of Ni-toxicity to plants cannot be ruled out when industrial or municipal wastes with high Ni concentrations are applied to agriculture lands. Nevertheless, like Zn and Cu, phototoxicity of Ni appears to provide an effective barrier against Ni toxicity to human population and animals.

Mercury
Mercury and its compounds are among the strongest toxic substances and are also extremely dangerous to humans and animals. Its concentration in world soils of uncontaminated region is very low and usually ranges from 0.02 to 0.4 ppm. In normal rhizospheric environment, it exists in soil solution mainly as Hg^{+2}, though its reduced form (Hg^{+}) may also be present at a very low concentration under reduced condition. Even though the retention and accumulation of Hg in soils depends highly on its chemical form, the mobility may also be controlled by the transport of the mobile phase (soil water) and the soil properties (pH, CEC, mineral composition, soil texture, organic matter, etc.). Mercury ions can be retained by organic matter, oxides of iron-aluminium, and clay minerals; and may precipitate with carbonate and hydroxide to form a stable solid phase in alkaline soils. As mercury is strongly absorbed in the soils especially by organic matter, its uptake is somewhat less pH dependent in comparison to other heavy metals. Mercury may be reduced to its elemental form (Hg^{0}) by both chemical and biological processes, which is volatile and extremely toxic to organisms. Methylmercury ($[CH_3Hg]^{+}$ or $[CH_3Hg]^{0}$), another toxic compound, may be formed by anaerobic microorganisms under reduced condition. However, this process is highly competitive with the chemical transformation process involving precipitation as HgS, which makes prediction on its mobility/toxicity difficult. Strong interaction between humic acid

and Hg^0 through thiolate ligand-induced oxidative complexation has also been observed under anaerobic environment (Gu et al. 2011). Oxidized Hg^{+2} may be effectively reduced to Hg^0 in the presence of reduced humic acid (having more proton reactive site); whereas production of Hg^0 is inhibited by complexation as humic acid concentration increases. This dual role played by dissolved organic matter in the reduction and complexation of mercury is expected to significantly influence the mercury species transformations and its biological uptake that leads to the formation of toxic methylmercury. In aquatic environments, inorganic mercury is microbiologically transformed into methylmercury, which makes it more prone to biomagnification in food chains. Consequently, populations with traditionally high dietary intake of food originating from fresh or marine environment have highest dietary exposure to this highly toxic element.

Mercury uptake by plants is little because of its very low solubility in soil due to reasons described above. In the uncontaminated environment, Hg-content of plants tissues seldom exceeds 500 parts per billion (ppb). In naturally contaminated areas, its level can be as high as 3500 ppb. Mercurial fungicides constitute an important source of Hg for crop plants. Its content in the above-ground plant parts is generally low despite Hg seed treatment or its addition to soil. In general, Hg hazards are much less in intensive agriculture in comparison to aquatic food chain.

Copper

Average concentrations of Cu in unpolluted soils vary from 6 to 80 mg kg^{-1}, depending on parent materials, topography, climatic conditions and soil properties. It exists in soils almost exclusively as Cu^{+2}. Copper ions are held very tightly on both organic and inorganic exchange sites. The process controlling fixation by soil constituents are related to adsorption, occlusion, organic chelation and complexing and microbial fixation. All soil minerals are capable of absorbing Cu ions from solution and these properties depend on the surface charge carried by the adsorbents. The surface charge is strongly controlled by pH, and therefore, the absorption of Cu ions species can be presented as a function of pH. The greatest amount of adsorbed Cu had always been found for Fe and Mn oxides (hematite, geothite, birnessite), amorphous Fe and Al hydroxides, and clays (montmorillonite, vermiculite, imogolite). In reduced soil (as under paddy cultivation), Cu has very low mobility due to formation of insoluble sulfides, Cu_2S and CuS. Chelation and complexation are the key reactions governing Cu behavior in most soils. Many kinds of organic substances form from both soluble and insoluble complexes with Cu; and thus Cu-binding capacities of soils and Cu solubility are highly dependent on the kind and amount of organic matter in soils. Although Cu is one of the least mobile heavy metals in soil, this metal is abundant in soil solutions of all types of soils. Overall solubility of both cationic and anionic forms of Cu decreases at about pH 7–8. However, the most common forms of Cu in soil solution are soluble organic chelates of this metal. Although very little is known of the kinds of soluble organic Cu forms, about 80% of the soluble Cu forms have been estimated to be as organic chelates. Although there is increasing evidence of the active absorption of Cu by plants, passive absorption is likely to occur, especially in the toxic range of this metal in solutions.

Toxic metal pollution affects the composition and metal tolerance of soil bacterial communities. In highly contaminated soils with $CuSO_4$ of industrial origin about 85 years ago, the relative abundances of members of several phyla or candidate phyla, including the *Proteobacteria*, *Bacteroidetes*, *Verrumicrobia*, *Chloroflexi*, *WS3*, and *Planctomycetes* decreased with increasing bioavailable Cu, while members of the dominant phylum, the *Actinobacteria*, showed no response and members of the *Acidobacteria* showed a marked increase in abundance (Berg et al. 2012). Pollution-induced community tolerance was significantly correlated to bioavailable Cu, as determined by the results seen with a Cu-specific bioluminescent biosensor strain, demonstrating a specific community response to Cu.

Zinc

Worldwide mean Zn content in soils varies between 17 and 125 mg kg^{-1}. During the last two decades, Zn as a micronutrient is being used indiscriminately in agricultural and husbandry practices. The most common and mobile Zn in soil is Zn^{2+}; but several other ionic species may also occur in soils. In aerobic soils having acidic pH, Zn had medium mobility as it is predominantly held in exchange sites of clays and organic matter. Unlike Cu, it does not complex tightly with soil humus in acidic condition. Chemisorption on sesquioxides and clays and complexation with soil organic matter markedly reduce Zn mobility at neutral and higher pH. Formation of soluble Zn-organic complex and Zn-hydroxy anion [$Zn(OH)_4{}^{-2}$] at alkaline pH may increase its solubility. Under reducing environment (as in flooded rice cultivation), release of Zn^{+2} from dissolving iron and manganese oxides may temporarily increase its solubility; but may subsequently get precipitated as extremely insoluble ZnS.

Soluble forms of Zn are readily available to plants and uptake of Zn has been reported to be linear with concentration in the nutrient solution and in soils. The rate of Zn absorption differs greatly among plant species and growth media. Even though Zn is an essential element in our diet, but its too much intake can be harmful.

Selenium

Soil selenium is normally present in the range between 0.05 and 1.25 mg kg^{-1}, in various oxidation states. Selenites ($SeO_3{}^{-2}$) predominate under aerobic acidic soil, which have lower solubility due to strong chemisorptions by oxides and aluminosilicates. On the other side in aerobic neutral to alkaline soil, it exists mainly as selenates ($SeO_4{}^{-2}$), which bind weakly to soil minerals rendering these with higher mobility. Under reduced condition, microbially mediated methylation of Se may form volatile compounds making these more mobile. Selenium is essential for animals and probably for people, but not for plants. However, concentration of Se less than 4–5 mg kg^{-1} in the food is often suggested as the amount that will avoid Se-toxicity. Se in wastewater and sludge is generally present as elemental Se, selenities bound to hydrous iron oxides, or trimethylselenonium salts. These forms when applied to soil are not easily available to plants. Thus, chances for the occurrence of Se-hazards through the application of waste water or sludge containing Se are low. Some pockets of seleniferous soils have been identified in northwestern India.

Arsenic

Arsenic content ranges from 2.2 to 25 mg kg^{-1} in the soils of unpolluted area. In general, soils overlying sulphide ore deposits or derived from shales and granites and those surrounding geothermal activity, have high arsenic contents. Release from As-enriched minerals is the primary geogenic source of As in the environment.

Trivalent arsenite [As(III)] and pentavalent arsenate [As(V)] are the most predominant forms in soil environment. Under reducing condition, arsenites are predominantly found; while arsenates are stable in aerobic soils. Arsenates behave similar to phosphates and undergo various immobilization processes such as sorption to solid phases such as Fe-, Al-, and Mn- (hydro)oxides, and to a smaller extent by clay minerals (particularly kaolinite). Arsenic may also bind to organic matter in soils, with arsenate and arsenite having maximum adsorption on humic acids at pH 5.5 and 8.0, respectively; though contribution of organic matter is less compared to oxides and aluminosilicates in this process. Multivalent cations like Fe, Al and Mn, increases arsenic adsorption by humic substances by bridging complexes for arsenate on humic acids. Adsorption affinity is higher for As(V) at lower pH values and for As(III) at higher pH values. As a result, arsenate mobility in aerobic soils is quite low in acidic condition with high clay and oxides content (as in red and lateritic soils). Arsenic solubility increases under reduced environment due to dissolution of Fe and Mn oxides/hydroxides and solubilized As(V) is further reduced to As(III) compounds.

Certain higher eukaryotic organisms (e.g., fungi) along with specific archaea and aerobic eubacteria can convert As(V) to As(III), which is then methylated to form less toxic species, such as monomethylarsonous acid (MMAA), dimethylarsenic acid (DMA) or trimethylarsine (TMA). However, methylated species are usually not abundant in aqueous solutions compared to inorganic forms of arsenic. Generally, inorganic As species are more toxic and mobile than organo-arsenic species, and As(III) is more toxic and mobile than As(V). Soluble As species are mostly taken up by plants and organisms and are also responsible for groundwater contamination. Uptake of different forms of As by rice plants follows the order: As (III)>As(V)>MMAA>DMAA. Upon absorption, inorganic species and MMAA are mainly accumulated in roots; while DMAA is readily translocated to the shoots (Carbonell et al. 1998).

Arsenic is phytotoxic as well as a proven carcinogen in human causing skin cancer and other internal cancers. As a result, concern on the toxicity of As to humans and other living organisms is increasing. Different crop plants vary in their tendency to accumulate As from the soil. Lowland rice seems to be especially susceptible to As-toxicity.

4.3.2.2 Phytotoxicity of Heavy Metals

Although some heavy metals (Cu, Fe, Mn, Mo, Zn, Co and Ni) are either essential or beneficial to plant, these are phytotoxic at high concentration. Other heavy metals (like Cd, Cr, Hg, Pb) do not have any role in plant physiology and are toxic at concentrations beyond certain critical level. Some plants are able to regulate heavy metal uptake, while others are able to tolerate high levels in their tissue. As soon as their critical concentration within plant tissue is exceeded, heavy metals can adversely affect metabolic processes directly or indirectly (through substitution of another essential metal in metalloproteins). Further, toxicity of heavy metals varies greatly from element to element and from species to species of plants. Some of well established phytotoxic manifestations include generation of reactive oxygen species and reactive nitrogen species as well as replacement of enzyme cofactors and transcription factors, inhibitions of antioxidative enzymes, cellular redox imbalance, ionic transport imbalance, DNA damage and protein oxidation (Sytar et al. 2013). Genetic diversity among Bermuda grass populations from Cd contaminated sites was found higher than those from populations from the neighbouring heavy metal free areas Hunan province of China, which was attributed to a large number of tolerant individuals, successive colonization events, pollen flow from adjacent normal populations and human disturbance (Xie et al. 2016). However several other researchers reported either no change or decrease in genetic diversity among plant species in heavy metal polluted sites (Mengoni et al. 2000; Deng et al. 2007).

Induced Fe deficiency chlorosis due to decreased uptake of Fe and its translocation to leaves is most commonly observed impact of heavy metal pollution on plant; and severity of chlorosis has been found more with Cu and Ni as compared to other metals (Bergmann 1992; Pandey and Sharma 2002). Cadmium has been shown to interfere with the uptake, transport and use of several essential elements (Ca, Mg, P and K) and water by plants (Das et al. 1997). Roots are damaged first, where precipitation of proteins in the root and root hairs may lead to its complete death. Although a great variation between plant species exists, critical leaf tissue concentrations affecting growth in most species usually are observed in the range of 200–300 $\mu g\ g^{-1}$ for Zn, 15–20 $\mu g\ g^{-1}$ for Cu and 8–12 $\mu g\ g^{-1}$ for Cd on dry weight basis which indicates Cd is most phytotoxic followed by Cu (Påhlsson 1989). Exposure to each Co^{2+}, Ni^{2+} and Cd^{2+} decreased chlorophyll content ($Ni^{2+} > Cd^{2+} > Co^{2+}$), with concomitant decrease in the activities of the Fe enzymes – catalase and peroxidase, suggesting reduced availability of Fe for chlorophyll–heme biosynthesis (Pandey and Sharma 2002).

At a particular given ecological condition, each metal has its own phytotoxicity potential. Several studies demonstrated and reported the phytotoxicity sequence of the heavy metals based on the effect of the heavy metals on plant growth, yield, plant physiology, etc. Copper exerted the most adverse effects on seed germination and early growth of crop seedlings followed by Zn and Pb (Mahmood et al. 2007). Wheat and rice seedlings were more susceptible to metal toxicity than barley. Metal

Fig. 4.4 (*top-left*) Spinach crop with no visible symptom when grown with 0, 50 and 100 mg Cu kg^{-1}; (*top-right*) showing retarded growth and toxicity symptoms with application of 400, 500 and 600 mg Cu kg^{-1}; (*bottom-left*) showing chlorosis of leaf at 400 mg Cu kg^{-1}; (*bottom-right*) showing necrosis of leaf blade at 600 mg Cu kg^{-1}

toxicity to new root primordia in *Salix viminalis* followed the order Cd> Cr> Pb, whereas root length was affected more by Cr than by other heavy metals such as Cd and Pb (Prasad et al. 2001). In another experiment, Cd was the most toxic metal to affect plant growth and seed yield of maize (*Zea mays* L.) followed by Co, Hg, Mn, Pb, and Cr (Abdul Ghani 2010). Toxicity of heavy metals when assessed using root elongation of rye grass (*Lolium perenne*) followed the sequence; Cu> Ni>Mn> Ni>Pb> Cd> Zn> Al> Hg> Cr > Fe (Wong and Bradshaw 1982). Heavy metals affected root growth and photosynthetic pigments production of white mustard (*Sinapis alba* L.) and the order of toxicity followed Cu > Se > Cd > Zn >>Pb (Fargasova 2004). It can be inferred from above instances of phytotoxicity experiments that Cu and Cd are, by and large more toxic to plants while Pb exhibit less toxicity. In an experiment on acidic sandy loam soil, order of toxicity of heavy metals on growth of spinach (*Spinacia oleracea* L.) were observed as: Cd > Ni > Cr > Cu > Zn and the soil metal levels corresponding to 80% relative yields were found to be 17.8 mg kg^{-1} Cd, 176.4 mg kg^{-1} Cr, 409 mg kg^{-1} Cu, 153 mg kg^{-1} Ni, and 852 mg kg^{-1} Zn (Saha et al. 2013). Spiking the soil with Pb up to 150 mg kg^{-1} did not have any significant effect on the above-ground biomass of the plant.

Table 4.6 Major Impacts of different pollutants on plants in brief

Pollutant	Its impact on plant
As	Red brown necrotic spots on old leaves, yellow browning of roots, growth reduction
Cd	Restricted growth, brown margin to leaves, chlorosis: chlorotic patches in older leaves and severe chlorosis in younger leaves at high level of toxicity, necrosis, curved leaves, brown stunted roots, reddish veins and petioles, reduction in growth, purple coloration
Cr	Affect seed emergence, stunted plant growth and decreased dry matter production
Pb	Dark green leaves, stunted foliage, increased amount of shoots
Ni	Chlorosis similar to Fe deficiency, necrosis, stunting, inhibition of root growth, decrease in leaf area
Hg	Severe stunting of seedlings and roots, chlorosis, browning of leaf tips, reduction in growth, tumor-like proliferations in deformed carrots
Cu	Chlorosis, yellow coloration, purple colouration of the lower side of the midrib, less branched roots, inhibition of root growth
Se	Interveinal chlorosis, black spots,bleaching and yellowing of young leaves, pink spots on root

Among the heavy metals investigated, only excess Cu levels showed visible toxicity symptom on spinach (Fig. 4.4) (Table. 4.6).

4.3.3 Heavy Metals Toxicity in Soil Microorganisms

Heavy metals exhibit toxicity to soil microorganisms, and consequently affect different microbially mediated soil processes of agricultural importance. Manifestation of toxicity is mostly observed through inhibition of growth, morphology and activities of various groups of microorganisms including symbiotic N_2 fixers like *R. leguminosarum, Mesorhizobium ciceri, Rhizobium* sp. and *Bradyrhizobium* sp. (Vigna) and Sinorhizobium (McGrath et al. 1988; Khan and Scullion 2002; Shi et al. 2002; Bondarenko et al. 2010; Arora et al. 2010; Bianucci et al. 2011). Biological nitrogen fixation is severely affected in contaminated soils due to expression of heavy metals toxicity by Rhizobia (Barajas-Aceves and Dendooven 2001; Hernandez et al. 2003). In heavy metal polluted sites, the frequency of tolerant microbes increased with a rise in heavy metal concentrations (Xie et al. 2014). Using Biolog-Ecoplate method, functional diversity of soil microorganism was analyzed in Cd contaminated or uncontaminated soils from Hunan province of China (Xie et al. 2016). The study indicated that total bioactivity, richness, and microbial diversity decreased with increasing concentration of heavy metal.

Nutrient cycling in the soil involves biochemical, chemical, and physicochemical reactions, with biochemical processes being mediated mainly by microorganisms. All the biochemical reactions are catalyzed by enzymes (Tabatabai 1982), and depending on their nature and level of contamination, many heavy metals have been found to impede soil enzyme activities to different extents (Tyler 1981; Kuperman and Carreiro 1997; Welp 1999; Yang et al. 2006). Soil enzyme activities are reflection of collective metabolism of the soil microbial community and therefore, reflect the ecological health of terrestrial ecosystems under the influence of soil heavy metal pollution. Soil enzyme activities such as those of arylsulfatase, β-glucosidase, acid and alkaline phosphatase, protease, urease, cellulase, dehydrogenase, protease, and invertase are sensitive to the presence of heavy metals (Kunito et al. 2001; Effron et al. 2004; Oliveira and Pampulha 2006; Wang et al. 2008).

The activities of soil dehydrogenases, proteases, lipases, and esterases were seen to decrease significantly with increasing levels of all the heavy metals in the soil (Saha et al. 2013). The activity of acid phosphatases was, however, not affected significantly by Pb, whereas the activity of arylsulfatases was significantly decreased only by Ni and Cr (Fig. 4.5). The ecological dose corresponding to 20% inhibition of enzyme activity (ED_{20}) due to Zn, Cd, and Ni toxicity were low in case of dehydrogenase, indicating maximal toxicity of these metals. Order of toxicity of different heavy metals towards dehydrogenase activity followed the order: Cd >Pb> Cr > Ni > Cu > Zn. Copper and Cr showed maximal toxicity to proteases, lipases, and esterases, as indicated by the low ED_{20} levels determined using fluorescein diacetate substrate. In general, heavy metals showed relatively lower toxicity towards acid phosphatase and arylsulfatase activities (Saha et al. 2013). Soil levels of all the heavy metals corresponding to 20% decrease in biomass growth were considerably higher than the limiting levels (ED_{20}) determined for soil enzyme activities, indicating adverse effects on microbial activity to appear much earlier than their adverse effects on plant growth (Saha et al. 2013). Mercury inhibited soil microbial activities like dehydrogenase and nitrification rate. The critical mercury contents (corresponding to 10% inhibition of enzyme activity) were found to be less than the available safe limits for inorganic mercury which demonstrated inadequacy of existing guideline values (Mahbub et al. 2016).

In an experiment under controlled condition, toxicity of different heavy metals to plants and soil microorganisms were compared (Saha et al. 2013). Phytotoxic concentration (PT_{20}, resulting in 20% reduction in growth) in soil and plant tissue followed the order Cd>Ni>Cr>Cu>Zn and Cr>Ni=Cu>Cd>Zn, respectively (Table 4.7). Such difference in order of toxicity in plant tissue from soil can be ascribed to the differences among the heavy metals in transfer rate from soil to above ground plant biomass. Microbial toxicity (ED_{20}, concentration in soil at which 20% reduction in microbial activity) in soil followed the order Cd>Pb>Cr>Ni>Cu>Zn. Lower

4.3.4 Human Health Effects of Heavy Metals

Environmental pollution by heavy metals is of major health concerns all over the world even if it is at low concentrations due to their long-term cumulative health effects. Intake of heavy metal through food materials impairs the function of other

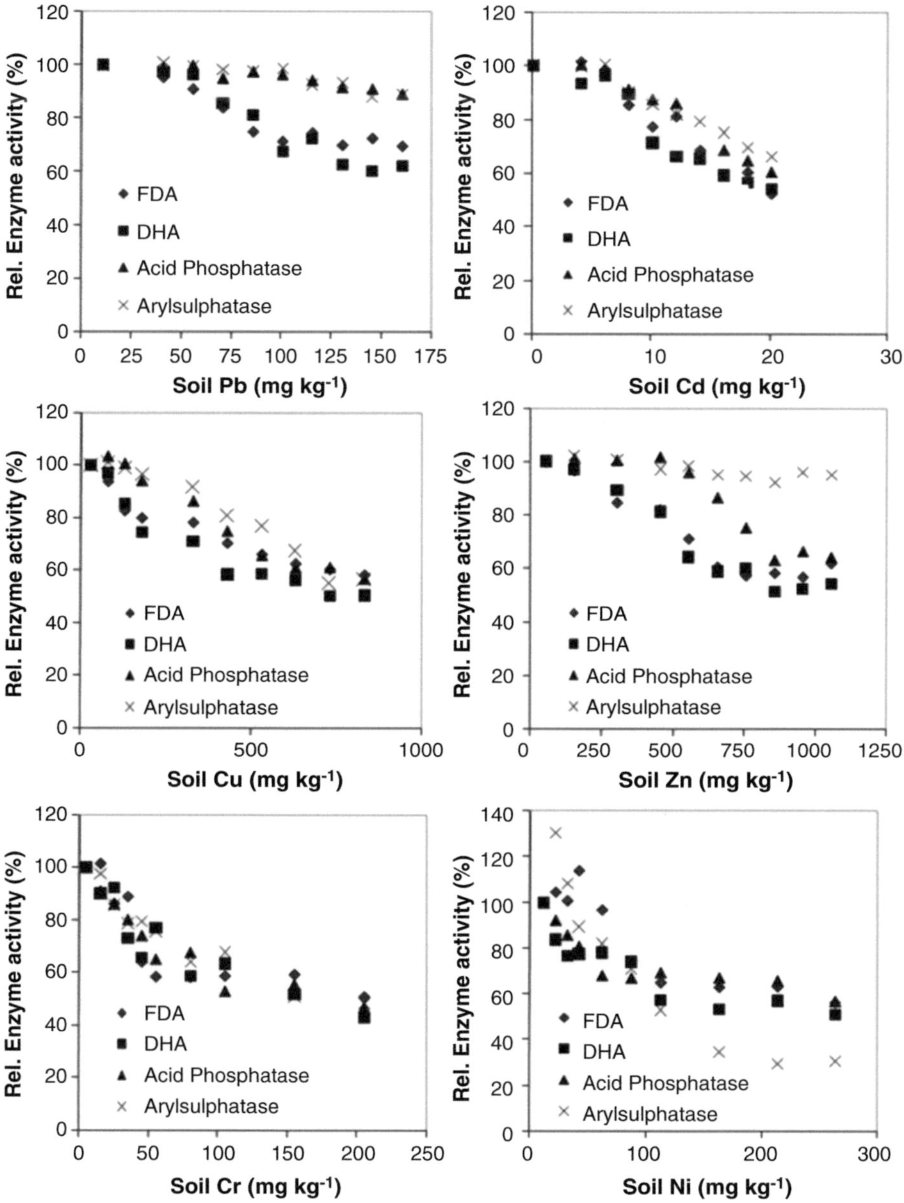

Fig. 4.5 Effects of heavy metal levels in soil on relative soil enzyme activities (FDA = activities of esterase + lipase + protease; DHA = dehydrogenase activity)

Table 4.7 Heavy metals toxicity to plant and soil microbial activity

	PT_{20} in soil	PT_{20} in spinach tissue	ED_{20} in soil
	$\mu M\ g^{-1}$		
Cd	0.16	1.17	0.08
Cr	3.39	0.58	0.6
Cu	6.44	1.03	2.82
Ni	2.61	1.02	0.87
Pb	–	–	0.39
Zn	13.03	37	5.99

Adapted from Saha et al. (2013)

metal ions and decreases the immunology of body, retardation of growth, poor mental development, poor psycho-social attitude, and a various type of disabilities in human beings (Young 1991, Iyengar and Nair 2000, Türkdogan et al. 2003, Nolan 2003, European Union 2002). Specific effects on human health due to high heavy metals intake has been indicated as below:

Arsenic

- It is an analogue of phosphate and interferes with the ATP synthesis pathway.
- It is categorized as highly toxic metalloid, inorganic in nature and causes carcinogenic diseases to skin, lungs, liver and bladder.
- Smaller level exposure caused vomiting and nausea, it impairs the function of blood cells, i.e. red and white; abnormal the heart beat and very high level results death.
- Long term exposure lead to dark color of skin and soles; failure of kidney and liver.

Cadmium

- It has carcinogenic properties.
- Inhalation of very high Cd concentration causes abnormal body biocycles, and leads to severely irritating stomach, vomiting and diarrhea.
- Long term contact with lesser Cd concentration causes damage to the kidneys and lungs, and makes the bones fragile.
- Disease caused by Cd poisoning in human has been termed as itai-itai disease which was detected in Japan in 1968. Disease is characterized by renal tubular dysfunction associated with atrophic kidney and osteomalacia (softening of the bones) and has been found among middle aged and elderly women with relatively frequent pregnancy. Survival rate of itai-itai disease patients is significantly lower.

Chromium

- Hexavalent Cr(VI) is more toxic to human health.

- Generally Cr(VI) is reduced to Cr(III) in soil in presence of oxidizing compounds, which also reduces its carcinogenicity.
- Intake of the high amount of Cr compounds causes irritation of skin, ulcers in nose; and its longer exposure causes damage to body organs, i.e. kidney, blood circulatory system, nervous system and also leads to cancer.

Lead

- It affects children more than adults.
- Higher concentration in children reduces the proper growth of body and mind. As a result it reduces intelligence; causes short memory, renal failure as well as poor development of cardiovascular organs.
- In adults, adverse effects like weakness in fingers, higher blood pressure and abnormal nervous system are observed.
- At high level of contamination, lead may cause death due to failure of multiple-organs.
- High level of Pb in human body reduces the fertility level of man and causes abnormal growth of the child during the pregnancy or miscarriage.

Mercury

- As per the EPA guidelines, mercury has been described as a human carcinogen.
- Among the body organs, nervous system is more sensitive to all types of Hg compounds
- Neurological syndrome caused by severe mercury poisoning is known as *minamata* disease. Symptoms include loss of physical control over bodily movements, numbness in the hands and feet, general muscle weakness, loss of peripheral vision, and damage to hearing and speech. In extreme cases, insanity, paralysis, coma, and death follow within weeks of the onset of symptoms. The disease was first detected in Japan in the year 1956.
- The toxicity can cause lasting damage to most of the body organs specially brain, kidneys, and developing fetuses.
- Long term exposure of lead results in complete failure of organs and may leads to death.
- Short term exposure causes irritation, diarrhea, abnormal blood pressure, vomiting and lung damage.

Selenium

- It is beneficial for human and animal in trace amounts as it plays valuable role in proper functioning of thyroid gland. However at higher level of contamination, it causes toxic effect and imparts abnormality in organs. Upper tolerable intake level of Se is 400 μg/day (ATSDR 2001).

- Short period oral intact causes vomiting, nausea and diarrhea.
- The higher level intake of Se causes selenosis in human and animals, which is characterized by loss of hair, weak nail. Neurological abnormalities are also observed. It also causes stomach pain due to heavy intake of Se.
- Short term high level of Se exposure causes respiratory abnormalities, i.e. difficulty in respiration, bronchitis pain or coughing.

4.3.5 Transfer of Heavy Metals in the Food Chain and Their Bioconcentration

Food chain is a linear network of links (organisms) in a food web starting from producer (such as plant) and ending at apex predator species (like tiger or vulture or killer whales). Metals are transferred across the food chain which begin with uptake by plant from soil in an agroecosystem and may end to animals and/or to human by various routes of food chain. Such transfer of heavy metals assumes severe human health implication in the polluted ecosystem due to its increased pace and biomagnification of metals in the tissue of organisms. Widely reported minamata disease in Japan due to Hg poisoning in human population is a classic example of biomagnifications of heavy metals. The level of different heavy metals (Hg, Cd, Pb) was studied in the producers (food plants) and consumers (four species of acridid grasshoppers) of a grassland of Greece (Devkota and Schmidt 2000). The solely herbivorous grasshoppers had a significantly higher Cd and Hg concentrations than in the food plants indicating their biomagnifications. On the contrary, Pb concentration in grasshoppers was always lower than in their food plants. The accumulation factor (concentration in grasshoppers/concentration in food plants) followed the order Cd>Hg>Pb. Transfer of heavy metals along a soil-plant-insect-chicken food chain at contaminated sites were investigated near lead-zinc mine in China (Zhuang et al. 2009). While concentrations of Zn, Pb and Cu increased slightly from plant to insect larva, Cd concentrations steadily declined with increasing trophic level. Major fraction of Pb, Zn, Cu, and Cd entered in the insect and chicken body is excreted through feces, avoiding metal bioaccumulation in insect and chicken body.

While monitoring Cu and Cd concentrations in a population of the grass hopper in and around a copper refinery, Hunter et al. (1987) observed Cu accumulation of up to 1600 μg/g among grasshoppers particularly in integuments indicating metal exclusion mechanism to protect vital organs.

Polar Eskimos in North West Greenland have high dietary access to methylmercury, though the area has no history of accidental mercury pollution (Hansen and Dasher 1997). Such example of biomagnification occurred because population in the area routinely consumes fish or a particular species of fish increasing the risk of methylmercury poisoning of aquatic origin. The transformation from the hydrophilic (Hg^{+2}) to the lipophilic state (methyl mercury) makes it more prone to biomagnification in aquatic food chain. Polar bears occupy a higher trophic level than bearded seals (*Erignathus barbatus*) and mercury increased significantly from seal to polar bear tissues indicating biomagnifications (Dehn et al. 2006). On the contrary, such biomagnification was not observed for Cd where concentrations of renal and hepatic Cd were significantly lower or not statistically different in polar bears compared to their primary prey ringed seals (*Phoca hispida*) and bearded seals (*Erignathus barbatus*).

Study on bioaccumulation in human tissue is limited. A study was conducted on the trends of concentration of persistent organic pollutants, Hg and Pb in umbilical cord blood of newly born infants belonging to indigenous inhabitants of northern Québec (Canada) (Dallaire et al. 2003). Those community people consume large quantities of marine food and therefore, are exposed to high doses of food chain contaminants. Organic pollutants PCBs, DDT and HCB were found decreasing from 1994 through 2001. A significant reduction of Pb and Hg concentrations were also observed, though the trend was not clear. Such decreases in pollutants concentrations in human tissue were suspected to be due to observed decrease in food contamination and/or to changes in dietary habits.

Potential health risks of heavy metals were assessed in Anhui province of Eastern China having high urban and industrial activity besides agricultural activity in some parts (Zhao et al. 2014). Although soils of the area contained heavy metals within safe limit, mean concentrations of Cr and Hg in grain exceeded the safety limits. The grain carcinogenic risk for Cr, Pb and As exceeded the accepted risk level and The target hazard quotient (THQ) of Cr, Pb, As and Hg exceeded the target value of 1, indicating a level of exposure sufficient to cause adverse non-carcinogenic health risks during a person's lifetime. Bioconcentration factor ($\frac{concentration\ in\ grain}{concentration\ in\ soil}$) values of heavy metals decreased in the sequence Hg > Se > Cd > Cr > Pb > As. Human gastric cancer correlated significantly with topsoil Pb concentration and human liver cancer correlated significantly with grain Hg. The study concluded that long-term low dose exposure of heavy metals may play a key role in cancer development in human.

References

Adhikari T, Biswas AK et al (2012) Heavy metal pollution in soil-plant system and its remediation, IISS Technical Bulletin. Indian Institute of Soil Science, Bhopal

Adviento-Borbe MAA, Doran JW, Drijber RA, Dobermann A (2006) Soil electrical conductivity and water content affect nitrous oxide and carbon dioxide emissions in intensively managed soils. J Environ Qual 35:1999–2010

Akerblom S, Baath E, Bringmark L, Bringmark E (2007) Experimentally induced effects of heavy metals on microbial activity and community structure of forest mor layers. Biol Fertil Soils 44:79–91

Apte AD, Verma S, Tare V et al (2005) Oxidation of Cr (III) in tannery sludge to Cr (VI): field observations and theoretical assessment. J Hazard Mater 121(1–3):215–222

Arora NK, Khare E, Singh S, Maheshwari DK (2010) Effect of Al and heavy metals on enzymes of nitrogen metabolism of fast and slow growing rhizobia under explanta conditions. World J Microbiol Biotechnol 26:811–816

ATSDR (2001) Toxicological profile for selenium. Agency for Toxic Substances and Disease Registry, U.S. Department of Health and Human Services, Public Health Service, Division of Toxicology/Toxicology Information Branch, Atlanta

ATSDR (2005) Toxicological profile for nickel. Agency for Toxic Substances and Disease Registry, U.S. Department of Health and Human Services, Public Health Service, Division of Toxicology/Toxicology Information Branch, Atlanta

Ayers RS, Westcot DW (1976) Water Quality for Agriculture. FAO Irrigation and Drainage Paper No, 29 (Rev 1), Food and Agriculture Organization of the United Nations

Barajas-Aceves M, Dendooven L (2001) Nitrogen, carbon and phosphorus mineralization in soils from semi-arid highlands of central Mexico amended with tannery sludge. Bioresour Technol 77:121–130

Berg J, Brandt KK, Al-Soud WA et al (2012) Selection for Cu-tolerant bacterial communities with altered composition, but unaltered richness, via long-term Cu exposure. Appl Environ Microbiol 78:7438–7446

Bergmann W (1992) Nutritional disorders of plants. Gustav Fischer, Jena/Stuttgart/New York

Bianucci E, Fabra A, Castro S (2011) Cadmium Accumulation and Tolerance in Bradyrhizobium spp. (Peanut Microsymbionts). Curr Microbiol 62:96–100

Bondarenko O, Rahman PKSM, Rahman TJ, Kahru A, Ivask A (2010) Effects of rhamnolipids from Pseudomonas aeruginosa DS10-129 on luminescent bacteria: toxicity and modulation of cadmium bioavailability. Microb Ecol 59:588–600

Brookes PC, McGrath SP (1984) The effect of metal toxicity on the size of soil microbial biomass. J Soil Sci 35:341–346

Brookes PC, McGrath SP et al (1986) Metal residues in soils previously treated with sewage-sludge and their effects on growth and nitrogen fixation by blue green algae. Soil Biol Biochem 18:345–353

Carbonell AA, Aarabi MA, DeLaune RD, Gambrell RP, Patrick WH (1998) Arsenic in wetland vegetation: availability, phytotoxicity, uptake and effects on plant growth and nutrition. Sci Total Environ 217:189–199

Chaney RL, Ryan JA (1994) Risk based standard for As, Pb, and Cd in urban soils. DECHEMA, Frankfurt, p 25

Chaney RL, Ryan JA, Brown SL (1999) Environmentally acceptable endpoints for soil metals. In: Anderson WC et al (eds) Environmental availability in soils: chlorinated organic, explosives, metals. American Academy of Environmental Engineers and Scientists, Annapolis, pp 111–154

Chaudri AM, McGrath SP, Giller KE, Rietz E, Sauerbeck DR (1993) Enumeration of indigenous Rhizobium leguminosarum biovar trifolii in soils previously treated with metal –contaminated sewage-sludge. Soil Biol Biochem 25:301–309

Connor JJ (1990) Bedrock geochemistry and the environment. In: Doe BR (ed) Proceedings of a U.S. Geological Survey Workshop on Environmental Geochemistry, US Geological Survey circular 1033. U.S. Department of The Interior, Washington DC, pp 61–62

Coors A, De Meester L (2008) Synergistic, antagonistic and additive effects of multiple stressors: predation threat, parasitism and pesticide exposure in *Daphnia magna*. J Appl Biol 45:1820–1828

CPCB (2010) Protocol for performance evaluation and monitoring of the common hazardous waste treatment Storage and disposal facilities including common hazardous waste incinerators, Hazardous waste management series. Central Pollution Control Board, New Delhi

Dallaire F, Dewailly E, Muckle G, Ayotte P (2003) Time trends of persistent organic pollutants and heavy metals in umbilical cord blood of inuit infants born in Nunavik (Québec, Canada) between 1994 and 2001. Environ Health Perspect 111:1660–1664

Das P, Samantaray S, Rout GR (1997) Studies on cadmium toxicity in plants: a review. Environ Pollut 98:29–36

Dehn LA, Follmann EH, Thomas DL et al (2006) Trophic relationships in an Arctic food web and implications for trace metal transfer. Sci Total Environ 362:103–123

Deng J, Liao B, Ye M, Deng D, Lan C, Shu W (2007) The effects of heavy metal pollution on genetic diversity in zinc/cadmium hyperaccumulator *Sedum alfredii* populations. Plant Soil 297:83–92

Devkota B, Schmidt GH (2000) Accumulation of heavy metals in food plants and grasshoppers from the Taigetos Mountains, Greece. Agric Ecosyst Environ 78:85–91

Diagboya PN, Olu-Owolabi BI, Adebowale KO (2015) Effects of time, soil organic matter, and iron oxides on the relative retention and redistribution of lead, cadmium, and copper on soils. Environ Sci Pollut Res 22:10331–10339

Effron D, de la Horra AM, Defrieri RL, Fontanive V, Palma PM (2004) Effect of cadmium, copper, and lead on different enzyme activities in a native forest soil. Commun Soil Sci Plant Anal 35:1309–1321

European Union (2002) Heavy metals in wastes, European Commission on Environment. http://ec.europa.eu/environment/waste/studies/pdf/heavy_metalsreport.pdf. Accessed 16 Sept 2016

Fargasova A (2004) Toxicity comparison of some possible toxic metals (Cd, Cu, Pb, Se, Zn) on young seedlings of *Sinapis alba* L. Plant Soil Environ 50:33–38

Ghani A (2010) Toxic effects of heavy metals on plant growth and metal accumulation in maize. Iranian J Toxicol 3:325–334

Gu B, Bian Y, Miller CL et al (2011) Mercury reduction and complexation by natural organic matter in anoxic environments. Proc Natl Acad Sci U S A 108:1479–1483

Hansen JC, Dasher G (1997) Organic mercury: an environmental threat to the health of dietary exposed societies? Rev Environ Health 12:107–116

Hanson B, Grattan SR, Fulton A (1999) Agricultural salinity and drainage. University of California irrigation program. University of California, Davis

Hernandez L, Probst A, Probst JL, Ulrich E (2003) Heavy metal distribution in some French forest soils: evidence for atmospheric contamination. Sci Total Environ 312:195–219

Hope BK (2006) An examination of ecological risk assessment and management practices. Environ Int 32(8):983–995

Hunter BA, Hunter LM, Johnson MS, Thompson DJ (1987) Dynamics of metal accumulation in the grasshopper Chorthippus brunneus in contaminated grasslands. Arch Environ Contam Toxicol 16:711–716

Iyengar V, Nair P (2000) Global outlook on nutrition and the environment: meeting the challenges of the next millennium. Sci Total Environ 249:331–346

Khan M, Scullion J (2002) Effects of metal (Cd, Cu, Ni, Pb or Zn) enrichment of sewage-sludge on soil micro-organisms and their activities. Appl Soil Ecol 20:145–155

Kunito T, Saeki K, Nagaoka K, Oyaizu H, Matsumoto S (2001) Characterization of copper-resistant bacterial community in rhizosphere of highly copper-contaminated soil. Eur J Soil Biol 37:95–102

Kuperman RG, Carreiro MM (1997) Soil heavy metal concentrations, microbial biomass and enzyme activities in a contaminated grassland ecosystem. Soil Biol Biochem 29:179–190

Liu WX, Shen LF, Liu JW (2007) Uptake of toxic heavy metals by rice (*Oryza sativa* L) cultivated in the agricultural soil near Zhengzhou City, People's Republic of China. Bull Environ Contam Toxicol 79:209–213

Mahbub KR, Krishnan K, Megharaj M, Naidu R (2016) Mercury Inhibits Soil Enzyme Activity in a Lower Concentration than the Guideline Value. Bull Environ Contam Toxicol 96:76–82

Mahmood T, Islam KR, Muhammad S (2007) Toxic effects of heavy metals on early growth and tolerance of cereal crops. Pak J Bot 39:451–462

Maliwal GL, Somani LL (2010) Nature, properties and management of saline and alkali soils. Agrotech Publishing Academy, Udaipur

Markus J, Mcbratney AB (2001) A review of the contamination of soil with lead II. Spatial distribution and risk assessment of soil lead. Environ Int 27:399–411

Martensson AM, Witter E (1990) Influence of various soil amendments on nitrogen fixing soil organisms in a long-term field experiments, with special reference to sewage sludge. Soil Biol Biochem 22:977–982

McBride MB (1994) Environmental chemistry of soils. Oxford University Press Inc, New York

McGrath SP (1993) Soil quality in relation to agricultural use. In: HJP E, Hamers T (eds) Integrated soil and sediment research: a basis for proper protection. Kluwer Academic Publishers, Dordrecht, pp 187–200

McGrath SP, Brookes PC, Giller KE (1988) Effects of potential toxic metals in soil derived from past applications of sewage sludge on nitrogen fixation by *Trifolium repens* L. Soil Biol Biochem 20:415–424

Mengoni A, Gonnelli C, Galardi F et al (2000) Genetic diversity and heavy metal tolerance in populations of *Silene paradoxa* L. (Caryophyllaceae):a random amplified polymorphic DNA analysis. Mol Ecol 9:1319–1324

Metternicht GI, Zinck JA (2003) Remote sensing of soil salinity: potentials and constraints. Remote Sens Environ 85:1–20

Munns Jr WR (2006) Assessing risks to wildlife populations from mulitple stressors: Overview of the problem and research needs. Ecol Soc 11(1):23. http://www.ecologyandsociety.org/vol11/iss1/art23/

NHMRC (1992) Australian and New Zealand guidelines for the assessment and management of contaminated sites. Australian and New Zealand Environment and Conservation Council (ANZECC) and National Health and Medical Research Council (NHMRC), Gov. of Australia. https://www.nhmrc.gov.au/guidelines-publications/eh17. Accessed 16 Sept 2016

Nolan K (2003) Copper toxicity syndrome. J Orthomol Psych 12:270–282

NRC (1983) Risk assessment in the federal government: managing the process. National Research Council, National Academy Press, Washington, DC

Oliveira A, Pampulha ME (2006) Effects of long-term heavy metal contamination on soil microbial characteristics. J Biosci Bioeng 102:157–161

Oster JD, Sposito G (1980) The gapon coefficient and the exchangeable sodium percentage-sodium adsorption ratio relation. Soil Sci Soc Am J 44:258–260

Pacyna JM, Pacyna EG (2001) An assessment of global and regional emissions of trace metals to the atmosphere from anthropogenic sources worldwide. Environ Rev 9:269–298

Påhlsson AB (1989) Toxicity of heavy metals (Zn, Cu, Cd, Pb) to vascular plants. Water Air Soil Pollut 47:287–319

Pandey N, Sharma C (2002) Effect of heavy metals Co^{2+}, Ni^{2+} and Cd^{2+} on growth and metabolism of cabbage. Plant Sci 163:753–758

Pathak H, Rao DLN (1998) Carbon and nitrogen mineralization from added organic matter in saline and alkali soils. Soil Biol Biochem 30:695–702

Prasad MNV, Greger M, Landberg T (2001) *Acacia nilotica* bark removes toxic elements from solution: corroboration from toxicity bioassay using *Salix viminalis* in hydroponic system. Int J Phytoremed 3:289–300

Rengasamy P, Churchman GJ (1999) Cation exchange capacity, exchangeable cations and sodicity. In: Peverill K et al (eds) Soil analysis and interpretation manual. CSIRO Publishing, Melbourne, pp 35–50

Richard FC, Bourg ACM (1991) Aqueous geochemistry of chromium: a review. Wat Res 25(7):807–816

Rooney CP, Zhao F, McGrath SP (2007) Phytotoxicity of nickel in a range of European soils: influence of soil properties, Ni solubility and speciation. Environ Pollut 145:596–605

Ryan JA, Chaney RL (1997) Issues of risk assessment and its utility in development of soil standards: the 503 methodology as an example. In: Prost R (ed) Contaminated Soils: Proceedings of the third international symposium on biogeochemistry of trace elements, Paris, France, 15–19 May 1995. Colloque No. 85. INRA Editions, Paris. pp 393–414

Saha JK (2005) Changes in salinity and sodicity of soils with continuous application of contaminated water near Industrial area. J Indian Soc Soil Sci 53:612–617

Saha JK, Sharma AK (2006) Impact of the use of polluted irrigation water on soil quality and crop productivity near Ratlam and Nagda industrial area, IISS Bulletin. Indian Institute of Soil Science, Bhopal, p 26

Saha JK, Panwar N et al (2010a) An assessment of municipal solid waste compost quality produced in different cities of India in the perspective of developing quality control indices. Waste Manag 30:192–201

Saha JK, Panwar N, Singh MV (2010b) Determination of lead and cadmium concentration limits in agricultural soil and municipal solid waste compost through an approach of zero tolerance to food contamination. Environ Monit Assess 168:397–406

Saha JK, Panwar N et al (2013) Risk assessment of heavy metals in soil of a susceptible agro-ecological system amended with municipal solid waste compost. J Indian Soc Soil Sci 61:15–22

Sauerbeck DR (1986) Effects of agricultural practices on the physical, chemical and biological properties of soils: part II, use of sewage sludge and agricultural wastes. In: Barth H, Hermite PL (eds) Scientific basis for soil protection in the European community. Elsevier, London, pp 181–210

Sauvé S, McBride M, Hendershot W (1997) Speciation of lead in contaminated soils. Environ Pollut 98:149–155

Segner H, Schmitt-Jansen M, Sabater S (2014) Assessing the impact of multiple stressors on aquatic biota: the receptor's side matters. Environ Sci Technol 48:7690–7696

Shanker AK, Cervantes C, Loza-Tavera H, Avudainayagam S (2005) Chromium toxicity in plants. Environ Int 31:739–753

Shi W, Bischoff M, Turco R, Konopka A (2002) Long-term effects of chromium and lead upon the activity of soil microbial communities. Appl Soil Ecol 21:169–177

Silva CMMS, Fay EF (2012) Effect of Salinity on Soil Microorganisms, Soil Health and Land Use Management, Dr. Maria C. Hernandez Soriano (Ed.), ISBN: 978–953–307-614-0, InTech, Available from: http://www.intechopen.com/books/soil-health-and-land-use-management/effect-of-salinity-on-soil-microorganisms

Smith JL, Doran JW (1996) Measurement and use of pH and electrical conductivity for soil quality analysis. In: Doran JW, Jones AJ (eds) Methods for assessing soil quality. Soil Science Society of America Spec Publ 49, SSSA, Madison, pp 169–185

Sullivan DM, Miller RO (2001) Compost quality attributes, measurements, and variability. In: Stofella PJ, Kahn BA (eds) Compost utilization in horticultural cropping systems. Lewis Publishers, New York, pp 97–120

Sytar O, Kumar A, Latowski D et al (2013) Heavy metal-induced oxidative damage, defense reactions, and detoxification mechanisms in plants. Acta Physiol Plant 35:985–999

Tabatabai MA (1982) Soil enzymes. In: Page AL, Miller RH, Keeney DR (eds) Methods of soil analysis, part 2, chemical and microbiological properties, Monograph no. 9. ASA-SSSA, Madison, pp 903–947

Theelen RMC (1997) Concepts in the Netherlands of risk assessment of soil contamination. Int J Toxicol 16:509–518

Tripathi S, Chakrabarty A et al (2007) Enzyme activities and microbial biomass in coastal soils of India. Soil Biol Biochem 39:2840–2848

Türkdogan MK, Fevzi K, Kazim K, Ilyas T, Ismail U (2003) Heavy metals in soil, vegetables and fruits in the endemic upper gastrointestinal cancer region of Turkey. Environ Toxicol Pharmacol 13:175–179

Tyler G (1981) Heavy metals in soil biology and biochemistry. In: Paul EA, Ladd JN (eds) Soil biochemistry. Marcel Dekker Inc., New York, pp 371–414

Wang Y, Li Q, Shi J et al (2008) Assessment of microbial activity and bacterial community composition in the rhizosphere of a copper accumulator and a non-accumulator. Soil Biol Biochem 40:1167–1177

Welp G (1999) Inhibitory effect of the total watersoluble concentrations of nine different metals on the dehydrogenase activity of a loess soil. Biol Fertil Soils 30:132–139
Wong MH, Bradshaw AD (1982) A comparison of the toxicity of heavy metals, Using root elongation of Rye Grass, *Lolium perenne*. New Phytol 91:255–261
Xie Y, Luo H, Du Z, Hu L, Fu J (2014) Identification of cadmium- resistant fungi related to Cd transportation in bermuda grass [*Cynodon dactylon* (L.) Pers.] Chemosphere 117:786–792
Xie Y, Fan J, Zhu W et al (2016) Effect of heavy metals pollution on soil microbial diversity and bermudagrass genetic variation. Front Plant Sci 7(755):1–12. doi:10.3389/fpls.2016.00755
Yang Z, Liu S, Zheng D, Feng S (2006) Effects of cadmium, zinc, and lead on soil enzyme activities. J Environ Sci 18:1135–1141
Young RA (1991) Toxicity profiles: toxicity summary for cadmium, Risk Assessment Information System, University of Tennessee. https://rais.ornl.gov/tox/profiles/cadmiu_c.html. Accessed 21 Mar 2016
Zhao Q, Wang Y, Cao Y et al (2014) Potential health risks of heavy metals in cultivated topsoil and grain, including correlations with human primary liver, lung and gastric cancer, in Anhui province, Eastern China. Sci Total Environ 470–471:340–347
Zhuang P, Zou H, Shu W (2009) Biotransfer of heavy metals along a soil-plant-insect-chicken food chain: field study. J Environ Sci 21:849–853

Chapter 5
Organic Pollutants

Abstract Concerns on organic pollutants as threat to environment is increasingly perceived since later half of the previous century. Although initially focus was on some of the pesticides used in controlling insects in agriculture and disease spreading vectors in residential area, concerns on several other types of organic pollutants increased afterwards with activities of chemical and petrochemical industries and use of xenobiotics in many lifestyle products. Threat from organic pollutants arises many a times due to their extreme toxicity at very low level of exposure as well as their high persistency in the soil. Due to advancements in detection capabilities, researchers have generated considerable information on the source, routes of contamination to soil, interactions with soil constituents, degradability by and impact on soil microorganisms, uptake by plants and contamination of food etc. The impacts on agroecosystem are indicated as multidimensional like food contamination, reduced soil enzymes activity leading to disruption in nutrient transformation processes, diminished biodiversity, mutagenicity among microorganisms etc. Most common groups of organic pollutants in soil including polycyclic aromatic hydrocarbons (PAHs), polychlorinated biphenyls (PCBs), pesticides, dye pollutants, antibiotics are discussed for their sources, fate and impact on the ecosystem.

Keywords Soil • Food • Water • Pollution • Polycyclic aromatic hydrocarbons • Polychlorinated biphenyls • Persistent organic pollutants • Dye • Antibiotics

The soil quality is immediate concern for most people, because we all depend on this natural resource on a daily basis for food, fodder, fibre and fuel. Due to intense pressure on land resource from industrial and urban related activities, soil gets contaminated or polluted; as a result its quality is at greater risk to sustain plant, animal and human life on long run. Soil is contaminated with various organic pollutants from various sources like industrial effluents discharges containing toxic pollutants, improper disposal of urban and industrial solid waste, and accidental spillage of contaminants during transportation or handling of hazardous materials and indiscriminate use of pesticide and fertilizers. This chapter attempts to discuss in details about various potential organic pollutants particularly those, which contribute considerably to environmental pollution and their adverse impact on soil, plant and microbes. Organic pollutants can be both naturally occurring and

J.K. Saha et al., *Soil Pollution - An Emerging Threat to Agriculture*, Environmental Chemistry for a Sustainable World 10, DOI 10.1007/978-981-10-4274-4_5

synthesised compounds and may pollute agro-ecosystem either deliberately through crop management activities or accidentally through entry of industrial/urban wastes. Their behaviour and fate in the soil depend mainly on their chemical structure, which determines interactions with mineral particulates, organic matter, water, gas and biota of the soil.

Most common organic pollutants are pesticides, polycyclic aromatic hydrocarbons (PAHs), polychlorinated biphenyls (PCBs), BTEX (benzene, toluene, ethylbenzene, and xylenes), other persistent organic pollutants (POPs) trichloroethylene (TCE), and perchloroethylene, nitro-aromatic compounds, organophosphorus compounds, petroleum oil, etc. These organic pollutants must be investigated thoroughly and controlled as they cause life /life-style threatening diseases such as cataracts, kidney and liver damage, jaundice and cancer; as well as degrade immune system, reproductive system, nervous system, endocrine system of human beings. Beyond the potentially harmful effects on humans, these chemicals affect the biogeochemical cycle and the ecosystem conservation. These organic pollutants even at low concentration in soil may inhibit metabolic activities and growth of soil-associated microbes (Moore et al. 2006). The advanced research has also proved that the few azo-dyes cause bladder cancer, hepatocarcinomas and chromosomal aberrations in mammalian cells in human and animal system.

5.1 Organic Pollutants Entry in Soil, Its Uptake by Plant and Food Chain Contamination

Organic pollutants can reach the soil either by purpose, such as pesticides and other agricultural chemicals, or incidentally through a variety of materials commonly used in agricultural practices, e.g., fertilizers and amendments. Once on the soil surface, the organic pollutant can be partially photo-decomposed and/or volatilized, and partially enter the soil or be transported to surface aquatic bodies by runoff and/or erosion. When in the soil, the pollutant can be subjected to partial or total chemical decomposition and/or biodegradation. The original pollutant and possibly, its breakdown products may be adsorbed to soil organic and inorganic constituents, taken-up by plant roots, and/or leached through the unsaturated zone eventually reaching the groundwater. All these processes are controlled by several factors including the physical and chemical properties of the pollutant and its breakdown products, the nature and thickness of the soil, the amount of water applied to the soil, and the type and extent of interactions between the pollutant and soil components.

Different models for uptake of organic chemicals have been proposed (Trapp 2004; Dettenmaier 2008). Organic chemicals are generally taken up by plant roots via vapor or water phases of soil through active (phenoxy acid herbicides), passive and diffusive process (Bromilow and Chamberlain 1995) and transported to the above ground biomass. Uptake of organic compounds through plant transpiration stream is significant for slightly hydrophobic compounds, but may be severely restricted for soil-bound highly hydrophobic compounds in many higher plants (Bromilow and Chamberlain 1995; Briggs et al. 1982; Ryan et al. 1988). Recent evidence indicates that plants can also translocate some hydrophilic compounds. Such organic compounds likely to be translocated by plants more closely resemble those that can cross the blood – brain barrier as compared to the intestine (Limmer and Burken 2014). Experimental evidences have shown that PAHs, chlorobenzenes, PCBs, and polychlorinated dibenzo-p-dioxins/dibenzofurans (PCDD/Fs) are found accumulated significantly in plant roots (Duarte-Davidson and Jones 1996). In addition to the chemical properties of the contaminant, plant root anatomy plays a major role in uptake mechanism. The higher uptake of organic contaminants by barley plant root over beans root is probably due to higher lipid content in the root of former plant (Bromilow and Chamberlain 1995). Similar observation on organic contaminants uptake based on lipid content in roots has been reported by Fismes et al. (2002) and Wild and Jones (1992). Specific surface area and lipid content were found the most crucial factors for phenanthrene (a PAH compound) uptake with contribution of former being greater than that of lipid content (Zhan et al. 2013). However, no correlation was found between phenanthrene uptake and total root length, and root-water. Several pharmaceuticals compounds have also been found to be taken up by plants. For example diclofenac which has been held responsible for collapse of vulture population in south Asia, was found to accumulate in plant through root uptake (Carter et al. 2014). These evidences indicate that organic pollutants may be taken up by plants from soil and accumulate in above ground plant biomass. Another model was proposed for plant contamination with pollutants based on aerial deposition, soil volatilization, and systemic translocation pathway (Collins and Finnegan 2010). However, this pathway could be significant only in locations with pollutant concentrations >100 mg kg^{-1} in soil. If such organic contaminant accumulated biomass is used for feeding livestock, it consequently increases the probability of exposure of pollutants to humans. In the following sections of this chapter, some of the commonly found organic pollutants, their sources and its effect on soil, plant and human life have been discussed in details.

5.1.1 Pesticides

Pesticides are substances or mixtures of substances used in various terms like herbicides, insecticides, fungicides, rodenticides, molluscicides, and nematicides to prevent, destroy or control any pest, and these are widely used mainly in agriculture and in domestic sector. Based on the chemical constituents, pesticides are classified as organochlorine, organophosphate, carbamate, synthetic pyrethroides and inorganic pesticides. There is another class of pesticides which is produced from natural materials called as biopesticide. These various pesticide formulations have been extensively used for decades and have significantly increased food production.

Global insecticide use in 2007 has been estimated at 404000 t of active ingredient (Grube et al. 2011). The agricultural sector is the primary user of pesticides, consuming over four million tons (Mt) of pesticides annually. However, according to an estimate by Abhilash and Singh (2009), about 2 Mt. of pesticides are consumed every year throughout the world; and out of these 24% is consumed in USA, 45% in Europe and 25% in rest of the world. Among the Asian countries, pesticide consumption is the highest in China followed by Korea, Japan and India. The usage of pesticide in India is about 0.5 kg/ha, of which major contribution is from organochlorine pesticides. In global level, the consumption of herbicide is the highest, whereas in India and other tropical countries, the consumption of insecticides is more. This is due to increased insect pest attack caused mainly by the prevailing warm humid climatic condition.

Some of the pesticides were banned several decades before due to longer persistence in the environment. Even after prohibition on use in most of the countries around the world, organochlorines are still found in agricultural soil, sediment and groundwater (Singh and Singh 2006, Matin et al. 1998; Singh 2001; Manirakiza et al. 2003; Gonçalves and Alpendurada 2005; Hilber et al. 2008). The organochlorine insecticides such as dichlorodiphenyl-trichloroethane (DDT) and hexachlorocyclohexane (HCH) were used widely previously in India due to their effectiveness in controlling malaria, locust and wide range of insects effectively. Presently, many organochlorine pesticides have been banned since 2003, except for DDT (permitted only for restricted use in disease spreading vector control) (Singh et al. 2007). However, some of these pesticides are still preferred by the small farmers because they are cost effective, easily available, and display a wide spectrum of bioactivity (Devi and Raha 2013). Intense and frequent use of these chemicals has led to accumulation of pesticides in soil due to lack of their speedy degradation. Low solubility and high structural stability of the pesticide limit their degradation in soil via chemical and biochemical processes by plants and microbes. Most of the hydrophobic pesticides are adsorbed to soil surface or to organic matter and get sequestrated into tiny pores of soil matrix, becoming less bioavailable. Soil microorganism generally degrades pesticides and converts them into less toxic form; but many a times, end products become more toxic than the original pesticide. Some of the factors that influence their persistency are microbial diversity, rainfall, soil temperature, exposure to sunlight, application rate as well as their solubility and mobility in soil.

5.1.1.1 Pesticides on Food Chain Contamination and Human Health

Generally less than 1% applied pesticides are used for the control of insect pest in crops (Arias-Estevez et al. 2008) and remaining unused pesticide contaminates soil, water and crop produce. Further, indiscriminate use of pesticides by the farmers' due to lack of awareness also results in their unwanted entry into soil with consequent adverse effects on the environment (Igbedioh 1991; Jeyaratnam 1985; Saiyed et al. 2003; Karunakaran 1958). Thus continued application of pesticides increases their concentration in soils and waters, which have adverse effect on different strata of organisms. The extent of these effects however, depends on the toxicity of the pesticides and degree of sensitivity of the organisms. This negative impact of pesticides is mainly due to the highly toxic, stable, less soluble active ingredients of pesticide. Among the various pesticides used, organochlorine pesticides pose serious health concerns. This group includes atrazine, DDT, benzene hexachloride (BHC), Lindane and Endosulfan. Humans are exposed to pesticides (found in environmental media such as soil, water, air and food) by different routes such as inhalation, ingestion and dermal contact, and such exposure results in acute and chronic health problems. Some of the chronic and acute toxicological effects of pesticides are chronic liver damage, endocrine and reproductive disorders, immune-suppression, various cancer, inhibition of choline esterases, Parkinson's and Alzheimer's diseases etc. (Ceron et al. 1995). Death and chronic illness caused by pesticide poisoning numbers about 1 million per year (Lorenz 2006).The World Health Organization has estimated that about 200,000 people are killed by pesticides worldwide every year. Even after ban, residues of organochlorine pesticides had been detected in edible crops, fruit, milk samples because of their transfer from soil through food chain (Jensen 1983; Banerjee et al. 1997). The incidence of pesticides contamination in the food chain is being reported regularly and is expected to increase in the future due to heavy dosage of pesticide use (Devi and Raha 2013). Organophosphate pesticides are generally regarded as safer as compared to organochlorine pesticides for use on crops and animals due to their relatively fast degradation rates.

Dieldrin and endrin residues in agricultural fields cause contamination of not only the aquatic environment but also of crops grown in contaminated soil. High levels of these pesticides have been detected in soils and in a variety of crops around the world. In spite of ban or restriction in use of organochlorine pesticides, their concentrations have been reported in soil of different countries (Table 5.1). Although use of pesticides in general is quite low in Africa, the presence of organochlorine pesticides in food crops have been reported from many countries in that continent. In Togo, West Africa, dieldrin and endrin residue levels of 39.50 and 13.16 ng g^{-1}, respectively, were found in cowpea and a dieldrin residue level of 18.09 ng g^{-1} was found in maize (Mawussi et al. 2009). In Nigeria, dieldrin residues of 6–80 ng g^{-1} were found in tubers (Adeyeye and Osibanjo 1999). Researchers from other countries have also reported residues of these pesticides in food; for example in cucumbers in Japan (Hashimoto 2005), winter squash in the USA (Johgenson 2001), in wheat in Serbia (Škrbić 2007) and vegetables such as spinach, garlic leaf, and pumpkin in China (Gao et al. 2005).

DDT and HCH concentrations in soils from India were lower and comparable with other locations of the world. Using the ratio of DDT/(DDD + DDE) for the contents in soils, Kumar et al. (2014) concluded that source of DDT in soils of Korba (India) may be from both ongoing and past applications. Also source of HCH in soil may be both lindane and HCH as indicated by ratio of α-HCH/γ-HCH. Cancer risk (ILCR) and non-cancer health hazard (HQ) were however lower than

Table 5.1 Levels of DDT and HCH in soils (μg kg^{-1}) reported after 2006

Country	ΣDDT	ΣHCH
China	ND – 663	ND – 654
Spain	126 – 316	1.8
Romania	20 – 50	–
Mexico	ND – 26980	–
Vietnam	BDL – 1538	BDL – 20.57
Pakistan	BDL – 1538	BDL – 119
India	BDL – 315	BDL – 104

Adapted from Kumar et al. (2014)

acceptable guideline values, suggesting low risk for adults and children residing in the area of study.

5.1.1.2 Effect of Pesticides on Biodiversity

Accumulation of pesticides may reduce number of macro- and micro-flora and fauna species like arthropods, earthworms, fungi, bacteria, protozoa, and other organisms in soil, which are responsible for several ecosystem functions related to soil fertility and soil formation. In a comprehensive review, Pelosi et al. (2014) concluded that pesticides had adverse impact on earthworms in soil at all organisation levels as these agro-chemicals disrupt enzymatic activities, increase individual mortality, decrease fecundity and growth, change individual behaviour such as feeding rate and decrease the overall community biomass and density of earthworms. Pesticides also affect other invertebrates in soil; although the impact varies with their types and accumulation levels as well as species of invertebrates (Frampton et al. 2006; Desneux et al. 2007). Exposure of birds to such toxic pesticides reduces the reproductive rate and biochemical reaction in the body, or even kills them directly at high doses (Wilson and Tisdell 2001).

Soil microbial diversity is important for maintenance of soil quality as well as for optimum ecosystem functions. The diverse groups of microorganisms commonly found in soil are mostly unicellular of prokaryotic or eukaryotic origin which include bacteria (eubacteria and archaebacteria), cyanobacteria, actinomycetes, fungi and algae. These soil microbes perform a variety of activities required for the proper functioning of the soil as a dynamic system. While soil enzymes released by soil microbes are responsible for many C and nutrient transformation processes important in crop production, microbial diversity plays important role in ecosystem multi-functionality, degradation of organic pollutants and resistance to invasion by pathogens (Kennedy and Smith 1995; He et al. 2009; van Elsas et al. 2012). Any change in their number or proportional distribution can potentially prohibit/enhance one or other soil biochemical processes important for soil fertility. Pesticides have been found to affect both soil enzyme activities and microbial diversity adversely and the effect depends on type of pesticides, their doses and time elapsed after application (Jacobsen and Hjelmsø 2014). Generally, impact on microbial diversity is more due to fumigant type of agro-chemicals like dithiocarbamate, organosulfur, organobromide etc. Glyphosate, an herbicide was found to impact negatively on the growth promoting bacteria in the rhizosphere (Zobiole et al. 2011).

Some pesticides (particularly non-organochlorine group) stimulate the growth of microorganisms, but other pesticides have either depressive effect or no effect. Organophosphorus pesticides are found less toxic to soil microorganisms (Digrak and Kazanaki 2001; Das et al. 2005). Soil microorganisms generally adapt initially to these pesticides and therefore, biodegradation is faster in soils that have had repeated applications of OPs compared to control soils which have never had OP applications (Ragnarsdottir 2000). Although half-life of OPs are generally 1–2 weeks, these may be detected in trace amount in soil years after application, probably due to their sorption on soil particles rendering these unavailable for microbial degradation (Das et al. 2005). Carbofuran (a carbamate compound) stimulated the population of *Azospirillum* and other anaerobic nitrogen fixers in flooded and non-flooded soil, but butachlor (acetanilide group) reduced the population of *Azospirillum* and aerobic nitrogen fixers in non-flooded soil (Lo 2010). Application of phorate (an organophosphorus compound) and carbofuran induced growth and development of several groups of microorganisms including N_2-fixing bacteria and phosphate solubilizing microorganisms (Das et al. 2005). In rice field agro-ecosystems of India, endosulfan and dichlorvos application at normal dose (@ 1 mL L^{-1}) did not show any adverse effect on soil micro flora; but showed a negative effect at double dose (Bhagabati and Sarma 2011). On the other hand, carbofuran showed an adverse effect on the growth of actinomycetes, bacteria and fungi even at recommended dose of application. Some pesticide compounds can severely limit the nitrogen-fixing capacities of blue-green algae, thereby affecting the overall nitrogen economy of soils in general (DaSilva et al. 1975). Application of aldrin, HCH, carbofuran, 2,4-D, and machette in higher dose lead to the reduction in the nitrifying bacteria population and in the concentrations of nitrate and nitrite in soil (Pandey and Rai 1993). Acid and alkaline phosphatase activities (responsible for phosphate mineralization) as well as the nitrification process (conversion of NH_4^+ to NO_3^-) were inhibited due to herbicide napropamide application (Cycoń et al. 2013). Pesticide application in paddy field soil decreased soil dehydrogenase activity with increasing doses and toxicity increased in the order insecticide > fungicide > herbicide (Subhani et al. 2002).

5.1.2 PAHs

The polycyclic aromatic hydrocarbons (PAHs) include more than 200 compounds with two or more fused benzene rings, the most significant of them are phenanthrene, anthracene, pyrene, fluoranthene, chrysene, benzo(a)pyrene, and benzo(a) anthracene. The differences in ring configurations lead to differences in chemical, physical, and toxicological characteristics of PAHs compounds. These are categorized as low molecular weight (LMW) and high molecular weight (HMW) PAHs based on molecular structure. The LMW PAHs include two and three rings structure while HMW PAHs comprise four and more rings structure and the aqueous solubility of PAHs decrease almost linearly with increase in molecular mass (Heitkamp and Cerniglia 1989).

5.1.2.1 PAH and Its Toxicity

The PAHs in environment are known to possess mutagenic, teratogenic, or carcinogenic properties and are widely associated with adverse effects on public health (Rost et al. 2002; Cerniglia 1993). A long back during 1791, the physician John Hill first linked nasal cancer with PAHs (Cerniglia 1984). The PAHs toxicity occurs mainly due to intercalation of the PAH aromatic ring system into the DNA molecule and this biochemical process is responsible for mutagenicity and carcinogenicity by this organic pollutant. The carcinogenicity of PAHs increased with increasing molecular weight. Human exposure to PAHs has been associated with an increased risk of developing cancer in the variety of organs (such as lung, bladder, stomach, skin, larynx, scrotum, breast, oesophageal, prostate, kidney and pancreas). Furthermore, they are known to suppress the immune system and are suspected of being endocrine disrupters. There are various industrial workplaces at which a significant increase of certain cancer diseases has been found that may be attributed to an unusually high exposure to PAHs. For instance, PAHs exposure is high in coke plants, coal tar manufacturing industries, asphalt, bitumen, aluminium plants, iron and steel industries, creosote, rubber, mineral oil, soot and carbon black-producing or manufacturing companies. As highly exposed occupational groups, chimney sweepers, road-men (involved in pavement-tarring) and roofers (involved in roof-tarring) are also under increased risk from these pollutants (Jacob and Seidel 2002). In a study, dermal absorption was found a major route of PAH entry in the body of coke-oven workers; where 75% of the total absorbed amount of PAHs (specifically pyrene) entered the body through the skin (VanRooij et al. 1993). United States Environmental Protection Agency has identified sixteen PAHs as priority pollutants due to their adverse effect on human health.

5.1.2.2 Sources of PAHs in Soil and Their Concentration Level

PAHs in soils may be generated from natural processes like biosynthesis (bacterial and algal synthesis, decomposition of vegetative liter fall) or geochemical reactions (forest fires or volcanic activity, erosion of sedimentary rocks containing petroleum hydrocarbons, fossil-fuel and mineral production). It is also believed that PAHs are produced during combustion of plant biomass, where these are synthesized during the chemical transformation of monomeric lignin components present in plant biomass (Medeiros and Simoneit 2008; Nussbaumer 2003). The species composition of burning biomass can influence the amount and nature of PAH compounds generated. Combustion of coniferous wood in fire place generated the maximum concentrations of hydrocarbons (Schauer et al. 2001). Also combustion of both pine and deciduous wood produced naphthalenes as the predominant PAH; however the amount being relatively smaller in case of deciduous wood. Combustion of herbaceous plants generated predominantly phenanthrene, pyrene and fluorene (Masclet et al. 1995). However, the amount of hydrocarbons produced in such case was comparable to that generated from the combustion of deciduous species, but smaller than during combustion of coniferous plants. These organic pollutants may also originate from anthropogenic sources such as burning of fossil fuels and biomass, industrial emissions, motor vehicle exhausts, abrasion of road surfaces and tires and petroleum processing (Masih and Taneja 2006). Contribution of anthropogenic sources in PAH introduction into the environment is much higher in comparison to their natural sources (Nam et al. 2003). Other potential sources of PAHs in soil include disposal from public sewage treatment, irrigation with coke oven effluent, and use of MSW compost (Santodonato et al. 1981; White and Lee 1980). In Brazil, Grossi et al. (1998) found increased concentrations of PAHs, PCBs, and PCDD/Fs in composts prepared from unsegregated MSW from metropolitan areas.

Concentration of total PAHs in soil may vary from 1 μg kg^{-1} to 300 g kg^{-1} soil, depending on the sources of contamination like combustion of fossil fuels, gasification and liquefaction of coal, incineration of wastes, and wood treatment processes (Bamforth and Singleton 2005). The high concentration of PAHs in the surface soil layer is observed especially in areas of municipal–industrial agglomerations, around roads/highways and fuel processing; and concentration of individual PAHs varies from below limit of detection to up to several hundred μg per kg of soil (Gąsecka et al. 2015). Coal-tar has been found a major source of PAHs in the environment particularly along roadside soils and water bodies (van Metre et al. 2009; Mahler et al. 2012). Median ΣPAH concentrations in dust from sealcoated pavement were 2200 mg kg^{-1}, which was almost 80 times higher than the similar concentration in dust samples received from unsealcoated roads (Van Metre et al. 2009). Such dust particles may be carried by air or storm water to contaminate nearby land. In India, most of the highways and city roads use coal-tar extensively as seal-coat and therefore, contamination of soils in the agricultural land alongside roads and highways with PAHs and their transfer to plant & food need to be investigated. High concentration of PAHs was observed in the atmosphere of Durg-Bhilai (India) city due to heavy vehicular traffic (Pandey et al. 1999). In an experiment at Jalandhar (Punjab), PAHs contamination in soil along the roadside was quantified. The average total PAHs concentration was found 4.04 μg g^{-1}, while 16 individual PAH concentration ranged between 0.008 and 28.4 μg g^{-1} (Kumar and Kothiyal 2011). The concentration of these pollutants was the highest within 1 m distance from road side. The ratio of noncarcinogenic and carcinogenic PAHs in all the soil samples was 1:2.95. In the most of the city locations, total carcinogenic PAHs concentration ranged from 60% to 80% of the total PAHs. The combined status of PAHs and metals in surface (0–5 cm) soils of three different land-use regions (industrial, flood-plain and a reference site) in Delhi was studied over a period of 1 year (Khillare et al. 2014). Mean Σ16PAH concentrations at the industrial and flood-plain sites (10.9 ± 2.8 and 3.1 ± 0.9 mg kg^{-1}, respectively) were ~15 and ~4 times respectively, higher than reference levels. Toxicity potentials (Benz[a]pyrene–equivalent concentrations, as a measure of carcinogenic potencies) of industrial and flood-plain soils were ~88 and ~8 times higher than reference levels.

$$\text{Benz[a]pyrene} - \text{equivalent concentrations} = \sum_{i}^{n} (C_i \times TEF_i)$$

where, TEF_i is Toxic equivalency factor of individual PAH relative to Benz[a] pyrene, as given by Tsai et al. (2004); and C_i is the concentration of an individual PAH. The values of TEF ranged from 0.001 (for Naphthalene, Acenaphthylene, Acenaphthene, Fluorene, Phenanthrene, Fluoranthene, Pyrene, Perylene, Coronene) to 1.0 (for Benzo(a)pyearene, Dibenzo(a,h)anthracene, Dibenzo(a,e) pyrene).

The identified sources of PAHs in the study were coal and wood combustion, vehicular and industrial emissions, and atmospheric transport. Health risk assessment (incremental lifetime cancer risk and hazard index) showed that flood-plain soils have potential high risk due to PAHs while industrial soils have potential risks due to both PAHs and Cr (Khillare et al. 2014). In another study to assess the health risk from PAH in Delhi, median concentrations of benzo(a)pyrene in soil was 0.029 ($\pm$0.002) mg/kg (Kumar et al. 2013a). Health risk from soil accumulated benzo(a) pyrene was low, as computed incremental life time cancer risk (ILCR) and index of additive cancer risk (IACR) values were much lower than the guideline values of $10^{-6} - 10^{-4}$ (ILCR) and <1 (IACR), respectively.

These organic contaminants have high environmental stability due to generally being resistant to microbial decomposition (Harrison et al. 1996; Zhang et al. 2009). Biochemical persistence of PAHs arises from dense clouds of π-electrons on both sides of the ring structures, making them resistant to nucleophilic attack. Their physical properties include low aqueous solubility and high solid-water distribution ratios, which decrease bioavailability for microbial utilization and promote their accumulation in the solid phases of soil mass. Bioavailability of the PAHs decreases almost logarithmically with increasing molecular mass. As a consequence of their low water-solubility, low vapour pressure and high degree of association with soil organic matter, PAHs tend to adsorb strongly onto soil particles (Kipopoulou et al. 1999; Orecchio 2010). In fact, most of the environmental burden of PAHs is considered to be due to their presence in soil (approximately 95%), as opposed to air (approximately 0.2%) (Smith et al. 1995). Once such PAHs seep into the soil, these get accumulated in horizons rich in organic matter and are likely to be retained for many years due to their persistence, hydrophobicity and slow degradation rate (Krauss et al. 2000).

5.1.2.3 Impact of PAHs on Soil and Crop Quality

While contaminated air and water are main routes of direct PAHs entry in human, their entry through contaminated soil is suspected only indirectly to food plant contamination or through grazing animals (soil ingestion). Soil system seems to be the important long-term repository for PAHs and is considered to be a steady indicator of the state of environmental pollution. PAHs with three rings or more tend to be very strongly adsorbed to the soil matrices (Knox et al. 1993), preferentially to small aggregates (<50 μm), which also contain the most humified organic matter (Quantin et al. 2005). Strong adsorption coupled with very low water solubility render PAH loss by leaching insignificant. The investigations have showed that assimilation of PAHs from soil into plants can be treated as an equilibrium process, where the accumulated PAHs are in equilibrium with available PAHs in soils. Available PAHs are only part of the total PAHs concentrations in soil. Therefore, the degree of bioaccumulation in plants, besides the physicochemical properties of organic pollutants and the plants' sensitivity to PAHs uptake, is closely related to the bioavailability of PAHs (van Bohemen and van de Laak 2003).

Absorption of PAHs from soil through roots and their transport into aboveground parts of the plant has been found possible mainly for low molecular weight (containing 2- and 3-ring) hydrocarbons due to their high solubility; whereas larger molecules (having >3 ring) of these are accumulated on the surface of the roots (Oleszczuk and Baran 2004). From a number of experiments, it has been concluded that contamination of crop plants via root absorption with PAHs is negligible (Ellwardt 1977; O'Connor 1996). Tao et al. (2006) determined the relationship among the contents of 16 PAHs (listed by EPA) present in cabbage, in air as well as in soil. The concentration of PAHs in cabbage was positively correlated with the concentration in the air, whereas dependence of the level of soil contamination was statistically insignificant indicating air pollution being major concern in respect food contamination. In another study, contamination of plants (trees, grasses, grain) including vegetables and fruits with PAHs depended on their contents in both soil and air (Liste and Alexander 2000). Only limited data are available on the toxicity of PAHs for plants; although it is generally opined that these pollutants do not produce any perceptible toxic effects in vegetation. PAHs did not have any adverse effect on the germination of several crops even at higher concentration (Ghanem et al. 2010). A study compared PAHs uptake by different plants in the vicinity of highway (Kluska 2005) and indicated that cabbage recorded the highest concentration (358 ng kg^{-1}) of total PAHs among studied vegetables, followed by cucumbers (71.3 ng kg^{-1}) and root of parsley (29 ng kg^{-1}). The concentrations in soil and plant decreased with distance away from highway. Aerial deposition of PAHs on leaves and their subsequent entry-translocation within plant are important mechanisms of their contamination of food. Leaf morphological features (cuticular wax, surface area, hairs, and stomata), leaf components (lipids) and contaminant molecular structure (low and high molecular weight PAHs) play important role in

PAH uptake and translocation by plant (Howsam et al. 2000). Many plants with relatively large surface area have greater potential to accumulate PAH from the polluted atmosphere through particle deposition on waxy leaf or by uptake in the gas phase through stomata (Kipopoulou et al. 1999; Lehndorff and Schwark 2004).

Changes in soil microbial activity and biodiversity have been reported due to accumulation of persistent and toxic organic pollutants. In the presence of PAHs, biomass-C, respiration, protease activity and heterotrophic counts were significantly enhanced, while urease activity was depressed. Mineralization of organic N in soil was initially, but reversibly inhibited in the presence of PAHs (Margesin et al. 2000). Adverse effect of these organic pollutants on enzymatic activity (dehydrogenase, phosphatase, urease and protease) was found influenced by soil physicochemical properties such as total organic carbon content and pH (Baran et al. 2004). Soil enrichment with organic matter considerably decreases the negative influence of PAH's on enzymatic activity (Kucharski et al. 2000; Baran et al. 2004). PAHs affected nodulation of alfalfa significantly and the effect was more in absence of humic matter (Wetzel and Werner 1995). In some of the soil samples, stimulating effect of PAH on dehydrogenase activity was also observed, which was attributed to an adaptation of the soil microflora and the use of the pollutant as a C and energy source (Baran et al. 2004).

5.1.2.4 Food Chain Contamination with PAHs

Accumulation of PAHs in soils and atmospheric pollution may lead to food chain contamination (Kipopoulou et al. 1999; Mueller and Shann 2006) and subsequently, cause direct or indirect exposure (ingestion, inhalation, and dermal contact) to human. Food appears to be the main source of PAH intake for humans who are not occupationally exposed to PAH. Raw foods should not contain high levels of PAH. In areas remote from urban or industrial activities, levels of PAH found in unprocessed foods reflect the background contamination. Generally, the average background values are in the range of 0.01–1 μg kg^{-1} in uncooked foods. Such PAH usually originates from long distance airborne transportation of contaminated particles as well as natural emissions from volcanoes and forest fires. In the neighbourhood of industrial areas or along highways, the contamination of vegetation can be ten-fold higher than in rural areas. The waxy surface of vegetables and fruits can concentrate low molecular mass of PAH mainly through surface adsorption. PAH concentrations are generally greater on plant surface (peel, outer leaves) than on internal tissue. Careful washing may remove up to 50% of the total PAH. Particle bound PAH are easily washed off the surface; whereas those in the waxy layer are less efficiently removed. Sediments of lakes and sea also accumulate organic pollutants originated from anthropogenic activities like biomass & fossil fuel combustion, road construction, oil spill etc. (Magi et al. 2002; Perra et al. 2009). The main route of human exposure to sedimentary PAHs is through marine lives which make up part of the human diet. PAHs contained in sediments and pore water are taken in by benthic organisms such as lobsters, mussels, and clams which are either consumed directly by humans or are consumed by higher predators such as squid and fish which are also a part of the human diet. Therefore, knowledge on contamination in soil, groundwater, marine and atmosphere with PAHs is needed to avoid any food production risk and to restrict the deleterious effect of these contaminants on human.

5.1.3 Other Persistent Organic Pollutants (POPs)

Persistent organic pollutants (POPs) are toxic chemicals that adversely affect human health and the environment around the world. Some of the most well-known POPs are PCBs, DDT, and dioxins. PCBs and DDT were being produced chemically due to their usefulness. PCBs were extensively used as dielectric and coolant fluids in transformers and capacitors, lubricants, plasticizers, paints, etc. Dioxins result from some industrial processes and from waste combustion (for example, municipal and medical waste incineration and backyard burning of trash). A study on assessment of air quality of different potential sources in different cities of India indicated that the concentrations of both polychlorinated dibenzo-p-dioxins and dibenzofurans (PCDDs/DFs) and dioxin-like PCBs from hazardous waste incinerators are much higher than those from some thermal processes (Thacker et al. 2013). A monitoring and toxicological study by Center for Marine Environmental Studies, Ehime University (Japan) revealed that municipal dumping sites were the major reservoirs and sources of terrestrial dioxins and related POPs, particularly PCDDs/DFs (Subramanian and Tanabe 2007). They also indicated that PCDD/DF homologue profiles of the dumpsite soils from India synchronized with the profiles reported for the soils representing environmental sources of these chemicals (municipal waste incinerator emissions) from United States. A decade after, the same group revisited the status of POPs in environmental samples of India and concluded that organochlorines such as DDTs and HCHs showed decreasing trend due to ban on the use of such chemicals, though their levels were still higher (Subramanian et al. 2015). Dumping site at Chennai (India) was estimated to receive 1400 mg of PCDD/DF annually (Minh et al. 2003). Use of polychlorinated biphenyls (PCBs) is more in Indian industries, but the production is totally contributed by other countries. Several experiments with increased sewage sludge application rates also demonstrated an accumulation of PCBs in the treated soils (Folch et al. 1996; Delschen 1999). Since 1966, PCBs and other organohalide compounds were identified in soil system and the sources of contribution were industrial effluent, sewage-sludge incinerators and biomass combustions. The discharged PCBs are adsorbed by the soil particles; affect the soil properties and persist for longer time. Exposure to these compounds is recognized as an important environmental risk factor for humans and animals because of the following effects: cancer; nervous system damages; reproductive and immune system impairments; and hormonal imbalance (Darnerud et al. 2001; Safe 2004). Presence of these organic contaminants however, has generally not been found to affect soil microorganisms

Table 5.2 Average concentrations of POPs in different environmental samples from various regions of India

Nature of sample	ΣDDT	ΣHCH/BHC	Period of sampling/ reporting
Surface water (ng L^{-1})	ND – 16,367,000	0.1 – 17,430,000	1988–2011
Groundwater (ng L^{-1})	ND – 75,000	88.43 – 400,000	2001–2011
Soil (ng g^{-1})	13.81 – 934	1.65 – 404,520.9	1996–2012
Sediments (ng g^{-1})	<0.01–128,600	<0.01 – –234,000	1998–2012
Aquatic organisms (ng g^{-1})	0.38 – 19,970	0.87 – 32,200	1995–2010
Human blood (μg L^{-1})	8 – 32,200	6 – 2633.7	1983–2012
Human milk (ng g^{-1} lipid wt.)	175.5 – 3210	127.25 – 4500	1981–2006

Adapted from Sharma et al. (2014)

and associated biochemical processes (Wild et al. 1994). However, their extreme persistence in the environment and ability to bioconcentrate in the food chain makes them great environmental and human health risks that need remedial action (Cousins and Jones 1998; Hickey 1999). Due to their high toxicity, many industrialized countries imposed regulations and bans on the production, use and discharge of these compounds. However developing countries are yet to initiate such actions due to the availability of limited baseline data.

Several studies reported elevated levels of persistent organic pollutants (POPs) in various environmental and biotic samples from India (Kumari et al. 2008; Devanathan et al. 2009; Someya et al. 2010; Sharma et al. 2014) (Table 5.2). In a study, large number of environmental samples (fish, dust and human milk) from different locations of India were analyzed for PCBs and other organohalide compounds (Devanathan et al. 2012). High levels of PCBs were found in human milk from surrounding municipal waste dumping site indicating higher risk for infants living near these sites. Significantly higher levels of organohalide compounds were found in dust samples from e-waste recycling locations, suggesting that the crude e-waste recycling/dismantling activities are the major emission sources of these contaminants. The hazard quotients values of PCBs were above one for the infants and toddlers living near the municipal dumping and e-waste recycling areas indicating high risk, which warrants regular monitoring and regulations to reduce the pollution levels. Dietary intake values of PCBs and polybrominated diphenyl ethers (PBDEs) through farm-raised fish were far lower than the guideline values indicating less risk through fish consumption. Soil samples from agricultural land near pesticide manufacturing industry at Lucknow contained very high concentrations of HCH (Jit et al. 2011; Abhilash and Singh 2009). Concentrations of DDTs and HCHs were generally lower in urban areas than in agricultural soils indicating higher use of these POPs as pesticides in agriculture (Sharma et al. 2014). Meta-analysis on time trend of data did not indicate a general decline in DDT and HCH residues in the environment and human body in India in spite of ban on their use; whereas, such samples from China showed a tendency towards decline in POP contamination (Sharma et al. 2014). In urban soils of Kurukshetra city, average of 28 PCBs and dioxin-like PCBs were 11.57 ± 2.00 μg kg^{-1} and 2.58 ± 0.34 μg kg^{-1}, respectively. However, analysis of incremental life time cancer risk indicated that these contaminated soils were not likely to pose any health risk to adults and children (Kumar et al. 2013b).

5.1.3.1 Impact of POPs on Soil and Produce Quality

Persistent organic pollutants are lipophilic /hydrophobic in nature and hence, these are strongly adsorbed to soil organic matter restricting their uptake by plant root. Availability of organic pollutants for plant uptake is governed its solubility in water which are generally determined by octanol-water partition coefficient (K_{ow}). Concentration of organic pollutants in soil and that in aboveground plant parts were found to be related by the equation below (Travis and Arms 1988):

$$\log B_v = 1.588 - 0.578 \log K_{ow}$$

where, B_v is the ratio of conc. of organic in aboveground plant to that in soil.

Contaminants with values of K_{ow} ranging between 0.5 and 3 are considered moderately hydrophobic and are most likely to be taken up by plants (Ryan et al. 1988; Wenzel et al. 1999). Hawker and Connell (1988) determined K_{ow} values for 13 PCB congeners and found its strong relationship with total surface area. Transfer of PCBs from contaminated soil to plant root is generally low; and higher chlorinated congeners are further less preferentially absorbed (Suzuki et al. 1977; McCrady et al. 1990; Wild et al. 1994). Several researches have also concluded that uptake and translocation of PCBs to aboveground biomass is very small with bioaccumulation factor of about 0.0045 (Webber et al. 1990; Gan and Berthouex 1994). However, some PCBs (like 2,3,7,8-Tetrachlorodibenzodioxin, TCDD) are likely to be absorbed by plant leaves via volatilization from soil surface (McCrady et al. 1990). In a study on a highly contaminated soil (0.6–200 μg g^{-1} total PCBs in terms of Aroclor 1254/1260), sedge (*Carex normalis)* showed significant accumulation of PCBs in shoot with bioaccumulation factors up to 0.29 indicating its potential for phytoremediation (Whitfield Aslund et al. 2007). Turrio-Baldassarri et al. (2007) found significant level of contamination in forage crop nearby PCB manufacturing plant in Brescia, Italy which was assumed to be from air-borne PCBs.

POPs are highly stable chemicals that resist the natural processes of degradation. Once introduced into the environment, they can persist for a long time. These are readily absorbed in fatty tissue and accumulate in the body fat of living organisms; and become more concentrated as they move up the food chain, especially into larger, longer-living organisms. Long-term consumption of contaminated forage plants led to their build-up in the body of bovines and human population residing in the contaminated area (Turrio-Baldassarri et al. 2008, 2009). Soil contamination with PCB (Aroclor1248; tetrachlorbiphenyl) resulted significant reduction in root growth of several plants (Chekol et al. 2004). This group of organic pollutant, particularly higher-chlorinated congeners has considerable adverse effect on soil microorganisms (Cámara et al. 2004; Correa et al. 2010). Soil ingestion during grazing has been found to play key role in toxic POPs transfer into livestock which in turn increases risk of their exposure to human (Wild et al. 1994).

5.1.4 Dye Pollutants

Several textile, pharmaceutical and printing industries use large volume of azo-dyes, which is a group of synthetic organic colorants (generally derivatives of benzene, toluene, naphthalene, phenol and aniline) containing N as azo group –N=N. Due to the extensive use of these dyes in industries, they have become an integral part of industrial effluent. A considerable fraction (about one-tenth) of azo-dyes used by textile industries does not bind the fabrics, and therefore is released through wastewater (Puvaneswari et al. 2006). In fact, of the one million ton of organic dyes annually produced worldwide, more than 11% is lost in effluents during manufacture and application processes (Forgas et al. 2004). The presence of dyes in water imparts color to it which can block sunlight penetration and oxygen dissolution, both of which are essential ingredients for aquatic life (Pereira and Alves 2012). Thus, there is a considerable need to treat these colored effluents before discharging them to various water bodies. Further, some azo dyes are bonded with heavy metals like Cr and Cu in order to impart shades and resistance to washing. As a result, wastewaters from textile industries, unless treated properly, pose considerable threat to environment from both organic and metal pollutants. Groundwater of several villages near industrial area of Ratlam and Bichhri (India) has been turned red due to contamination from dye industry effluents (DTE 1999; Saha and Sharma 2006) (Fig. 5.1). Several types of azo-dyes and their breakdown products are toxic (lethal, mutagenic and carcinogenic) to terrestrial and aquatic organisms including mammals (Brown and De Vito 1993; Puvaneswari et al. 2006). Many of these dyes are resistant to degradation and remediation under natural conditions and through conventional treatment methods (Tahir et al. 2016). Reported half-life of hydrolyzed Reactive-Blue 19 is 46 years (Hao et al. 2000).

5.1.4.1 Impact of Dye Pollutants on Soil Quality

Dyes show different degree of stability in soils from days to several weeks depending upon the nature of chemical compounds. Direct-Red 81 was more resistant to microbial degradation as compared to Reactive-Black 5 and Acid-Yellow 19 (Imran et al. 2015). As a result of irrigation with effluent from dye industries, considerable accumulation of total organic dyes in cultivated soil had

Fig. 5.1 Groundwater near Ratlam industrial area has turned *red*, suspected to be due to contamination with azo-dyes

been observed and such accumulation has also been transported in plant tissue (Zhou 2001). Phytotoxicity of dyes has been reported by some workers (Kalyani et al. 2008; Ayed et al. 2011). There are evidences of plant uptake of dye compounds by plants (Uera et al. 2007; Muthunarayanan et al. 2011). Different azo dyes have expressed toxicity to soil microorganisms and toxicity is more on Gram-negative bacteria than Gram-positive bacteria (Imran et al. 2015). Certain azo-dye had been found to affect adversely on growth of atmospheric N fixing cyanobacterium *Anabaena* sp. (Hu and Wu 2001). Sulfonated azo dye had significant adverse effect on urease activity, ammonification and nitrification rates in soil and therefore reduces nitrogen use efficiency in crop production (Topac et al. 2009). Ammonia oxidizing bacteria population and rate of nitrification process was depressed considerably in presence of reactive-black 5 dye (Batool et al. 2015). These results show dye pollutants show considerable toxicity to soil microorganisms and agriculturally important nutrient cycling processes.

5.1.5 Antibiotics Contamination in Soil and Their Impact

Antibiotics inhibit the activity of microorganisms, viruses, and eukaryotic cells and are commonly used in human and veterinary (bovines, swines and birds) medicines for treating infections. Their benefits have been recognized in agriculture, aquaculture, beekeeping, and livestock as growth promoters. A significant quantity of these is also routinely used as a feed supplement to promote growth of food and milch animals to inhibit subclinical infections and to reduce growth-depressing microbial metabolites (Kumar et al. 2005a, b). There are several different kinds of antibiotics and these can be classified based on their chemical structure, action mechanism, action spectrum, and the route of administration. Although antibiotics are designed to perform effectively at very low concentration, they are administered in the body at larger dose because of lower efficacy in reaching the targeted pathogens. Most antibiotics administered in human or fed to animals are poorly absorbed in the gut and as a result a substantial amount of these are excreted through urine and faeces, which in turn end up in municipal sewage effluent and manure.

5.1.5.1 Antibiotics Release into Environment and Their Entry into Agroecosystem

It is estimated that as much as 90% of some antibiotics may be excreted as parent compound, thereby contaminating the environment (Phillips et al. 2004). Tetracyclines (oxytetracycline and chlortetracycline), tylosin, sulfamethazine, amprolium, monensin, virginiamycin, penicillin, and nicarbazine are the most common antibiotics present in swine, beef, and poultry/turkey manures. The concentration of these antibiotics varies from traces to as high as 216 (mg L^{-1}) of manure slurry (NAAS 2010). Soil, sediment, sludge, groundwater, wastewater, tap water, surface water (lakes, streams, rivers, sea), plants, and aquatic animals have been reported for contamination with antibiotics by several researchers (Hirsch et al. 1999; Hamscher et al. 2002; Fick et al. 2009). Pharmaceutical industries in Hyderabad (India) have been reported to contaminate surface, groundwater and drinking water severely with antibiotics like ciprofloxacin, ofloxacin, cetirizine. Wastewater from a common effluent treatment plant receiving process water from approximately 90 bulk drug manufacturers contained several antibiotics in very high concentration (Fick et al. 2009). Mutiyar and Mittal (2013) found concentration of amoxicillin in the untreated raw sewage in Delhi (India) in the range from "not detected" (ND) to 172.6 ng L^{-1}. In another study by this group, significant concentrations of ampicillin (23.5–263.3 μg l^{-1}), ciprofloxacin (9.7–45.4 μg l^{-1}), sparfloxacin (ND to 63.2 μg l^{-1}) and gatifloxacin (ND to 8.4 μg l^{-1}) had been found in wastewater influents from Delhi (Mutiyar and Mittal 2014). Removal efficiencies of the STP varied between 55 and 99%. Thus significant amounts of antibiotics in the effluents were discharged in Yamuna River. The Yamuna River also receives effluents from 17 STPs and also carries sewage and industrial discharges from 17 stormwater drains. The concentration of the antibiotics in the river varied from ND to 13.75 μg l^{-1} ampicillin, ND to 1.44 μg l^{-1} for ciprofloxacin, ND to 0.48 μg l^{-1} for gatifloxacin, and ND to 2.09 μg l^{-1} for sparfloxacin. Similarly, significant levels of antibiotics had been detected in different river-water in different countries (Munoz et al. 2009; Feitosa-Felizzola and Chiron 2009; Watkinson et al. 2009; Lin and Tsai 2009). As use of treated and untreated sewage effluent for irrigation is quite prevalent in urban and peri-urban areas, soil contamination with antibiotics may be widespread.

Animal derived manure, an important soil amendment in Indian agriculture is also considered as another source of antibiotic contamination in soil. Most of the antibiotics fed to animals are poorly absorbed in the animal gut and consequently, a substantial amount of these is excreted with the urine and faeces within a few days of medication (Aust et al. 2008). In general, 72% of the active ingredients initially dispensed in animals were recovered in the animal wastes within 2 days of application (Winckler and Grafe 2001). Tetracyclines, tylosin, sulfamethazine, amprolium, monensin, virginiamycin, penicillin, and nicarbazine are commonly observed antibiotics in pig, cattle, and turkey manures (Webb and Fontenot 1975; De Liguoro et al. 2003; Kumar et al. 2005a). Antibiotics were found to degrade during composting of animal dung and the extent of degradation is highly variable among different antibiotics (Dolliver et al. 2008). At the end of the composting period, degradation of chlortetracycline was >99%; whereas monensin and tylosin degraded to the extent of 54–76%. Sulfamethazine, however, didn't degrade significantly during composting.

Composting methods (passively aerated composting or frequently turned aerobic composting, or vessel composting) did not have significant impact on degradation of antibiotics (Mitchell et al. 2015). Thermophilic stage (with $\geq$60 °C) during composting process has been found to enhance degradation kinetics of antibiotics, with greater degradation in cattle dung manure as compared to biosolids from wastewater treatment plant. However, most of the composting in India is not properly managed to impart thermophilic treatment for whole feedstock due to infrequent turning (Saha et al. 2010). Therefore certain antibiotics can still persist in the manure being prepared in the country. A review by Kumar et al. (2005a) indicated that antibiotic concentrations in manure ranged from trace levels to >200 mg kg^{-1}, with typical concentrations in the 1–10 mg kg^{-1} range. Although there is very few data available on veterinary antibiotics use in various food-producing animals in developing countries such as India (WHO 2001), their amount may be considerable due to huge livestock population (more than 33% of the world) and lack of regulatory mechanism with antimicrobial drug use in animals intended for food. Antibiotics have been detected in soils amended with livestock manure, with periods of detection ranging from days to months after application, depending on the antibiotics type and their degradation kinetics (Hamscher et al. 2002; De Liguoro et al. 2003). Sustained elevated levels of externally added antibiotics in the environment may contribute to the development of antibiotic-resistant microbial populations; and even very low quantities of antibiotics encourage the selection of antibiotic-resistant bacteria (Boxall et al. 2003). Contact of such antibiotic-resistant bacteria in the soil and water with human-associated microbiota has been suspected to be a cause in development of multi-drug-resistant human pathogens, a concern among medical practitioners (Baquero et al. 2008).

For sustainable agriculture, maintenance of soil organic matter is of paramount importance under different land use management systems and this necessitates regular incorporation of organic matter into the rhizosphere from external sources. Indian soils are generally deficient in organic matter due to high mineralization rate under prevalent climatic condition and therefore, frequent application of animal manure in crop land is advocated for improving and sustaining soil quality. Data from long-term experiments have revealed that additional yields of different crops can be realized over and above soil test based optimum NPK rates, only if 10–15 tonnes of FYM per hectare are annually supplemented (Swarup and Wanjari 2001). Similarly scarcity of irrigation water for agriculture is prevalent in India due to erratic rainfall and stiff competition for this resource with industry and municipality. As a result, farmers in the peri-urban area who live nearby such water sources divert this untreated wastewater for irrigation whenever needed as it proves to be low-cost alternative to conventional irrigation water. It supports livelihood and generates considerable value in urban and peri-urban agriculture of India, despite the health and environmental risks associated with this practice. Conclusions based on the considerable research worldwide, limited studies in Indian context and available related secondary information indicates that presence of antibiotics in manure and wastewater (both being important inputs for agriculture) may be widespread across India and their regular use may be contaminating the soil.

5.1.5.2 Antibiotics Persistence and Its Effects on Soil Properties

Antibiotics concentration in soil has been reported by researchers across the world to vary widely in the range of ND to 1534.4 μg kg^{-1} (Du and Liu 2012). Impact of antibiotics in soil depends on their persistence and solubility. Distribution coefficients (K_d), a measure of contaminant activity retardation, for the adsorption to soil constituents of different groups of antibiotics vary from 0.6 to 4.9 for sulfonamides, from 290 to 1620 for tetracyclines and from 310 to 6310 for fluoroquinolones (Thiele-Bruhn 2003). Sorption of antibiotics to soil minerals is weaker than to soil organic matter and is governed by pH, ionic strength, and type of ions in the medium. Adsorption of most of the antibiotics is fast and reduces their antibiotic potency; though the process does not completely eliminate their antimicrobial activity. Most of the antibiotics have low solubility and therefore, their mobility in soil is mainly facilitated by their association with mobile soil colloids and dissolved organic matter (Thiele-Bruhn 2003).

Soil microorganisms play important role in several nutrient transformation processes in soil like nitrogen fixation, nitrification, ammonification, phosphate solubilisation etc. As antibiotics are meant to kill mainly microorganisms, their entry is expected to have harmful effects on agriculturally important soil microorganisms. Repeated contamination through livestock manure application and wastewater irrigation during cropping can accumulate antibiotics in the soil, which may reach beyond the level of the threshold inhibitory concentrations for agriculturally important microorganisms in the soil ecosystem. A comprehensive review by Thiele-Bruhn (2003) indicated that effects of antibiotics in soil are diverse and are generally microorganism and dose specific, which tend to cause shift in microbial community. Antibiotics on the soil have been found to reduce microbial population including soil enzyme activities (Kotzerke et al. 2008; Akimenko et al. 2015). For example, sulfonamides affected activities of dehydrogenase (responsible for organic matter mineralization) and urease (responsible for hydrolysis of urea) as well as structural diversity of the soil microbial community (decrease in bacterial biomass and increase in proportion of fungal biomass) even at a relatively low concentration (Gutierrez et al. 2010). In another study, six antibiotics (chlortetracycline, tetracycline and tylosin; sulfamethoxazole, sulfamethazine and trimethoprim) inhibited soil phosphatase activity during the 22 days' incubation; while sulfonamides (sulfamethoxazole and sulfamethazine) inhibited total microbial activity as indicated by reduced soil respiration (Liu et al. 2009). The effective concentration doses (EC_{10}) for soil respiration in the first 2 days were 7 mg kg^{-1} for sulfamethoxazole, 13 mg kg^{-1} for sulfamethazine and 20 mg kg^{-1} for trimethoprim.

5.1.5.3 Effect of Antibiotics on Plant

The presence of antibiotics in surface water, groundwater, seawater, soil and sludge opened their entry into biota. The contaminant transfer usually takes place from contaminated soils to the plants, via retention by root surfaces, root uptake, translocation and foliar uptake. There are some reports available which indicate adverse effects of antibiotics on plant growth and yield; although their toxic doses are several folds higher in plants as compared to soil microorganisms. Phytotoxicity depended on plant species and antibiotic types (Batchelder 1982; Jjemba 2002; Farkas et al. 2007; Liu et al. 2009). For example, significant gentamicin (but not streptomycin) uptake was observed in radish, while carrot roots or lettuce leaves didn't uptake these two antibiotics (Bassil et al. 2013). Experimentally determined phytotoxicity doses for different antibiotics are generally much higher than the reported range of antibiotics in different contaminated sites, which indicates that phytotoxicity due to antibiotic contamination is rare in agroecosystem (Thiele-Bruhn 2003). High-level accumulation of antibiotics in food crops may raise potential human health concerns through food chain. A recent review indicated absorption of antibiotics by crops from contaminated soils and detected concentrations in different plant parts ranged between 0.9 and 6.1 mg/kg (Du and Liu 2012). A greenhouse studies with plants grown in antibiotic contaminated manure-applied soil indicated that corn, green onion and cabbage absorbed chlortetracycline but not tylosin antibiotics present in manure (Kumar et al. 2005a). The concentrations of chlortetracycline in plant tissues were small (2–17 ng g^{-1} fresh weight), but these concentrations increased with increasing amount of antibiotics present in the manure. Antibiotics with large molecular weight are generally not taken up easily by plants (Kumar et al. 2005b). Although the maximum residue value for antibiotics in animal-based products has been established, the limit for antibiotics in plant-based products is absent. Antibiotics present in plant materials ingested by humans may provide resistance to human pathogens thus resulting in illnesses that may be difficult to cure with currently available antibiotics. This can be a serious threat, as has been proved in many animal studies.

5.1.5.4 Antibiotic Resistance in Microbes

The most obvious risk of antibiotic release into the environment is related to the development of antibiotic-resistant microorganisms which has resulted in the reduction of therapeutic potential against human and animal pathogens. Residues of antibiotics can induce resistance in pathogens either directly or indirectly by transfer of plasmids from non-pathogens to pathogenic microorganisms. The detection of antibiotics residues (even at sub-nanogram per liter) is alarming for ecosystem sustainability. These compounds are specially engineered to show their effect at trace levels. There are no reports of direct effect on human beings from contaminated water but effects on other organisms have been documented. Kummerer (2004) reported the development of various resistant bacterial strains associated with discharges of antibiotics. A study on archived long-term soil-series from different locations in The Netherlands showed a significant accumulation of antibiotic resistance genes in agricultural soils sampled since 1940, postulated to be due to decades of increasing use of antibiotics (Knapp et al. 2010). Three Indian soil samples irrigated with wastewater (from Aligarh, Kanpur and Ghaziabad) contained the antibiotic resistance genes (*ampC)*; while it was absent in the samples from groundwater irrigated soils (Malik et al. 2008). Similarly, *E. coli* resistant to cefotaxime, ciprofloxacin, and cefoxitin was present in treated effluent (originated from hospital) samples (Galvin et al. 2010). These observations indicate that antibiotic resistant genes might have been transferred from the sewage system to soil. Therefore, anthropogenic activities such as manure/biosolid application, wastewater irrigation, and agricultural application of antibiotic compounds are supposed to be responsible in transmitting antibiotic resistance to the environment.

Studies have also found that highly diverse and abundant levels of antibiotic resistant bacteria are already present in soils. Rhizosphere comprises large populations of different microorganisms and some of these naturally produce antibiotics which are lethal to susceptible rhizobial populations in the soil. Therefore, intrinsic resistance to antibiotics is a desirable trait for the rhizobial population. It increases the rhizobia's chances of growth, multiplication and persistence in the soil (Naamala et al. 2016). Hollowell et al. (2015) genotyped more than 400 isolates from plant roots and soils across California and assayed 98 of them for resistance traits against 17 clinically relevant antibiotics. However, resistance to certain group of antibiotics had been found to reduce symbiotic effectiveness of *Rhizobium sp.* and that affect biological nitrogen fixation (Pankhurst 1977; Turco et al. 1986). Thus, there is a need for detailed investigations on assessing impacts antibiotics of anthropogenic origin on soil biodiversity, development of multidrug resistant organic organisms and transfer to human through food chain contamination.

References

Abhilash PC, Singh N (2009) Pesticide use and application: an Indian scenario. J Hazard Mater 165:1–12

Adeyeye A, Osibanjo O (1999) Residues of organochlorine pesticides in fruits, vegetables and tubers from Nigerian markets. Sci Total Environ 231:227–233

Akimenko YV, Kazeev KS, Kolesnikov SI (2015) Impact Assessment of soil contamination with antibiotics (for example, an ordinary chernozem). Am J Appl Sci 12:80–88

Arias-Estevez M, Lopez-Periago E, Martínez-Carballo E et al (2008) The mobility and degradation of pesticides in soils and the pollution of groundwater resources. Agric Ecosyst Environ 123:247–260

Aust MO, Godlinski F, Travis GR et al (2008) Distribution of sulfamethazine, chlortetracycline and tylosin in manure and soil of Canadian feedlots after subtherapeutic use in cattle. Environ Pollut 156:1243–1251

Ayed L, Mahdhi A et al (2011) Decolorization and degradation of azo dye methyl red by an isolated *Sphingomonas paucimobilis*: biotoxicity and metabolites characterization. Desalination 274:272–277

Bamforth SM, Singleton I (2005) Bioremediation of polycyclic aromatic hydrocarbons: current knowledge and future directions. J Chem Technol Biotechnol 80(7):723–736

Banerjee BD, Zaidi SSA, Pasha ST et al (1997) Levels of HCH residues in human milk samples from Delhi, India. Bull Environ Contam Toxicol 59:403–403

Baquero F, Martinez JL, Canton R (2008) Antibiotics and antibiotic resistance in water environments. Curr Opin Biotechnol 19:260–265

Baran S, Bielińska JE et al (2004) Enzymatic activity in an airfield soil polluted with polycyclic aromatic hydrocarbons. Geoderma 118:221–232

Bassil RJ, Bashour II, Sleiman FT, Abou-Jawdeh YA (2013) Antibiotic uptake by plants from manure-amended soils. J Environ Sci Health B 48:570–574

Batchelder AR (1982) Chlortetracycline and oxytetracycline effects on plant growth and development in soil systems. J Environ Qual 11:675–678

Batool S, Khalid A, Jalal KCA et al (2015) Effect of azo dye on ammonium oxidation process and ammonia-oxidizing bacteria (AOB) in soil. RSC Adv 5:34812–34820

Bhagabati KS, Sarma H (2011) Differential effects of pesticides on soil microflora in cultivated soil of Indian rice field agro-ecosystems. J Appl Nat Sci 3:277–279

Boxall ABA, Kolpin DW, Halling-Srensen B et al (2003) Are veterinary medicines causing environmental risks? Environ Sci Technol 37:286–294

Briggs GG, Bromilow RH, Evans AA (1982) Relationships between lipophilicity and root uptake and translocation of non-ionised chemicals by barley. Pest Manag Sci 13:495–504

Bromilow RH, Chamberlain K (1995) Principles governing uptake and transport of chemicals. In: Trapp S, McFarlane JC (eds) Plant contamination: modeling and simulation of organic chemical processes. Lewis Publishers, Boca Raton, pp 37–68

Brown MA, De Vito SC (1993) Predicting azo dye toxicity. Crit Rev Environ Sci Technol 23:249–324

Cámara B, Herrera C, Gonzalez M et al (2004) From PCBs to highly toxic metabolites by the biphenyl pathway. Environ Microbiol 6:842–850

Carter LJ, Harris E, Williams M et al Boxall (2014) Fate and uptake of pharmaceuticals in soil–plant systems. J Agric Food Chem 62 (4):816-825

Cerniglia CE (1984) Microbial metabolism of polycyclic aromatic hydrocarbons. Adv Appl Microbiol 30:31–71

Cerniglia CE (1993) Biodegradation of polycyclic aromatic hydrocarbons. Curr Opin Biotechnol 4:331–338

Ceron JJ, Gutierrez-Panizo C, Barba A, Camara MA (1995) Endosulfan isomers and metabolite residue degradation in carnation (*Dianthus caryophyllus*) by product under different environmental conditions. J Environ Sci Health A 30:221–232

Chekol T, Vough LR et al (2004) Phytoremediation of polychlorinated biphenyl-contaminated soils: the rhizosphere effect. Environ Int 30:799–804
Collins CD, Finnegan E (2010) Modeling the plant uptake of organic chemicals, including the soil−air−plant pathway. Environ Sci Technol 44:998–1003
Correa PA, Lin L et al (2010) The effects of individual PCB congeners on the soil bacterial community structure and the abundance of biphenyl dioxygenase genes. Environ Int 36:901–906
Cousins IT, Jones KC (1998) Air soil exchange of semi-volatile organic compounds (SOCs) in the UK. Environ Pollut 102:105–118
Cycoń M, Wójcik M, Borymski S, Piotrowska-Seget Z (2013) Short-term effects of the herbicide napropamide on the activity and structure of the soil microbial community assessed by the multi-approach analysis. Appl Soil Ecol 66:8–18
Darnerud PO, Eriksen GS, Johannesson T et al (2001) Polybrominated diphenyl ethers: occurrence, dietary exposure, and toxicology. Environ Health Perspect 109:49–68
Das AC, Chakravarty A, Sen G et al (2005) A comparative study on the dissipation and microbial metabolism of organophosphate and carbamate insecticides in Orchaqualf and Fluvaquent soils of West Bengal. Chemosphere 58:579–584
DaSilva EJ, Henriksson LE, Henriksson E (1975) Effect of pesticides on blue-green algae and nitrogen-fixation. Arch Environ Contam Toxicol 3:193–204
De Liguoro M, Cibin V, Capolongo F et al (2003) Use of oxytetracycline and tylosin in intensive calf farming: evaluation of transfer to manure and soil. Chemosphere 52:203–212
Delschen T (1999) Impacts of long-term application of organic fertilizers on soil quality parameters in reclaimed loess soils of the Rhineland lignite mining area. Plant Soil 213:43–54
Desneux N, Decourtye A, Delpuech JM (2007) The sublethal effects of pesticides on beneficial arthropods. Annu Rev Entomol 52:81–106
Dettenmaier E (2008) Measuring and modelling of plant root uptake of organic chemicals. Dissertation, Utah State University, Logan, Utah
Devanathan G, Subramanian A, Someya M et al (2009) Persistent organochlorines in human breast milk from major metropolitan cities in India. Environ Pollut 157:148–154
Devanathan G, Isobe T, Subramanian A et al (2012) Contamination status of polychlorinated biphenyls and brominated flame retardants in environmental and biota samples from India. In: Kawaguchi M, Misaki K, Sato H et al (eds) Interdisciplinary studies on environmental chemistry. Environmental pollution and ecotoxicology, pp 269–277
Devi NL, Raha P (2013) Contamination of Organochlorine Pesticides (OCPs) in India. Bull Environ Sci Res 2:9–14
Digrak M, Kazanaki F (2001) Effect of some organophosphorus insecticides on soil microorganisms. Turk J Biol 25:51–58
Dolliver H, Gupta S, Noll S (2008) Antibiotic degradation during manure composting. J Environ Qual 37:1245–1253
DTE (1999) Bloody waters. Down To Earth, Issue Apr 1–15, Society for Environmental Communications, New Delhi, India. http://www.downtoearth.org.in/node/19679
Du L, Liu W (2012) Occurrence, fate, and ecotoxicity of antibiotics in agro-ecosystems. A review. Agron Sustain Dev 32:309–327
Duarte-Davidson R, Jones KC (1996) Screening the environmental fate of organic contaminants in sewage sludge applied to agricultural soils: II. The potential for transfers to plants and grazing animal. Sci Total Environ 185:59–70
Ellwardt P (1977) Variation in content of polycyclic aromatic hydrocarbons in soil and plants by using municipal waste compost in agriculture. In: Proceedings of symposium on soil organic matter studies. Vol. II. International Atomic Energy Agency, Vienna. p 291
Farkas MH, Berry JO, Aga DS (2007) Chlortetracycline detoxification in maize via induction of glutathione S-transferases after antibiotic exposure. Environ Sci Technol 41:1450–1456
Feitosa-Felizzola J, Chiron S (2009) Occurrence and distribution of selected antibiotics in a small Mediterranean stream (Arc River, Southern France). J Hydrol 364:50–57

Fick J, Söderström H, Lindberg RH et al (2009) Contamination of surface, ground, and drinking water from pharmaceutical production. Environ Toxicol Chem 28:2522–2527
Fismes J, Perrin GC, Empereur BP et al (2002) Soil-to-root transfer and translocation of polycyclic aromatic hydrocarbons by vegetables grown on industrial contaminated soils. J Environ Qual 31:1649–1656
Folch I, Vaquero MT et al (1996) Extraction and clean-up methods for improvement of the chromatographic determination of polychlorinated biphenyls in sewage sludge-amended soils: elimination of lipids and sulphur. J Chromatogr A 719:121–130
Forgas E, Cserhati T, Oros G (2004) Removal of synthetic dyes wastewaters: a review. Environ Int 30:953–971
Frampton GK, Jaensch S, Scott-Fordsmand JJ et al (2006) Effects of pesticides on soil invertebrates in laboratory studies: a review and analysis using species sensitivity distributions. Environ Toxicol Chem 25:2480–2489
Galvin S, Boyle F, Hickey P (2010) Enumeration and characterization of antimicrobial-resistant *Escherichia coli* bacteria in effluent from municipal, hospital, and secondary treatment facility sources. Appl Environ Microbiol 76:4772–4779
Gan R, Berthouex P (1994) Disappearance and crop uptake of PCBs from sludge amended farming. Water Environ Res 66:54–69
Gao HJ, Jiang X, Wang F, Wang DZ et al (2005) Residual levels and bioaccumulation of chlorinated persistent organic pollutants (POPs) in vegetables from suburb of Nanjing, People's Republic of China. Bull Environ Contam Toxicol 74:673–680
Gąsecka M, Włodarczyk-Makuła M et al (2015) Phytoremediation of PAH- contaminated areas. In: Ansari AA et al (eds) Phytoremediation: management of environmental contaminants, Volume 1. Springer International Publishing, Cham, pp 295–308
Ghanem A, D'Orazio V et al (2010) Phytotoxicity assay of selected plants to Pyrene contaminated soil. 19th World congress of soil science, soil solutions for a changing world, 1–6 August 2010, Brisbane, Australia. Published on DVD, pp 74–77
Gonçalves C, Alpendurada MF (2005) Assessment of pesticide contamination in soil samples from an intensive horticulture area, using ultrasonic extraction and gas chromatography-mass spectrometry. Talanta 65:1179–1189
Grossi G, Lichtig J, Kraub P (1998) PCDD/F, PCB, and PAH content of Brazilian compost. Chemosphere 37:213–216
Grube A, Donaldson D, Kiely T et al (2011) Pesticides industry sales and usage: 2006 and 2007 market estimates. U.S. Environmental Protection Agency, Office of Chemical Safety and Pollution Prevention, Washington, DC
Gutierrez IR, Watanabe N, Harter T et al (2010) Effect of sulfonamide antibiotics on microbial diversity and activity in a Californian mollic haploxeralf. J Soils Sediments 10:537–544
Hamscher G, Sczesny S, Hoper H et al (2002) Determination of persistent tetracycline residues in soil fertilized with liquid manure by high-performance liquid chromatography with electrospray ionization tandem mass spectrometry. Anal Chem 74:1509–1518
Hao OJ, Kim H, Chang PC (2000) Decolorization of wastewater. Crit Rev Environ Sci Technol 30:449–505
Harrison R, Smith DJT, Luhana L (1996) Source apportionment of atmospheric polycyclic aromatic hydrocarbons collected from an urban location in Birmingham, UK. Environ Sci Technol 30:825–832
Hashimoto Y (2005) Dieldrin residue in the soil and cucumber from agricultural field in Tokyo. J Pestic Sci 30:397–402
Hawker DW, Connell DW (1988) Octanol-water partition coefficients of polychlorinated biphenyl congeners. Environ Sci Technol 22:382–387
He JZ, Ge Y, Xu Z, Chen C (2009) Linking soil bacterial diversity to ecosystem multifunctionality using backward-elimination boosted trees analysis. J Soils Sediments 9:547–554

Heitkamp MA, Cerniglia CE (1989) Polycyclic aromatic hydrocarbon degradation by a Mycobacterium sp. in microcosms containing sediment and water from a pristine ecosystem. Appl Environ Microbiol 55:1968–1973
Hickey WJ (1999) Transformation and fate of polychlorinated biphenyls in soil and sediment. In: Adriano DC (ed) Bioremediation of contaminated soils. ASA, CSSA and SSSA, Madison, pp 213–237
Hilber I, Mäder P, Schulin R et al (2008) Survey of organochlorine pesticides in horticultural soils and there grown *Cucurbitaceae*. Chemosphere 73:954–961
Hirsch R, Temes T, Haberer K et al (1999) Occurrence of antibiotics in the aquatic environment. Sci Total Environ 225:109–118
Hollowell AC, Gano KA, Lopez G et al (2015) Native California soils are selective reservoirs for multidrug-resistant bacteria. Environ Microbiol Rep 7(3):442–449
Howsam M, Jones KC, Ineson P (2000) PAHs associated with the leaves of three deciduous tree species. I – concentrations and pro- files. Environ Pollut 108:413–424
Hu T, Wu SC (2001) Assessment of the effect of azo dye RP2B on the growth of a nitrogen fixing cyanobacterium – *Anabaena* sp. Bioresour Technol 77:93–95
Igbedioh SO (1991) Effects of agricultural pesticides on humans, animals and higher plants in developing countries. Arch Environ Health 46:218–223
Imran M, Shaharoona B, Crowley DE et al (2015) The stability of textile azo dyes in soil and their impact on microbial phospholipid fatty acid profiles. Ecotoxicol Environ Saf 120:163–168
Jacob J, Seidel A (2002) Biomonitoring of polycyclic aromatic hydrocarbons in human urine. J Chromatogr B 778(1–2):31–47
Jacobsen CS, Hjelmsø MH (2014) Agricultural soils, pesticides and microbial diversity. Curr Opin Biotechnol 27:15–20
Jemba PK (2002) The effect of chloroquine, quinacrine, and metronidazole on both soybean plants and soil microbiota. Chemosphere 46:1019–1025
Jensen AA (1983) Chemical contaminants in human milk. Residue Rev 89:1–12
Jeyaratnam J (1985) Health problems of pesticide usage in the third world. Brit J Ind Med 42:505–506
Jit S, Dadhwal M, Kumari H et al (2011) Evaluation of hexachlorocyclohexane contamination from the last lindane production plant operating in India. Environ Sci Pollut Res Int 18 (4):586–597
Johgenson JE (2001) Aldrin and dieldrin: a review of research on their production, environmental deposition and fate, bioaccumulation, toxicology, and epidemiology in the United States. Environ Health Perspect 109:113–139
Kalyani DC, Patil PS et al (2008) Biodegradation of reactive textile dye Red BLI by an isolated bacterium Pseudomonas sp. SUK1. Bioresour Technol 99:4635–4641
Karunakaran CO (1958) The Kerala food poisoning. J Indian Med Assoc 31:204–207
Kennedy AC, Smith KL (1995) Soil microbial diversity and the sustainability of agricultural soils. Plant Soil 170:75–86
Khillare PS, Hasan A, Sarkar S (2014) Accumulation and risks of polycyclic aromatic hydrocarbons and trace metals in tropical urban soils. Environ Monit Assess 186:2907–2923
Kipopoulou AM, Manoli E, Samara C (1999) Bioconcentration of polycyclic aromatic hydrocarbons in vegetables grown in an industrial area. Environ Pollut 106:369–380
Kluska M (2005) Polycyclic aromatic hydrocarbons uptake by selected plants in the vicinity of highway. Archiwum Ochrony Środowiska 31:61–70
Knapp CW, Dolfing J, Ehlert PAI, Graham DW (2010) Evidence of increasing antibiotic resistance gene abundances in archived soils since 1940. Environ Sci Technol 44:580–587
Knox RC, Sabatini DA, Canter LW (1993) Subsurface transport and fate processes. Lewis Publishing, Boca Raton
Kotzerke A, Sharma S, Schauss K et al (2008) Alterations in soil microbial activity and N-transformation processes due to sulfadiazine loads in pig-manure. Environ Pollut 153:315–322

Krauss M, Wilcke W, Wolfgang Z (2000) Availability of polycyclic aromatic hydrocarbons (PAHs) and polychlorinated biphenyls (PCBs) to earthworms in urban soils. Environ Sci Technol 34:4335–4340

Kucharski J, Jastrzębska E, Wyszkowska J et al (2000) Effect of pollution with diesel oil and leaded petrol on enzymatic activity of the soil. Zesz Probl Postep Nauk Rol 472:457–464

Kumar V, Kothiyal NC (2011) Distribution behavior of polycyclic aromatic hydrocarbons in roadside soil at traffic intercepts within developing cities. Int J Environ Sci Technol 8:63–72

Kumar K, Gupta SC, Baidoo SK et al (2005a) Antibiotic uptake by plants from soil fertilized with animal manure. J Environ Qual 34:2082–2085

Kumar K, Gupta SC, Chander Y, Singh AK (2005b) Antibiotic use in agriculture and their impact on terrestrial environment. Adv Agron 87:1–54

Kumar B, Gaur R, Kumar S, Sharma CS (2013a) Environmental and human health risk assessment of benzo[a]pyrene levels in agricultural soils from the national capital region Delhi, India. Hum Ecol Risk Assess 19:118–125

Kumar B, Verma VK, Kumar S, Sharma CS (2013b) Probabilistic health risk assessment of polycyclic aromatic hydrocarbons and polychlorinated biphenyls in urban soils from a tropical city of India. J Environ Sci Health, Part A: Toxic Hazard Subst Environ Eng 48:1253–1263

Kumar B, Verma VK, Mishra M et al (2014) DDT and HCH (Organochlorine Pesticides) in residential soils and health assessment for human populations in Korba, India. Hum Ecol Risk Assess 20:1538–1549

Kumari B, Madan VK, Kathpal TS (2008) Status of insecticide contamination of soil and water in Haryana, India. Environ Monit Assess 136:239–244

Kummerer K (2004) Resistance in the environment. J Antimicrob Chemother 54:311–320

Lehndorff E, Schwark L (2004) Biomonitoring of air quality in the cologne conurbation using pine needles as a passive sampler – part II: polycyclic aromatic hydrocarbons (PAH). Atmos Environ 38(23):3793–3808

Limmer MA, Burken JG (2014) Plant translocation of organic compounds: molecular and physicochemical predictors. Environ Sci Technol Lett 1:156–161

Lin AYC, Tsai YT (2009) Occurrence of pharmaceuticals in Taiwan's surface waters: impact of waste streams from hospitals and pharmaceutical production facilities. Sci Total Environ 407:3793–3802

Liste HH, Alexander M (2000) Accumulation of phenanthrene and pyrene in rhizosphere soil. Chemosphere 40:11–14

Liu F, Ying GG, Tao R et al (2009) Effects of six selected antibiotics on plant growth and soil microbial and enzymatic activities. Environ Pollut 157:1636–1642

Lo CC (2010) Effect of pesticides on soil microbial community. J Environ Sci Health B 45:348–359

Lorenz ES (2006) Potential health effects of pesticides. College of Agricultural Science Penn State University, University Park

Magi E, Bianco R, Ianni C, Di Carro M (2002) Distribution of polycyclic aromatic hydrocarbons in the sediments of the Adriatic Sea. Environ Pollut 119:91–98

Mahler BJ, Metre PCV, Crane JL et al (2012) Coal-tar-based pavement sealcoat and PAHs: implications for the environment, human health, and stormwater management. Environ Sci Technol 46(6):3039–3045

Malik A, Celik EK, Bohn C et al (2008) Detection of conjugative plasmids and antibiotic resistance genes in anthropogenic soils from Germany and India. FEMS Microbiol Lett 279:207–216

Manirakiza P, Akinbamijo O, Covaci A, Pitonzo R, Schepens P (2003) Assessment of organochlorine pesticide residues in west African city farms: Banjul and Dakar case study. Arch Environ Contam Toxicol 44:171–179

Margesin R, Walder G et al (2000) The impact of hydrocarbon remediation (diesel oil and polycyclic aromatic hydrocarbons) on enzyme activities and microbial properties of soil. Acta Biotechnol 20:313–333

Masclet P, Cachier H, Liousse C et al (1995) Emissions of PAH by savannah fires. J Atmos Chem 22:41–45
Masih A, Taneja A (2006) Polycyclic aromatic hydrocar-bons (PAHs) concentrations and related carcinogenic po-tencies in soil at a semi-arid region of India. Chemosphere 65:449–456
Matin MA, Malek MA, Amin MR, Rahman S, Khatoon J, Rahman M, Aminuddin M, Mian AJ (1998) Organochlorine insecticide residues in surface and underground water from different regions of Bangladesh. Agric Ecosyst Environ 69:11–15
Mawussi G, Sanda K, Merlina G, Pinelli E (2009) Assessment of average exposure to organochlorine pesticides in southern Togo from water, maize (*Zea mays*) and cowpea (*Vigna unguiculata*). Food Addit Contam Part A Chem Anal Control Expo Risk Assess 26:348–354
McCrady JK, McFarlane C, Gander LK (1990) The transport and fate of 2,3,7,8-TCDD in soybean and corn. Chemosphere 21:359–376
Medeiros PM, Simoneit BRT (2008) Source profiles of organic compounds emitted upon combustion of green vegetation from temperate climate forests. Environ Sci Technol 42:8310–8316
Minh NH, Minh TB, Watanabe M et al (2003) Open dumping site in Asian developing countries: a potential source of polychlorinated dibenzo-p-dioxins and polychlorinated dibenzofurans. Environ Sci Technol 37:1493–1502
Mitchell SM, Ullman JL, Bary A et al (2015) Antibiotic degradation during thermophilic composting. Water Air Soil Pollut 226:1. doi:10.1007/s11270-014-2288-z
Moore FP, Barac T, Borremans B et al (2006) Endophytic bacterial diversity in poplar trees growing on a BTEX-contaminated site: the characterization of isolates with potential to enhance phytoremediation. Syst Appl Microbiol 29:539–556
Mueller KE, Shann JR (2006) PAH dissipation in spiked soil: impacts of bioavailability, microbial activity, and trees. Chemosphere 64:1006–1014
Munoz I, Gomez-Ramos MJ, Aguera A et al (2009) Chemical evaluation of contaminants in wastewater effluents and the environmental risk of reusing effluents in agriculture. Trends Anal Chem 28:676–694
Muthunarayanan V, Santhiya M, Swabna V et al (2011) Phytodegradation of textile dyes by Water Hyacinth (*Eichhornia crassipes*) from aqueous dye solutions. Int J Environ Sci 1:1702–1717
Mutiyar PK, Mittal AK (2013) Occurrences and fate of an antibiotic amoxicillin in extended aeration-based sewage treatment plant in Delhi, India: a case study of emerging pollutant. Desalin Water Treat 51:6158–6164
Mutiyar PK, Mittal AK (2014) Occurrences and fate of selected human antibiotics in influents and effluents of sewage treatment plant and effluent-receiving river Yamuna in Delhi (India). Environ Monit Assess 186:541–557
Naamala J, Jaiswal SK, Dakora FD (2016) Antibiotics resistance in rhizobium: type, process, mechanism and benefit for agriculture. Curr Microbiol 72:804. doi:10.1007/s00284-016-1005-0
NAAS (2010) Antibiotics in manure and soil- a grave threat to human and animal health, Policy paper no 43. National Academy of Agricultural Sciences, New Delhi, p 20
Nam JJ, Song BH, Eom KC, Lee SH et al (2003) Distribution of polycyclic aromatic hydrocarbons in agricultural soils in South Korea. Chemosphere 50:1281–1289
Nussbaumer T (2003) Combustion and co-combustion of biomass: fundamentals, technologies, and primary measures for emission reduction. Energy Fuel 17:1510–1521
O'Connor GA (1996) Organic compounds in sludge-amended soils and their potential uptake by crop plants. Sci Total Environ 185:71–81
Oleszczuk P, Baran S (2004) Concentration of PAHs in plants growing on soils amended with various doses of sewage sludge. Arch Environ Prot 30:35–50
Orecchio S (2010) Contamination from polycyclic aromatic hydrocarbons (PAHs) in the soil of a botanic garden localized next to a former manufacturing gas plant in Palermo (Italy). J Hazard Mater 180(1–3):590–601
Pandey RK, Rai SN (1993) Effect of varying doses of pesticides on nitrifying bacteria and nitrogen transformations. Indian J Agric Chem 26(2/3):123–132

Pandey PK, Patel KS, Lenicek J (1999) Polycyclic aromatic hydrocarbons: need for assessment of health risks in india? Study of an urban-industrial Location in India. Environ Monit Assess 59:287–319

Pankhurst CE (1977) Symbiotic effectiveness of antibiotic-resistant mutants of fast- and slow-growing strains of Rhizobium nodulating Lotus species. Can J Microbiol 23:1026–1033

Pelosi C, Barot S, Capowiez Y, Hedde M, Vandenbulcke F (2014) Pesticides and earthworms- a review. Agron Sustain Dev 34:199–228

Pereira L, Alves M (2012) Dyes-environmental impact and remediation. In: Malik A, Grohmann E (eds) Environmental protection strategies for sustainable development, Strategies for sustainability. Springer Science+Business Media B,V, Dordrecht, pp 111–162

Perra G, Renzi M, Guerranti C et al (2009) Polycyclic aromatic hydrocarbons pollution in sediments: distribution and sources in a lagoon system (Orbetello, Central Italy). Transit Waters Bull 3:45–58

Phillips I, Casewell M, Cox T et al (2004) Does the use of antibiotics in food animals pose a risk to human health? A critical review of published data. J Antimicrob Chemother 53:28–52

Puvaneswari N, Muthukrishnan J et al (2006) Toxicity assessment and microbial degradation of azo dyes. Indian J Exp Biol 44:618–626

Quantin C, Joner EJ, Portal JM, Berthelin J (2005) PAH dissipation in a contaminated river sediment under oxic and anoxic conditions. Environ Pollut 134:315–322

Ragnarsdottir KV (2000) Environmental fate and toxicology of organophosphate pesticides. J Geol Soc 157:859–876

Rost H, Loibner AP, Hasinger M, Braun R, Szolar OHJ (2002) Behavior of PAHs during cold storage of historically contaminated soil samples. Chemosphere 49:1239–1246

Ryan JA, Bell RM, Davidson JM, O'Connor GA (1988) Plant uptake of non-ionic organic chemicals from soils. Chemosphere 17:2299–2323

Safe S (2004) Endocrine disruptors and human health: is there a problem. Toxicology 205:3–10

Saha JK, Sharma AK (2006) Impact of the use of polluted irrigation water on soil quality and crop productivity near Ratlam and Nagda industrial area. IISS Bulletin, Indian Institute of Soil Science, Bhopal, p 26

Saha JK, Panwar N et al (2010) An assessment of municipal solid waste compost quality produced in different cities of India in the perspective of developing quality control indices. Waste Manag 30:192–201

Saiyed HN, Bhatnagar VK, Kashyap R (2003) Impact of pesticide use in India. Asia Pacific Newsletter: 1999–2003. http://www.ttl.fi/Internet/Eng-lish/Infotion/Electronic+journals/Asian-Pacific+News-let-ter/1999-03/05.htm

Santodonato J, Howard P et al (1981) Health and ecological assessment of polynuclear aromatic hydrocarbons. J Environ Pathol Toxicol Oncol 5:1–364

Schauer JJ, Kleeman MJ, Cass GR et al (2001) Measurement of emissions from air pollution sources, 3. C_1–C_{29} organic compounds from fireplace combustion of wood. Environ Sci Technol 35(9):1716–1728

Sharma BM, Bharat GK, Tayal S et al (2014) Environment and human exposure to persistent organic pollutants (POPs) in India: a systematic review of recent and historical data. Environ Int 66:48–64

Singh RP (2001) Comparison of organochlorine pesticide levels in soil and groundwater of Agra, India. Bull Environ Contam Toxicol 67:126–132

Singh AN, Singh JS (2006) Experiments on ecological restoration of coal mine spoil using native trees in a dry tropical environment, India: a synthesis. New For 31:25–39

Singh KP, Malik A, Sinha S (2007) Persistent organochlorine pesticide residues in soil and surface water of northern Indo-Gangetic alluvial plains. Environ Monit Assess 125:147–155

Škrbić B (2007) Organochlorine and organophosphate pesticide residues in wheat varieties from Serbia. Food Addit Contam Part A Chem Anal Control Expo Risk Assess 24:695–703

Smith DJT, Edelhauser EC, Harrison RM (1995) Polycyclic aromatic hydrocarbon concentrations in road dust and soil samples collected in the United Kingdom and Pakistan. Environ Technol 16:45–53

Someya M, Ohtake M, Kunisue T et al (2010) Persistent organic pollutants in breast milk of mothers residing around an open dumping site in Kolkata, India: specific dioxin-like PCB levels and fish as a potential source. Environ Int 36:27–35

Subhani A, Min L, Yong HC et al (2002) Alteration of certain soil microbiological indices of a paddy soil under anthropogenic stress. J Zhejiang Univ Sci A 3:467–474

Subramanian A, Tanabe S (2007) Persistent toxic substances in India. In: Li A, Tanabe S, Jiang G et al (eds) Developments in environmental science, vol 7, pp 433–485. doi:10.1016/S1474-8177(07)07009-X

Subramanian A, Kunisue T, Tanabe S (2015) Recent status of organohalogens, heavy metals and PAHs pollution in specific locations in India. Chemosphere 137:122–134

Suzuki M, Aizawa N, Okano G et al (1977) Translocation of polychlorobiphenyls in soil into plants: a study by a method of culture of soybean sprouts. Arch Environ Contam Toxicol 5:343–352

Swarup A, Wanjari RH (2001) Lessons from long-term fertility experiments. All India coordinated research project on long-term fertilizer experiments to study changes in soil quality, crop productivity and sustainability. Indian Institute of Soil Science, Bhopal

Tahir U, Yasmin A, Khan UH (2016) Phytoremediation: potential flora for synthetic dyestuff metabolism. J King Saud Univ Sci 28:119–130

Tao S, Jiao XC, Xu FL et al (2006) Uptake of vapor and particulate polycyclic aromatic hydrocarbons by cabbage. Environ Pollut 140:13–15

Thacker N, Sheikh J, Tamane SM et al (2013) Emissions of polychlorinated dibenzo-p-dioxins (PCDDs), dibenzofurans (PCDFs), and dioxin-like polychlorinated biphenyls (PCBs) to air from waste incinerators and high thermal processes in India. Environ Monit Assess 185:425–429

Thiele-Bruhn S (2003) Pharmaceutical antibiotic compounds in soils-a review. J Plant Nutr Soil Sci 166:145–167

Topac FO, Dindar E, Uçaroglu S, Başkaya HS (2009) Effect of a sulfonated azo dye and sulfanilic acid on nitrogen transformation processes in soil. J Hazard Mater 170:1006–1013

Trapp S (2004) Plant uptake and transport models for neutral and ionic chemicals. ESPR – Environ Sci Pollut Res 11(1):33–39

Travis CC, Arms AD (1988) Bioconcentration of organics in beef, milk, and vegetation. Environ Sci Technol 22:271–274

Tsai PJ, Shih TS, Chen HL et al (2004) Assessing and predicting the exposures of polycyclic aromatic hydrocarbons (PAHs) and their carcinogenic potencies from vehicle engine exhausts to highway toll station workers. Atmos Environ 38:333–343

Turco RF, Moorman TB, Bezdicek DF (1986) Effectiveness and competitiveness of spontaneous antibiotic-resistant mutants of *Rhizobium leguminosarum* and *Rhizobium japonicum*. Soil Biol Biochem 18:259–262

Turrio-Baldassarri L, Alivernini S, Battistelli CL et al (2007) A study on PCB, PCDF/PCDF industrial contamination in an mixed urban/agricultural area significantly affecting the food chain and the human exposure. Part I: soil and feed. Chemosphere 67:1822–1830

Turrio-Baldassarri L, Abate V, Battistelli CL et al (2008) PCDF/PCDF and PCB in human serum of differently exposed population groups of an Italian city. Chemosphere 73:228–234

Turrio-Baldassarri L, Alivernini S, Carasi S et al (2009) PCB, PCDD and PCDF contamination of food of animal origin as the effect of soil pollution and the cause of human exposure in Brescia. Chemosphere 76(2):278–285

Uera RB, Paz-Alberto AM, Sigua GC (2007) Phytoremediation potentials of selected tropical plants for ethidium bromide. Environ Sci Pollut Res 14:505–509

van Bohemen HD, van de Laak WHJ (2003) The influence of road infrastructure and traffic on soil, water, and air quality. Environ Manag 31:50–68

van Elsas JD, Chiurazzi M, Mallon CA et al (2012) Microbial diversity determines the invasion of soil by a bacterial pathogen. Proc Natl Acad Sci U S A 109:1159–1164
Van Metre PC, Mahler BJ, Wilson JT (2009) PAHs underfoot: contaminated dust from coal-tar sealcoated pavement is widespread in the United States. Environ Sci Technol 43(1):20–25
VanRooij JGM, Bodelierbade MM, Jongeneelen FJ (1993) Estimation of individual dermal and respiratory uptake of polycyclic aromatic-hydrocarbons in 12 coke-oven workers. Brit J Ind Med 50:623–632
Watkinson AJ, Murby EJ, Kolpin DW et al (2009) The occurrence of antibiotics in an urban watershed: from wastewater to drinking water. Sci Total Environ 407:2711–2723
Webb KE, Fontenot JP (1975) Medicinal drug residues in broiler litter and tissues from cattle fed litter. J Anim Sci 41:1212–1217
Webber MD, Pietz RI et al (1990) Plant uptake of PCBs and other organic contaminants from sludge-treated coal refuse. J Environ Qual 23:1019–1026
Wenzel WW, Adriano DC, Salt D, Smith R (1999) Phytoremediation: a plant-microbe-based remediation system. In: Bioremediation of contaminated soils, Agronomy monograph no. 37. SSSA, Madison, pp 457–508
Wetzel A, Werner D (1995) Ecotoxicological evaluation of contaminated soil using the legume root nodule symbiosis as effect parameter. Environ Toxicol Water Qual 10:127–133
White CM, Lee ML (1980) Identification and geochemical significance of some aromatic components of coal. Geochim Cosmochim Acta 44:1825–1832
Whitfield Aslund ML, Zeeb BA, Rutter A, Reimer KJ (2007) In situ phytoextraction of polychlorinated biphenyl – (PCB) contaminated soil. Sci Total Environ 374:1–12
WHO (2001) Bathing water quality and human health, HO/SDE/WSH/01.2, Geneva
Wild SR, Jones KC (1992) Polynuclear aromatic hydrocarbons uptake by carrots grown in sludge amended soil. J Environ Qual 21:217–225
Wild SR, Harrad SJ, Jones KC (1994) The influence of sewage sludge applications to agricultural land on human exposure to polychlorinated dibenzo-p-dioxins (PCDDs). Environ Pollut 83:357–369
Wilson C, Tisdell C (2001) Why farmers continue to use pesticides despite environmental, health and sustainability costs. Ecol Econ 39:449–462
Winckler C, Grafe A (2001) Use of veterinary drugs in intensive animal production. J Soils Sediments 1:66–70
Zhan X, Liang X, Xu G, Zhou L (2013) Influence of plant root morphology and tissue composition on phenanthrene uptake: stepwise multiple linear regression analysis. Environ Pollut 179:294–300
Zhang Y, Tao S, Shen H, Ma J (2009) Inhalation exposure to ambient polycyclic aromatic hydrocarbons and lung cancer risk of Chinese population. Proc Natl Acad Sci U S A 106:21063–21067
Zhou Q (2001) Chemical pollution and transport of organic dyes in water soil crop system of the chinese coasts. Bull Environ Contam Toxicol 66:784–793
Zobiole LHS, Kremer RJ, Oliveira RS, Constantin J (2011) Glyphosate affects micro-organisms in rhizospheres of glyphosate- resistant soybeans. J Appl Microbiol 110:118–127

Chapter 6
Collection and Processing of Polluted Soil for Analysis

Abstract Risk assessment of polluted land generally involves formulation of site specific remedial strategy, which further depends on history and type of polluting activities, contaminant migration pathways and their visible impact on ecosystem. Methods of soil sample collection and their processing before analysis in the laboratory become highly important for accurate interpretation of results and devising appropriate strategy for remediation of the site. Several statistical and non-statistical methods have been used for identifying sampling locations, each of which has advantages over other. Systematic and judgmental sampling methods are preferred where contaminant migration history is known and polluted area requires more intensive sampling due to high heterogeneity. On the other hand, statistical methods avoid sampling biasness and minimize uncertainty related errors in situations. Processing and storage of samples before analysis of samples depend on the type of contaminants. This chapter also discusses on selection of appropriate instruments for soil sampling, methods of processing and storage as well as precautionary measures for minimizing errors of analysis.

Keywords Soil • Polluted site • Collection • Sampling • Processing • Storage

Soil samples collection and their processing are the initial steps during impact assessment process of soil pollution cases. Although these are also integral part of soil fertility evaluation, collection and processing of soils demands extra care and precautions during EIA due to several following reasons.

(i) Stakeholders in this case are more than one (only farmer or researcher in the other case), which may include aggrieved party (land owner), government, judiciary, polluting agency, researcher etc. This demands satisfactory protocol for all the stakeholders.
(ii) In many cases, pollutants to be determined in samples are present in very small concentration requiring extra precision and accuracy of determination through minimization of error in sample handling.
(iii) In most cases, contaminants migration pathway (both horizontal and vertical) are not known on polluted sites, which makes impact assessment more complex and uncertain. On the contrary, agricultural land are relatively uniform

J.K. Saha et al., *Soil Pollution - An Emerging Threat to Agriculture*, Environmental Chemistry for a Sustainable World 10, DOI 10.1007/978-981-10-4274-4_6

within a limited boundary area due to similar cropping and soil management history which makes soil sampling for fertility evaluation is relatively simple.

This chapter briefly describes different sampling and processing methodology (including tools) usually followed under different pollution scenario with their advantages and uncertainties.

6.1 Environmental Risk Assessment

Environmental risk assessment of the soil pollution is a site based approach, which provides a decision support during formulating remedial strategy. The major objective of risk assessment is to assess whether a contaminated site poses any harmful effect on organisms residing in the ecosystem. In the perspective of human health, risk assessment generally involves following sequential steps (Malherbe 2001):

1. Site characterization
2. Exposure assessment
3. Toxicity assessment
4. Risk characterization

The 1st step of the risk assessment (i.e., site characterization) is very important as it influences the success of decision making process and remedial actions. Sampling of soils is an important part of site characterization. It may be carried out in one or more steps depending on the existing information on the site and the objective(s) of site characterization. Due care is needed to devise appropriate strategy as it has consequences on cost of remediation and public health. Following points should be given due care before initiating sampling process for achieving its objectives.

- Identifying site location
- Determining sample density
- Selection of sampling method
- Type of sample
- Depth of sample

In case of polluted land, soil samples are collected with one or more of following different objectives

1. To identify the exact locations having accumulated pollutants
2. To identify the nature of contaminants and quantify their concentrations
3. To assess quantitatively the area and volume of land/soil mass requiring remediation
4. To evaluate quantitatively the accumulation of toxic elements/components in the soil fractions which contribute more in the exposure pathway(s)

The first two objectives are most commonly adopted in India and other countries mostly to fulfill different legal obligations in reply to public interest litigations. Care and diligence should be exercised to ensure that the samples collected were truly representative of the environmental components to be studied. Different types sampling pattern have been used depending on the purpose, cost and availability of time to deliver outcome, which can be categorized into statistical and non-statistical approaches.

6.2 Sampling Pattern Selection

Several factors like history of contamination, nature of contaminants, and objective of sample collection will determine sampling pattern selection. The sampling cost, time, error and its efficacy is influenced by sampling pattern. If the concentration levels of a contaminant are known or are suspected to exhibit periodic spatial variations, the sampling pattern should be oriented in such a way that it will not be in or out of phase with the periodicity of the variations. Where periodic variations are known to exist but the orientation of the periodicity is not known, a 'stratified random' sampling pattern should be considered.

6.2.1 Sampling Based on Non-statistical Approach

6.2.1.1 Judgmental Sampling Pattern

For this method, sampling points are selected on the basis of the investigator's knowledge of the probable distribution of contaminants at the site. It is an efficient sampling method which makes use of the site history and field observations but has the disadvantage of being potentially biased. The quality of the sampling results depends on the experience of the investigator and the available site history information. This technique may be used during preliminary investigation to recognise the nature of contamination before conducting systematic sampling.

Sampling Location Identification The selection of sampling locations depends upon the study objectives, the data quality objectives, preliminary information, and on-site conditions (Paetz and Wilke 2005). While sampling from the identified locations the details of the following on-site conditions need to be considered:

(i) Local topography
(ii) Climatic conditions
(iii) Bedrock/Parent material type
(iv) Soil type and its physicochemical characteristics (texture, pH, organic matter content etc.)
(v) Ground cover

(vi) Water and its proximity to the location site
(vii) Location of contaminant source and its background level of contamination
(viii) Nature of contaminants
(ix) Contaminant pathway
(x) Ecological receptors

For the purpose of soil sample collection, the site should be divided into areas representing similar material types (like tailings, natural soil or rock), similar contaminants, and other site-specific features as indicated in the site history. Variation in contamination distribution in soil differs on depth as well as surface location. Finer sized particles and organic matter have relatively more contaminant than the coarse sized particle and its deposition on the site location (surface and depth) depends on flow characteristics (velocity) of a body of water. Hence, the soil with finer and high organic carbon content, the sampling is most likely from depositional zones. On the other hand, soils with low organic carbon and coarse texture have few pollutants and are mostly found in erosion zones. Therefore, sampling location selection greatly influences the analytical results.

6.2.1.2 Systematic Sampling

Systematic sampling is collection of samples from points at regular and even intervals. The site is divided into rectangular or triangular grids and each grid points are given numbers. If sampling is done from each of the grid points, it is termed as systematic sampling (see Fig. 6.1 as an example). This method of sampling ensures complete site coverage and a homogeneous distribution of samples, hence minimizing bias in the estimation of mean concentrations of the pollutant. Systematic sampling does not generate clusters of sampling points and is easier to use for identifying survey sampling locations than random sampling technique. A square grid is the most preferred type of systematic sampling pattern. Major advantage of the method is that results of analysis of the collected samples can be used to prepare geographical maps showing spatial variability in contaminants concentration using geostatistical tool.

6.2.2 *Statistical Approach*

In a statistically-based sampling strategy, samples should be collected in an unbiased manner, i.e, all locations within the sampling area should have an equal chance of being selected as a sampling point. Sampling patterns can be classified into following two basic types.

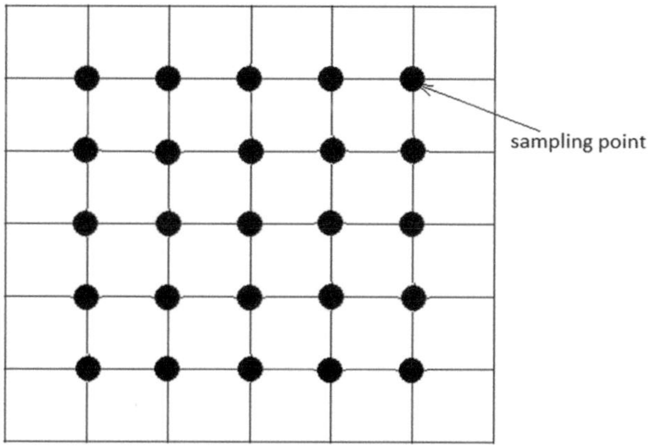

Fig. 6.1 Schematic layout showing 'systematic sampling' locations

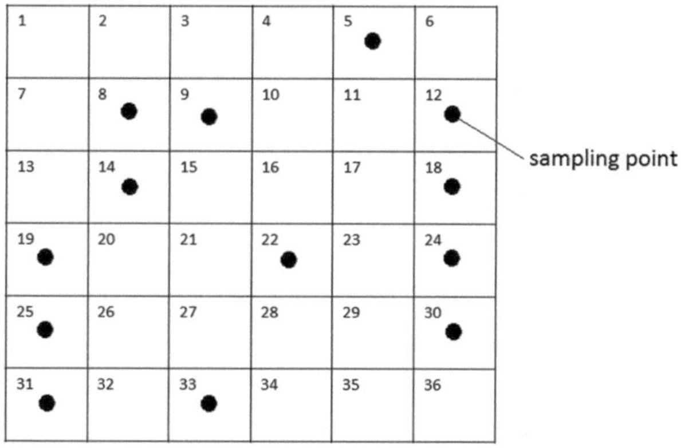

Fig. 6.2 Schematic layout showing 'random sampling' pattern

6.2.2.1 Random Sampling Pattern

For random sampling pattern, the site is divided into rectangular or triangular grids and each grid points are numbered (see Fig. 6.2 as an example). Sampling points are selected randomly but not arbitrarily. A legitimate 'random number generator' should be used to determine sampling point coordinates. This randomization process ensures that any location within the sampling area has an equal chance of being selected as a sampling point. While random sampling is statistically unbiased, sampling points by chance, can cluster together. Highly contaminated regions may be insufficiently sampled making it inefficient in detecting hot spots. Although, an overall picture of the spatial distribution of the contamination may be obtained, it has limited use for decision making during remediation process.

6.2.2.2 Stratified Sampling Pattern

At the first step, the site is divided into sub-areas according to geological and geographical features, nature of the contamination, previous usage pattern of the

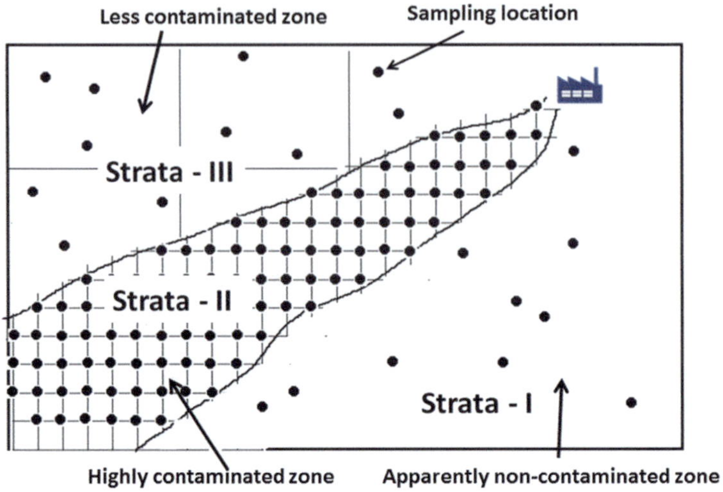

Fig. 6.3 Schematic layout showing 'stratified sampling' pattern

site, intended future use of the sub-area and other relevant factors. Each sub-area can then be treated as an individual site; and subsequently different sampling patterns and sampling densities can be applied (see Fig. 6.3 as an example). A stratified sampling pattern approach is best suited for investigations of large sites with complex contaminant distributions. This sampling pattern may require a more complex statistical analysis. However advantageously, it avoids the risk of insufficient sampling from highly contaminated area.

6.2.2.3 Stratified Random Sampling

Polluted site is divided into several blocks/strata through grids and subsequently, exact sampling points within each of the blocks are determined randomly (see Fig. 6.4 as an example). If minimization of number of sampling points is necessary in view of technical limitations, the blocks/strata for sampling may be selected statistically using random number and subsequently, sampling points may be selected randomly (see Fig. 6.5 as an example).

6.2.2.4 Multi-stage Sampling

Selection of appropriate sampling strategy is only possible if prior information related to site characteristics and soil contamination is available. In case such facts are not available, a two- or multi-stage sampling is often better for devising appropriate sampling strategy, which can be modified and improved at each step. However, time and cost of sampling increases with stage of sampling. Collection of information on nature of polluting activities, mode of contaminants spread, landscape, slope, present and past land use helps in reducing number of sampling stages.

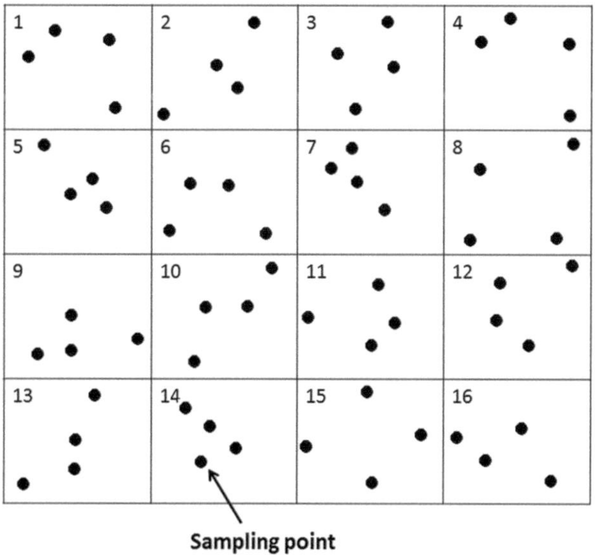

Fig. 6.4 Schematic layout showing 'stratified random sampling' pattern where randomised sampling is done for each block

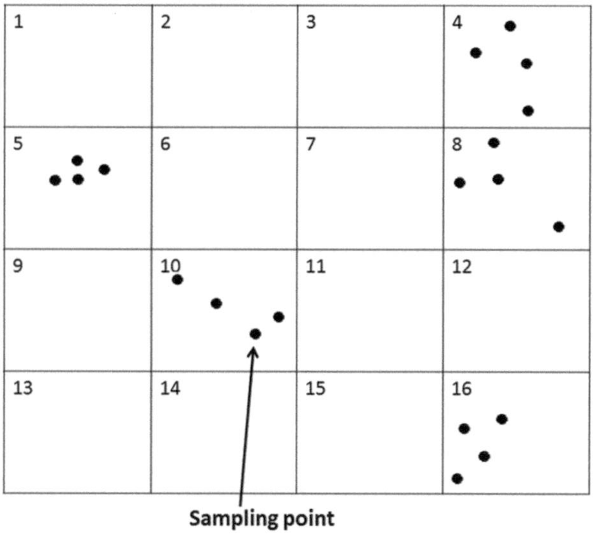

Fig. 6.5 Schematic layout showing 'stratified random sampling' pattern where randomised sampling is done for few blocks selected randomly

6.3 Sampling Uncertainty

One can never be 'certain' about an answer derived from sampling. Therefore, the uncertainty must be specified for a statistical statement to have meaning. In statistics, uncertainty is technically referred to as *confidence* level or *risk*. Confidence level is quantified as a percentage. A 95% confidence' level is equivalent to a 0.05 risk. For example, if a particular statistical statement is quoted as having a 95%

confidence level or a 0.05 risk, this implies that on average the statistical statement will be correct 95 out of 100 times.

In a set of normally distributed measurements, 'mean, μ' is used as true value. In environmental analysis, 'variance, σ^2' (indicating spread of the measurements from true value) of measurements is also used to represent 'measure of uncertainty' and calculated using the formula:

$$\sigma^2 = \frac{\sum (x_i - \mu)^2}{n - 1}$$

where, x_i is ith measurement and 'n' is the number of measurements.

Variance during measurement can arise from two sources; during sampling and during analysis of contaminants. Therefore, uncertainty in sampling can be derived as:

$$\sigma_s^2 = \sigma_T^2 - \sigma_a^2$$

where, σ_T^2 is total variance due to sampling and is estimated by collecting and analyzing several samples for a contaminant which are expected to produce identical results;

σ_a^2 is variance due to analytical measurements and is estimated from the several analysis of same sample for the contaminant; and

σ_s^2 is variance due to sampling.

6.3.1 Number of Samples to Be Collected

During EIA, 'average concentration' of a contaminant in a land area is most frequently sought parameter. In view for the inherent uncertainty in sampling, number of samples collected to determine 'average concentration' becomes an important determinant for the quantum of uncertainty or error associated the conclusion.

A simple approach to calculate the number of samples is to collect and analyze for estimating an overall standard deviation, σ_s from few samples through preliminary study (Mitra and Kebbekus 1998). Using Students-t distribution, the number of samples 'n' required for achieving a given confidence level is calculated as:

$$n = \left(\frac{t \times \sigma_s}{\sigma_s^*}\right)^2$$

where 't' is the t-statistic value selected for a given confidence level, and 'σ_s^*' is the acceptable level of uncertainty (i.e. standard deviation). If an experimental value of σ_s is not available, an estimate may be done from previous similar studies.

6.4 Sampling Instruments

Depending upon the type of samples required and soil type, different sampling methods and equipment's are used for soil sample collection. Soil samples are collected using devices like scoop, digger, auger and drills. Surface litters of non-decomposed plant materials should be removed before sampling so as to ensure that only soil sample is collected. The use of hand auger ensures the reproducibility of test results, and guarantees high quality data collection. Following sample collection, the soil has to be uniformly homogenized in a uncontaminated container and properly stored in a labeled sample containers. While using devices for sampling, due care must be taken to avoid devices plated with chrome or other metals. A stainless steel or lab spoon or plastic scoop suits better for most of sampling applications. Since most of the probable contamination occurs on the surface of the sample which comes in contact with the sampling tool, contamination can be reduced by collecting samples with high volume to surface ratio.

6.4.1 Sampling Surface Soils with a Trowel/Spade and Hand Scoop

While collecting soil samples with a scoop and trowel the following protocol has to be kept in mind:

1. Vegetation and stubbles should be removed from soil surface area. Thereafter, soil volume from the desired depth (usually in the zone of maximum root activity of annual crops) is loosened with trowel/spade and subsequently, the desired volume of soil can be collected from the sampling area with a scoop. Collected sample is thereafter, transferred into an appropriate samples container.
2. For obtaining a composite sample, bulk samples are collected from several points from a small area (usually $<1000\ m^2$) and are homogenized in a container made of stainless steel using the stainless steel spoon, pestle or trowel. If sample volume is required to be reduced, the homogenized sample is spread on plastic sheet in a circular shape and four quarters are marked. At each time any two opposite quadrants were selected and the rest quadrants were rejected. This process of homogenizing- quartering is repeated if further reduction of sample volume is required.
3. If the samples type is solid, chemical preservation is not recommended.
4. Sample containers need to be properly labeled and tagged. All the relevant information like depth, location, color, proximity to source/water bodies etc. should be recorded on the soil sample data sheets.

6.4.2 *Sampling Surface Soils with a Hand Auger*

The hand held auger has a series of extension rod with 'T' shaped handle along with a thin-walled sampler tube. While sampling for a desired depth, a hole is made with the help of auger rod and the rod has to be withdrawn from the hole. Thereafter the auger tip is replaced with a tube core sampler, is lowered down the bore hole, and subsequently is withdrawn along the sample. During this process The following precautions and steps should be followed while collecting soil samples with a hand auger:

1. The auger has be inserted at 0° to 45° angle from vertical which helps in minimizing spillage of the sample from the tube sampler. While extracting samples from the tube auger, tilting of the sampler may be required.
2. Make the auger to rotate once or twice so as to cut a core of material.
3. The auger should be slowly withdrawn, while ensuring the slot to face upward.

6.4.3 *Sampling Subsurface Soils with a Drill Rig*

A coring device, handle, and hammer are required for this sampling method. The procedure for collecting subsurface samples using a sample coring device is described below:

1. The sampler is placed in a position perpendicular to the land surface. In case of firm/hard soil, the sampler has to be used with the drive hammer, whereas in case of soft soil, "T" shaped handle is sufficient for sampling. While using the "T" handle in soft soil for sampling, downward pressure has to be applied until the device reached the desired soil depth.
2. Thereafter the sampler is withdrawn by pulling the hammer upward direction and subsequently removing the hammer from the sampler.
3. The sample thus collected has to be transferred to an uncontaminated stainless steel or plastic bucket and it should be mixed thoroughly so as to achieve a representative homogeneous sample.
4. If the sample type is volatile in nature, it has to be collected directly from the bucket before mixing the sample in order to minimize the volatilization loss of contaminants.

6.5 Preventing Sample Cross-Contamination

Sampling equipment should not be made of a material that will contaminate the sample (e.g., sampling devices should not be painted, plated, greased, or have some other type of chemically treated surface). Stainless steel is usually preferred, and inert plastic equipment may also suit in some cases while collecting metal-contaminated soils. Prevention of cross-contamination in the field is essential, and the level of effort required will depend upon the characteristics of the contaminant (s) of concern. Sampling devices should be cleaned between sampling locations and sites as well as at the end of sampling by:

- Wiping, washing, or scrubbing the equipment
- Rinsing the equipment with purified water
- Equipment can also be cleaned with non- phosphate containing soap, but the equipment must be thoroughly rinsed following a soap wash to ensure that the samples do not become contaminated with soap residue.
- If the samples are to be used to collect soil containing organic contaminants, rinsing the equipment should be done at least three times with a solvent (acetone or hexane) followed by air-drying.

6.6 Soil Sample Collection from Profile

If soil horizons are not distinct, soil samples should be collected at depth-wise interval. Depth-wise collection of soil samples is appropriate in many industrial, commercial, residential, or urban soils where the surface soil horizons have been mixed or disturbed either through construction, landfilling, dumping or capping. As a standard, soil sample collection at 0 to 15 cm, 0 to 30 cm, 15 to 30 cm, 15 to 60 cm, and 30 to 60 cm depths are followed (Mason 1992; Dalpé and Hamel 2008). If the soil is contaminated with atmospheric deposition, 0 to 5 cm depth soil sample collection is recommended. Similarly, sampling not deeper than 30 cm (where most of microbes habitat exists) is appropriate in case of any ecologically relevant studies (Spurgeon et al. 2002; ISO 2006). For any specific crop related studies, rooting depth is considered for sampling zone (Dalpé and Hamel 2008).

There are some challenges associated with sampling forest top soil (*Ah* horizon) as it can sometimes be difficult to identify the exact boundary between overlying organic matter (forest floor) and the mineral horizon. For comparison of certain soil characteristics (e.g., soil carbon pools) among forest management practices, the collection of soil samples by depth may be preferred (Belanger and Van Rees 2008).

6.7 Composite Sampling

Composite soil sampling is defined as a technique that combines a number of discrete samples collected from a body of material into a single homogenized sample so as to represent the average conditions in the sampled body of material for the purpose of analysis. In the contaminated site, composite sampling is used to reduce sampling and analytical costs. The principal limitation of composite sampling is the potential hotspots to remain undiscovered due to the inherent characteristics of the process. For example, a discrete subsample may contain a high concentration of contaminant that may be masked due to the dilution effect in the composite sampling procedure. Though Composite sampling may be useful for assessment of soil fertility of a piece of land with definite ownership boundary; it should not be used for site-specific health and ecological risk assessments. Composite sampling is not generally recommended to determine the extent of site contamination, for health and safety risk assessment related to remediation activities and also during validation purpose following remediation of contaminated site. Further composite soil sampling should not be practiced for soils contaminated with semi-volatile or volatile compounds such as polycyclic aromatic hydrocarbons and petroleum hydrocarbon.

6.8 Reference Soil Collection for Testing

While conducting a site-specific soil testing, one of the most important and sometimes the most challenging tasks is the collection of a well-matched, site-specific, field-collected reference soil(s). A site-specific field-collected reference soil is a one that has physical and chemical properties similar to those of the contaminated field-collected soil(s), but is not contaminated. The importance of including a reference soil as a control soil in a site-specific ecotoxicity assessment is that it enables the investigator to differentiate between the influence of the soil's physical and chemical properties on test organism's performance from that of soil contaminants. In many cases inherent soil properties such as texture, pH, electrical conductivity, fertility, and organic matter content strongly influence the growth and vigour of plants, the growth and reproduction of invertebrates, and the activity and diversity of soil microorganisms. In order to avoid a misinterpretation of the toxicity data, well-matched reference soils should therefore be included in any site-specific ecotoxicity assessment.

General guidance on the degree to which a difference in soil physicochemical properties between contaminated and reference soils is considered acceptable, is difficult to prescribe as it will vary with the type of soil, toxicity test being performed, test organisms chosen etc. Some of the most critical soil physicochemical properties that are normally considered to match for both the reference and contaminated soils are:

- Particle size distribution (Per cent distribution of clay, sand, and silt)
- Cation exchange capacity
- Fe and Al oxides content
- Content of silicates
- Type of clay minerals

Other soil properties may also be critical to match depending on the type of testing to be conducted. More than one reference soil might be required to match the contaminated soil samples, especially if there is a wide range of physicochemical properties among site soils.

6.9 Sample Identification/Coding

To ensure preservation of the integrity of the samples collected, sample identification/coding should be written on a permanent label mentioning specific details so as to ensure sample authenticity. Samples should be properly sealed, carrying labels with information such as, identification code or sample number (location of sample), date and time of sampling, type of preservation, sample description etc.

6.10 Sample Preservation, Containers, Handling, and Storage

For soil sampling, different containers have to be used for specific purpose like (i) aluminum foil for total hydrocarbons determination, and (ii) polythene bags/ containers for general physico-chemical and heavy metal analysis. Chemical preservation of solids is generally not recommended and therefore, cooling is usually the best approach, supplemented by the appropriate holding time (Zagury et al. 2002).

6.11 Preparation of Soil Samples

The extent and type of soil preparation before analysis is dependent on the type of biological testing to be conducted, soil type, and the condition of the soil sample upon receipt at the laboratory.

6.11.1 Preparation of Soil Samples for Microbial Testing

Due care must be taken while preparing unconsolidated soil samples for microbial tests, including microcosm tests for functional capability, biomass, or diversity. During soil preparation, visible plant materials, stubbles and fauna should be removed properly. Drying-rewetting procedure is commonly used while preparing soils. If possible air-drying should be avoided because it changes microbial composition (Topp et al. 2008). Also wetting-drying events can induce significant changes in microbial C and N dynamics. Pre-incubating soils by adding moisture can significantly influence respiration rates, substance utilization efficiency, nitrification potential, microbial biomass, nitrate and ammonium concentrations of soil samples, and these changes can persist for more than a month following the stress (Fierer and Schimel 2002). Re-wetting after drying also causes bursts of respiration and growth of distinct populations of bacteria (ISO 2006). Therefore testing soils as soon as possible after sample collection, and thereby avoiding the necessity of drying and rewetting, is recommended. If the field moisture of the soil is too high to sieve, the soil should be air-dried at ambient temperature to workable moisture content. Normally, the soil samples were sieved through 2 mm sieve and aerobic soil conditions has to be maintained. However, if the soil is high in organic matter, 5 mm mesh sieve is recommended and the sieved samples has to be homogenized thoroughly (Bailey et al. 2008).

6.11.2 Preparation of Soil Samples for Toxicity Testing

This section describes recommended procedures to prepare unconsolidated samples for testing other than microbial testing.

6.11.2.1 Drying

Drying of soil samples might be required if the sample is too moist to homogenize and sieve. Soil samples should not be dried, if the method or study objectives require soils to be tested at field moisture level. Soils should be air-dried at ambient temperature to "workable" moisture content. A "workable" moisture content is a soil moisture content usually between field moist and completely air- dried soil (completely air-dried soil has moisture in equilibrium with the humidity of the ambient air). A "workable" moisture content result in soil sufficiently dry to allow sieving without the production of dust and/or permits the soil to flow freely to facilitate thorough homogenization. Soils should not be completely air-dried or oven- dried during analysis for microbial parameters. Completely air-drying or oven- drying soil to any degree, causes a number of reactions in the soil: micro-fauna and flora die or become senescent; contaminants dissolved in soil pore water become concentrated, precipitate or coagulate; and, solid organic matter deforms, can become hydrophobic and/or expose underlying mineral surfaces (Sheppard and Addison 2008).

Method of Air-Drying of Soils

- Spread out the sample on a sheet of thick polyethylene or polypropylene plastic or clean stainless steel surface and air-dry the soil only until it is dry enough to sieve/homogenize; leave sufficient margins between the soil and the edge of the work surface to avoid contamination from other sources.
- The time required to dry might range from 1 hour to 5 days, depending on the moisture content of the soil, the texture of the soil, and the temperature and humidity of the room.

- Ensure that while drying, the soil sample is placed in a location safe from disturbance, dust and/or contamination.
- If there is too much standing water, decant the standing water before spreading out the soil and then dry the soil as described above.

6.11.2.2 Sieving

Soil sample contains coarser materials (gravel, stones), plant roots, macro flora and fauna which has to be removed before analysis by sieving. Depending upon the soil analysis method, different mesh sieves (stainless steel) like 2, 4, 6, 8 or 10 mm were used for sieving dry samples. Sometimes even smaller sized mesh sieves (0.25, 0.50, 0.75 and 1 mm) are also used for specific analysis. For example, soil sample sieved by 0.5 mm mesh is recommended for organic carbon estimation. In most of the cases, soil samples are sieved through 2 mm sieve which contains all soil separates like sand, silt and clay in proportionate amount. While sieving, minimum disturbance has to be maintained so as to ensure that soil sample aggregate remains closer to the field condition. If soil samples of larger in size > 2 mm is removed entirely by sieving, it may sometimes leads to bias during contaminant determination. For example, removing plant roots or other plant related materials having size greater than > 2 mm sieve size where contaminants are associated, may result in underestimation of pollutants.

6.11.2.3 Crushing or Grinding

Soil samples are subjected to grinding, crushing or milling when their homogenization is difficult to attain by normal sieving method. Further, if the soil gets harder while drying as in case of clayey soil, crushing/grinding is recommended. It also helps in proper homogenization of organic matter along with soil separates. However, grinding of soil samples should be avoided as much as possible as it destroys the soil aggregates that represent the field condition.

6.11.3 Preparation of Soil Samples for Plant Available Nutrient Testing

6.11.3.1 Drying of Samples

Moist samples received in the laboratory for plant nutrient analysis should be air dried by spreading the samples on a polyethylene or polypropylene or plastic or wooden trays. Proper care should be taken to maintain the identity of each sample at all stages of preparation either by numbering or tagging the sample trays. After labeling, the trays may be air dried or placed under racks in a hot air cabinet whose temperature should not exceed 35 °C and relative humidity should be between 30 and 60%. Soil sample drying either at high temperature or by oven drying is not recommended as it can cause profound changes in soil parameter. For example, the microbial population and nitrogen transformation is highly affected by high temperature drying and exposure time. With excessive drying, soil potassium may be released or fixed depending upon the original level of available potassium. Exchangeable potassium will increase, if its original level is less than 1 meq/100 g soil (1 cmol kg^{-1}) and *vice-versa*, but the effect depends upon the nature of clay minerals in the soil. In general, excessive drying, such as oven drying of soil, affects availability of most of the nutrients present in the sample and should be avoided and therefore, only air drying is recommended.

Nitrate, nitrite and ammonium determinations must be carried out on samples brought straight from the field (Drury et al. 2008). These samples should not be dried. However, the results are expressed on oven dry basis by separately estimating moisture content in the samples.

6.11.3.2 Post Drying Care

Air dried samples are ground with a wooden pestle and mortar so that the soil aggregate are crushed but the soil particles do not breakdown. After grinding, the soil sample is sieved though 2 mm mesh sieve. While sieving, the entire sample lot should be used without retaining/discarding sample subset. If some sample portion is discarded or not used for sieving, the rejected/discarded portion may have element of interest that may leads to erroneous result while analyzing the sample. Therefore, repeated sieving is required to pass the soil sample as much as possible through 2 mm sieve, leaving (retaining) only the coarser fractions (gravels, stones), organic matter and concretions on the sieve. If the soil is to be analyzed for trace elements, containers made of copper, zinc and brass must be avoided during grinding and handling. After sieving, the sample should be mixed thoroughly and placed in an uncontaminated sample container. The sample container has to be labeled properly and placed with numbered wooden/plastic tray in a separate room.

6.12 Sampling Plan Checklist

Sampling plan or check list is a key component of the study plan which contains detailed description of the procedures to be followed during sample collection and preparation stage. A sampling plan is an important supporting document which includes all of the information like site location, site manager contact details, sampling point identification, sampling devices and procedures, sample density, field observations, sample preparations and handlings (packaging, transportation and storing).

References

Bailey VL, Bolton JH, Smith JL (2008) Substrate-inducted respiration and selective inhibition as measures of microbial biomass in soils. In: Carter MR, Gregorich EG (eds) Soil sampling and methods of analysis, Canadian society of soil science, 2nd edn. Boca Raton, CRC Press Taylor and Francis Group, pp 515–526

Belanger N, Van Rees KJC (2008) Sampling forest soils. In: Carter MR, Gregorich EG (eds) Soil sampling and methods of analysis, Canadian society of soil science. CRC Press Taylor and Francis Group, Boca Raton, pp 15–24

Dalpe Y, Hamel C (2008) Vesicular-arbuscular mycorrhiza. In: Carter MR (ed) Soil sampling and methods of analysis. Canadian society of soil science, Lewis Publishers, Boca Raton, pp 287-302

Drury CF, Hart SC, Yan XM (2008) Nitrification techniques for soils. In: Carter MR, Gregorich EG (eds) Soil sampling and methods of analysis, Canadian society of soil science. Boca Raton, CRC Press, Taylor and Francis Group, pp 495–513

Fierer N, Schimel JP (2002) Effects of drying-rewetting frequency on soil carbon and nitrogen transformations. Soil Biol Biochem 34:777–787

ISO (2006) Draft 10381-6. Soil quality sampling part 6: Guidance on the collection, handling and storage of soil under aerobic conditions for the assessment of microbial processes, biomass and diversity in the laboratory. International Organization for Standards, Geneva, Switzerland

Malherbe L (2001) Designing a contaminated soil sampling strategy for human health risk assessment. Eurolab/Eurachem International Workshop "Sampling", Lisbonne, Portugal

Mason BJ (1992) Preparation of soil sampling protocols: sampling techniques and strategies. United State Environmental Protection Agency, Washington, DC

Mitra S, Kebbekus B (1998) Environmental chemical analysis. Chapman and Hall/CRC, Boca Raton

Paetz A, Wilke BM (2005) Soil sampling and storage. In: Margesin R, Schinner F (eds) Manual of soil analysis – monitoring and assessing soil bioremediation, soil biology, vol 5. Springer-Verlag, Berlin/Heidelberg, pp 1–45

Sheppard SC, Addison JA (2008) Soil sample handling and storage. In: Carter MR, Gregorich EG (eds) Soil sampling and methods of analysis, Canadian Soc Soil Sci. CRC Press Taylor and Francis Group, Boca Raton, pp 39–49

Spurgeon D, Svendson C, Hankard P et al (2002) Review of sublethal ecotoxicological tests for measuring harm in terrestrial ecosystems, Environment Agency science report P5-063/TR1. Environment Agency, Bristol, p 191

Topp E, Tien YC, Hartmann A (2008) Bacterial community analyses by denaturing gradient gel electrophoresis. In: Carter MR, Gregorich EG (eds) Soil sampling and methods of analysis, Canadian society of soil science. CRC Press Taylor and Francis Group, Boca Raton, pp 567–576

Zagury GJ, Dudal Y, Bureau J, Bastien C, Chasse R (2002) Sample handling and preparation for estimation of mobility, bioavailability and toxicity of contaminants in soils. In: Sunahara GI et al (eds) Environmental analysis of contaminated sites, Ecological and environmental toxicology series. John Wiley and Sons Ltd, West Sussex, pp 9–23

Chapter 7
Assessment of Heavy Metals Contamination in Soil

Abstract Several heavy metals are toxic to organisms including human and therefore their entry into food chain from polluted agricultural land is required to be controlled. Management and remedial measures of metal contaminated sites depend on the level and distribution of contamination. In view of high heterogeneity of soils in respect of heavy metals concentration, mere analysis of a sample may not confirm contamination due to polluting activities. This chapter describes various approaches for assessing heavy metals contamination of soil and expected degree of threat it may pose to the environment. Importance of baseline concentration in areas with high polluting activity is thoroughly discussed along with statistical significance. Degree of contamination and its threat has also been estimated by researchers through various models. Methods of assessment of risks to organisms due to heavy metals contamination in soil involving critical analysis of their translocation within plants and to human through various routes have also been described. Different countries have adopted regulatory limits to protect the land from contamination based on different approaches which resulted wide variations in their limiting values. Soil testing methods for rapid assessment of contaminated site has also been described in this chapter.

Keywords Heavy metals • Assessment • Contamination • Risks • Organisms • Critical limits

Toxic heavy metals and other pollutants are increasingly introduced into the environment by different anthropogenic activities. Precautionary measure and management of heavy metal contaminated sites are easier to adopt and cost effective when contamination levels do not cross threshold limit and adverse effect on environment is exhibited. However, identification of a contaminated site (level of which has not reached threshold limit) is very difficult unless information on pollutant levels before pollution activity in soils for the area is available. In most of the sites particularly having lower levels of polluting activities, contamination could not be confirmed due to large variations in the natural heavy metal contents in soils. Soils of different regions naturally contain variable concentrations of different heavy metals due to differences in parent material, climate, land topography, texture etc. Many fold variations in metal contents in soils of different regions of the world have been reported (Kabata-Pendias and Pendias 1992; McBride 1994). Hence, natural

J.K. Saha et al., *Soil Pollution - An Emerging Threat to Agriculture*, Environmental Chemistry for a Sustainable World 10, DOI 10.1007/978-981-10-4274-4_7

concentration of heavy metals determined for a particular region may not be applicable for other regions. While investigating heavy metals contamination in soils of a particular region, it is therefore important to establish background/ baseline concentrations. Reliable data on this aspect are also needed because if these data are incorrect, legislators may mistakenly mandate guideline levels lower than those are realistically attainable (Davies 1992). Therefore, proper legislation can only be enacted when databases exist with accurate background trace element concentrations of soil. Although several EU countries and USA has determined baseline concentrations in soil for different regions, very few studies have been undertaken in this respect in low and middle income countries including India.

India's agricultural land mass has been categorized under 20 agro-ecological zones covering 103 soil suborders which clearly indicates large variability in soil properties (Bhattacharyya et al. 2013). Recommended soil management also varies according to the soil type (Rao and Saha 2012). Nevertheless, it is important to assess the background concentration which is essential for assessing the degree of metal pollution affected by anthropogenic activities and inputs and also for regulation of metals in water/soil systems (Zhao et al. 2007; Marandi and Karro 2008; Vencelides et al. 2010). Due to large variations in parent materials, topography and climate, different soil types have been developed in India (Kabata-Pendias 2011). Heavy metals of lithogenic (inherited from lithosphere) and pedogenic (concentration and distribution of lithogenic metals change due to pedogenic processes) origin ideally constitute the natural background level in soils. However, anthropogenic heavy metals (entered due to direct or indirect human activities) into the soil bodies cannot be ruled out even in areas far away from point sources having significant polluting activities.

7.1 Presence of Heavy Metals in Natural Soil

Heavy metals occur naturally in the soil as these are derived due to weathering of rocks. Soil formation from parent material occurs in two stages: first, physical and chemical changes in primary mineral constituents of the parent rocks (weathering); and secondly, formation of a soil profile from the weathered rock material (pedogenesis). Weathering and pedogenesis are continuous processes occurring simultaneously at the same site. Weathering involves dissolution, hydration, hydrolysis, oxidation, reduction, and carbonation processes. Temperature, rainfall, vegetation and rock types of a geographical region determine the severity of each of these processes, resulting into formation of soil types. Though rock types generally determine trace elements (including heavy metals) concentration in the initial phases of soil development, other three factors are responsible for their transformation and movement within the soil profile or outside the system. All these factors and processes ultimately determine background concentration of elements in the surface soil layers.

Table 7.1 Content of various elements in soils

Metals	World average for Soils, mg kg^{-1}	Range of trace elements in world Soils, mg kg^{-1}	Reported range of contents in soils of polluted area in India (Panwar et al. 2010; Krishna and Govil 2007, 2008; Bhagure and Mirgane 2011; Parth et al. 2011; Purushotham et al. 2012)
As	6.83	<0.1–66.5	0.073–411
B	42	<1–210	
Cd	0.41	0.01–2.7	0.05–39
Cr	59.5	1–1100	43–480
Co	11.3	0.1–70	7.6–97
Cu	38.9	1–140	21.2–373
F	321	<10–1194	
Pb	27	1.5–176	1.1–1833
Mn	488	7–9200	
Hg	0.07	0.008–1.11	1.11–1.54
Mo	1.1	0.1–7.35	2.3–15
Ni	29	0.2–450	14.1–465
V	129	6.3–500	18.4–452
Zn	70	3.5–770	65–7120

Adopted from Kabata-Pendias (2011) and Murthy (2008)

Natural background concentration is defined as the ambient concentration of chemicals in soils without human influence (Kabata-Pendias and Pendias 1992; Gough et al. 1994). Mean heavy metal concentrations varies widely from 0.06–1.1 mg kg^{-1} for Cd, 1.6–21.5 mg kg^{-1} for Co, 7–221 mg kg^{-1} for Cr, 6–80 mg kg^{-1} for Cu, 0.02–0.41 mg kg^{-1} for Hg, 4–55 mg kg^{-1} for Ni, 10–84 mg kg^{-1} for Pb, 17–125 mg kg^{-1} for Zn in uncontaminated soils of the world (McBride 1994). The content of various elements in the soils is depicted in Table 7.1. Wide variations have been observed in the contents of several heavy metals among the soil types as well as within a soil type (Kabata-Pendias and Pendias 1992). This is due to differences in parent materials (from which soil has developed), climate, land-use history, soil composition etc., influencing soil forming processes (McBride 1994). Multivariate statistical approach using principal component analysis (PCA) and cluster analysis as well as GIS tools have been used to find out large scale variability, origin (pedogenic or anthropogenic) and possible non-point sources of contamination of heavy metals in cultivated soils (Ratha and Sahu 1993; Garcia et al. 1996; Facchinelli et al. 2001; da Silva et al. 2016). Variability of Cr, Co and Ni was mainly controlled by parent materials; while Pb, Cu and Zn were controlled by long-term anthropogenic activities as indicated by large scale anomalies in spatial distribution map in Piemonte region of northwestern Italy (Facchinelli et al. 2001). In order to distinguish natural background (geogenic) concentration from anthropogenic contribution in soils, geochemical mapping technique has been used with the help of multifractal IDW (Inverse Distance Weighted) interpolation method, spectral analysis (S-A) and the concentration–area fractal method (C-A) (Cheng et al. 1996, 2000; Lima et al. 2003). In volcanic soils of the metropolitan and provincial areas of Napoli (Italy), Cicchella et al. (2005) used this technique to identify baseline concentration values of heavy metals. Maps obtained by the S-A method indicated anthropogenic contribution for some metals (Pb, Zn, Sb, Hg, Cd, Cr, Cu) due to the long period of constant human activities in the study area. These maps also showed a high geogenic contribution most probably derived from hydrothermal fluids of nearby active volcanic fields. The maps obtained using the S-A method and implemented in GeoDAS, reflected both the geogenic and anthropogenic contribution. Concentration–area fractal method (C-A) was used to classify the pixel values of baseline maps for determining background values to be used as reference values for the environmental Italian regulatory Law. Multivariate analysis has been used to estimate degree of contamination in surface soils of Guangdong Province in China due to intense industrialization and urbanization (Hu et al. 2013). Mean Cd, Cu, Zn, and As concentrations were over two times higher than the background values, with Cd, Cu, and Zn clearly contributed by anthropogenic sources. In Spain, similar technique was used to identify the sources of heavy metals in soils under vegetable cultivation (Mico et al. 2006). The content of Co, Cr, Fe, Mn, Ni and Zn were associated with parent rocks and corresponded to the first principal component called the lithogenic component. A significant correlation was found between lithogenic metals and some soil properties such as soil organic matter, clay content, and carbonates, indicating an important interaction among them. On the other hand, elements such as Cd, Cu and Pb were related to anthropic activities and comprised the second (Cu and Pb) and third principal components (Cd), designated as anthropogenic components.

Due to large variation in their contents, it becomes very difficult to identify a possibly contaminated soil through mere chemical analysis of a particular site without the knowledge on background concentration of the area. It is almost impossible to establish the true natural background level of trace metals in soils because of long-range atmospheric transport and precipitation of contaminants along with various other anthropogenic impacts. However, contributions of these anthropogenic sources are very small and hence are generally ignored during baseline concentration determination in soils of particular region.

7.2 Concept of Baseline Concentrations of Metals in Soil

The information on baseline concentrations are needed to provide regulators for a scientific database in order to address the how clean is clean issues and to establish reasonable clean up criteria. Baseline concentration represents an expected range of element concentrations around a mean in a normal sample medium and is defined as 95 % of the expected range of background concentrations. This can be determined in both numerical and statistical ways as *a measure of central tendency (e.g. mean)± a measure of deviation from central tendency (e.g. standard deviation)*. Although 'range' and 'mean' of analytical data is widely used by researchers

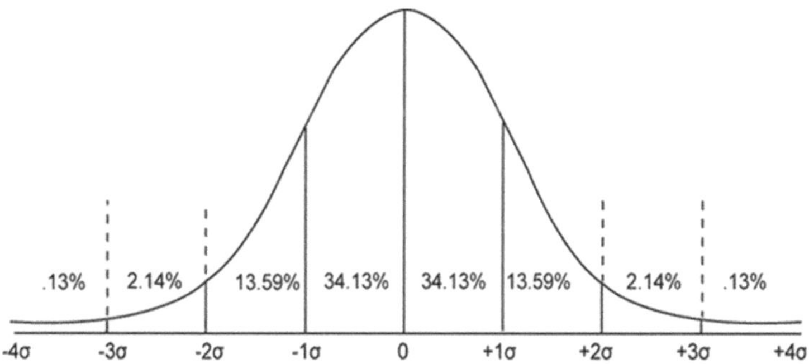

Fig. 7.1 Normal distribution curve (X-axis: sample value; Y-axis: frequency of occurrence)

to indicate true background values, these are not applicable to other area or to even other sets of data in the same area due to some exceptionally high or low analytical values, which may not be a result of predominant ongoing lithogenic and pedogenic processes. In many naturally occurring processes, population follow a normal distribution where about 68%, 95% and 99% of the population lies within $\mu \pm \sigma$, $\mu \pm 2 * \sigma$ and $\mu \pm 3 * \sigma$ range ($\mu =$ mean, and $\sigma =$ standard deviation), as indicated in Fig. 7.1 (Feller 1971). Geochemical baseline concentration has statistically been defined as concentration range in which 95% of samples are expected to belong (Gough et al. 1994). Several researchers computed baseline concentration values as '$\mu \pm \sigma$' (Campos et al. 2003). Quality reference values (similar to background concentration) of metals in soils of Cuba were determined based on 75th percentile of samples after removing anomalies through a boxplot (Alfaro et al. 2015).

However, many investigators found that heavy metal concentrations in naturally occurring unpolluted soils follow a log-normal distribution, as exemplified in Fig. 7.2 (Chen et al. 1999; Zhang et al. 2008). Log-transformation of data minimizes distorting effects of a few high concentrations. In view of this, baseline concentration of a region is generally represented using geometric mean (GM, a measure of central tendency) and geometric standard deviations (GSDs, a measure of variation of the data). Natural background concentrations of major heavy metals are, thus defined as the range between $GM/GSD^2$and $GM \times GSD^2$ which represents the varied concentrations between the 5th and the 95th percentiles. Upper baseline concentration of a heavy metal (i.e. $GM \times GSD^2$), thus includes 95% of the samples (Dudka et al. 1995; Chen et al. 1999, Zhang et al. 2008).

7.2.1 *Baseline Concentrations Determined for Different Countries in the World*

Background concentrations have been determined in various countries, for example, in United Kingdom (Thornton 1982), Belgium (De Temmerman et al. 1984), Italy (Bini et al. 1988), India (Kuhad et al. 1989), China (Chen et al. 1991), the

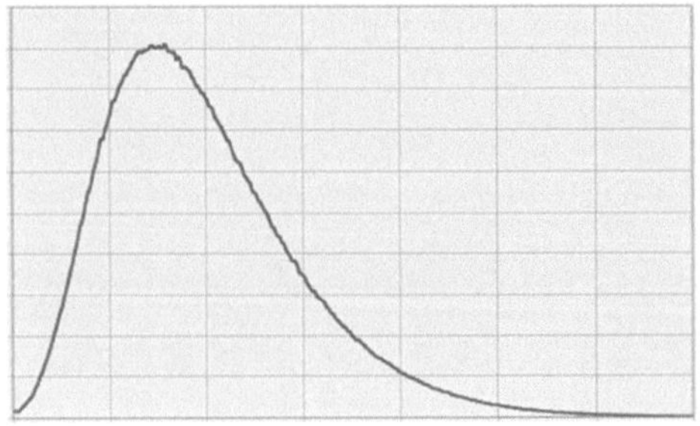

Fig. 7.2 Log-normal distribution curve (X-axis: sample value; Y-axis: frequency of occurrence)

Table 7.2 Upper baseline concentration of heavy metals reported from different countries

	Cuba[a]	Spain[b]	USA[c]	Poland[d]	USA[e]	USA[f]	USA[g]	China[h]	Spain[i]
Ag	1	–	2.5	–	–	–	–	–	–
Ba	111	617.9	112	409	–	380	1110	–	–
Cd	0.6	–	0.33	–	0.41	–	–	0.13	0.7
Co	25	–	–	–	–	–	–	–	11
Cr	153	120.2	80.7	75.3	17.2	29	392	87	36
Cu	83	111.6	21.9	18	16.3	20	77.8	28.7	28
Ni	170	118.6	48.5	27	9.4	4.6	207	23.5	31
Pb	50	217.5	42.0	50.1	18.1	15	48.8	57.6	28
Sb	6	–	0.79	–	–	–	–	–	–
V	86	101.2	–	84.2	–	–	–	–	–
Zn	86	326.8	29.6	154.7	20.3	12	–	77.8	83
Mo	0.1	–	6.76	–	–	–	–	–	–
As	19	–	7.01	–	37.0	–	–	–	–
Hg	0.1	–	0.04	–	0.028	–	–	0.15	–
Sr	–	128.8	–	138.5	–	–	–	–	–

Sources: [a]Alfaro et al. (2015), [b]Bech et al. (2005), [c]Chen et al. (1999), [d]Dudka (1992), [e]Ma et al. (1997), [f]Gough et al. (1994), [g]Bradford et al. (1996), [h]Zhang et al. (2008) and [i]Mico et al. (2007)

Netherlands (Denneman and Robbersen 1993), Western Europe (Angelone and Bini 1992), Poland (Andersen et al. 1994), Germany (Bachman et al. 1995), United States of America (Holmgren et al. 1993; Bradford et al. 1996; Ma et al. 1997), and Spain (Bech et al. 2005) (Table 7.2). All the reports indicated that the origin of heavy metals is parent materials and anthropogenic activities influenced their spatial distribution. Apart from these, climate, organic and inorganic components of soils as well as redox potential also affect the concentration level of trace metals in soils. In a study to determine spatial distribution of heavy metals in Guangdong province, Zhang et al. (2008) concluded that the primary source of heavy metals in soil is parent material and their concentrations are also influenced by pedogenesis. Anthropogenic activities were other secondary factor playing crucial role in their spread over the region.

Although several EU countries and USA have determined baseline concentrations in soil for different regions by various methods, very few studies have been undertaken in this respect in India. Geochemical distribution as well as baseline levels of 29 elements were determined in soils of Siddipet, Medak District, Andhra Pradesh (Dantu 2010). Regional parent materials and pedogenesis were the primary factors influencing the concentrations of trace elements while anthropogenic activities had secondary influence. For agricultural soils of Indo-Gangetic plains of Haryana, Kuhad et al. (1989) determined background levels of some heavy metals in physiographic zones and found no significant differences among them. However, investigators did not follow any statistical approach in their methods.

7.2.2 Assessment of Heavy Metal Contamination Around Industrial Congregations

While baseline concentration of heavy metals for a specified geographical area is a useful tool to detect occurrence of contamination due to anthropogenic activities with a certain degree of confidence, its determination requires systematic collection and analysis of a large number of samples. Hence, such information is not available for most of the locations. Several other techniques have also been proposed for assessing the extent and degree of contamination of heavy metal in and surrounding area of industrial sites.

7.2.2.1 Enrichment Factor

Enrichment factors (EFs) are commonly used as a means of identifying and quantifying human interference with global element cycles and are also used to differentiate trace metals originating from human activities and those from lithogenic source (parent material) (Blaser et al. 2000; Reimann and de Caritat 2005; Dantu 2009; Karageorgis et al. 2009; Namaghi et al. 2011). Enrichment factor is calculated using modified formula given by Krzysztof et al. (2004) based on the equation suggested by Buat-Menard and Cherselet (1979), which is double ratios of the target metal and a reference metal in the soil and earth's crust as indicated below. The reference metal must be one that is overwhelmingly derived from a single source. Usually, Al, Sc, Ti, Fe Mn and Sr are used as the reference as these are present in significantly larger concentrations in soil and are not significantly altered by anthropogenic activities (Reimann and de Caritat 2000; Sutherland 2000; Saur and Juste 1994). As an example, EFs can be calculated using Fe as the reference, with the following relationship:

$$EF = \frac{\left(\frac{X}{Fe}\right)_{sample}}{\left(\frac{X}{Fe}\right)_{crust}}$$

Where, $\left(\frac{X}{Fe}\right)_{sample}$ and $\left(\frac{X}{Fe}\right)_{crust}$ refer, respectively, to the ratios of mean concentrations (milligrams per kilogram, dry weight) of the target metal (X) and Fe in the sediments and continental crust. The EFs are calculated on the basis of earth's crust mean abundance of the elements (Lide 2005).

Five contamination categories are recognized on the basis of the enrichment factor (Sutherland 2000).

$EF =< 2$	Deficiency to mineral enrichment
$EF = 2-5$	Moderate enrichment
$EF = 5-20$	Significant enrichment
$EF = 20-40$	Very high enrichment
$EF => 40$	Extremely high enrichment

7.2.2.2 Enrichment Ratios

This is another approach used to characterize degree of anthropogenic pollution in soil. Originally, enrichment ratio (ER) based assessment approach was used by Chester et al. (1985) and is computed as below:

$$ER_{ext} = \frac{C_n\ Sample}{\mu_n Back + (2 \times \sigma)}$$

where 'ER_{ext}' is the enrichment ratio for a given extraction for a given element, 'C_n Sample' represents the concentration of a given element 'n' in sample under investigation, 'μ_n Back' represents the mean concentration of the element 'n' in the background soil samples, and 'σ' is standard deviation for concentration of the element 'n' in the background soil samples. The method of extraction may be for total content or for more labile fraction of metals by acid, EDTA etc.

Sutherland (2000) used 'Median' and 'Median absolute deviation (MAD)' in place of 'Mean' and 'standard deviation (σ)' for the background samples and computed enrichment ratio ER_{ext} as below:

$$ER_{ext} = \frac{C_n\ Sample}{Median_n Back + (2 \times MAD_n)}$$

Median absolute deviation (MAD) is calculated as:

$$MAD = median(|X_i - median(X)|)$$

Above equation had been indicated as more robust as it overcomes the problem of positive skewness commonly encountered with geochemical data. Values of $ER_{ext} > 1$ reflect a potential anthropogenic input above the background geogenic variation. Using this method, Sutherland et al. (2000) found enrichment ratios for EDTA-extractable Pb ≥ 11.5 in one-fourth of 169 roadside soil samples.

7.2.2.3 Contamination Index

Assessment of soil contamination is also performed by the contamination index (P_i) and integrated contamination index (P_c) as suggested by Huang (1987), and is expressed by the fuzzy functions given below:

$$P_i = \frac{C_i}{X_a} \ldots\ldots \text{ if } C_i \leq X_a$$
$$P_i = 1 + \frac{C_i - X_a}{X_b - X_a} \quad \ldots \text{ If } X_a < C_i \leq X_b$$
$$P_i = 2 + \frac{C_i - X_b}{X_c - X_b} \quad \ldots \text{ If } X_b < C_i \leq X_c$$
$$P_i = 3 + \frac{C_i - X_c}{X_c - X_b} \quad \ldots \text{ If } C_i > X_c$$

Where,

C_i is the observed concentration of heavy metal in the soil sample;
X_a is the no-polluted threshold concentration of heavy metal;
X_b is the lowly polluted threshold concentration of heavy metal;
X_c is the highly polluted threshold concentration of heavy metal.
X_a, X_b and X_c in the above functions have been defined according to Class I, Class II, and Class III criteria (Table 7.3), respectively, based on Chinese Environmental Quality Standard for Soils (GB 15618-1995) (SEPAC 1995). Class I criteria were suitable to keep natural background values, and Class II could be used as the threshold values for protecting human health, while Class III could be used as the threshold values for plant growth.

7.2.2.4 Integrated Contamination Index

Integrated contamination index (P_c) for each site is defined as the summation of the difference between the contamination index (P_i) for each metal and 1 (one) and is calculated by the following formula:

Table 7.3 Threshold values (mg kg^{-1}) for contamination index

Threshold concentrations	Zn	Cd	Pb	Ni	Cu	As
X_a	100	0.2	35	40	35	15
X_b	200	0.3	250	60	50	30
X_c	500	1.0	500	200	400	40

SEPAC (1995)

$$P_c = \sum_{n}^{i=1} (P_i - 1)$$

Average and standard deviation values of P_c were used to establish confidence intervals. Accordingly, a $P_c \times$ distance curve was adjusted using a log scale for real distance. Values of contamination index (P_i) and integrated contamination index (P_c) have been interpreted as below in the Table 7.4.

7.2.2.5 Geo-accumulation Index

Index of geo-accumulation (I_{geo}) enables the assessment of contamination by comparing current and pre-industrial concentrations. Background concentration of heavy metals in the earth's crust was used as a reference value implicating pre-industrial environment. This is calculated using below equation:

$$I_{geo} = \log_2 \frac{C_n}{1.5B_n}$$

where, C_n denoted the concentration of a given element in the soil tested, while B_n denoted the concentration of elements in the earth's crust (Taylor and McLennan 1995). The metal content accepted as background was multiplied each time by the constant 1.5 in order to take into account natural fluctuations of a given substance in the environment as well as very small anthropogenic influence.

Descriptive classification for the index of geo-accumulation (I_{geo}) is given in Table 7.5.

7.2.2.6 Nemerow's Pollution Index (P_n)

Geo-Accumulation Index is suitable for the evaluation of soil heavy metal contamination in industrial and mining affected areas. However, it is only applicable for assessing the impact of single heavy metal contaminant; and therefore, this index

Table 7.4 Limits for contamination indices

Contamination level	Contamination index (P_i)	Integrated contamination index (P_c)
No	$P_i \leq 1$	$P_c \leq 0$
Low	$1 \leq P_i \leq 2$	$0 < P_c \leq 7$
Moderate	$2 \leq P_i \leq 3$	$7 < P_c \leq 21$
High	$P_i > 3$	$P_c > 21$

Table 7.5 Interpretation of geo-accumulation index values

Class	I_{geo}value	Interpretation
0	$I_{geo} < 0$	practically uncontaminated
1	$0 < I_{geo} < 1$	uncontaminated to moderately contaminated
2	$1 < I_{geo} < 2$	moderately contaminated
3	$2 < I_{geo} < 3$	moderate to strongly contaminated
4	$3 < I_{geo} < 4$	strongly contaminated
5	$4 < I_{geo} < 5$	strongly to very strongly contaminated
6	$5 < I_{geo}$	very strongly contaminated

Forstner (1990)

Table 7.6 Classification of contaminated sites based on Nemerow's Pollution index

Class	Nemerow's pollution index	Interpretation
1	$P_n \leq 0.7$	Unpolluted
2	$0.7 < P_n \leq 1.0$	Marginally polluted
3	$1.0 < P_n \leq 2.0$	Slightly polluted
4	$2.0 < P_n \leq 3.0$	Moderately polluted
5	$P_n > 3.0$	Severely polluted

Jie-liang et al. (2007)

cannot provide a comprehensive description of the contamination status of the study area. Nemerow's pollution index was applied to assess degree of soil ecosystem pollution and integrative assessment of soil ecosystem quality (Jie-liang et al. 2007; Guan et al. 2014). The method can be described as below:

$$P_n = \sqrt{\left(MaxP_i^2 + AveP_i^2\right)/2}$$

where,

$$P_i = \frac{C_i}{S_i}$$

P_n is the Nemerow's pollution index, P_i is the pollution index of the ith heavy metal, C_i is the measured concentration of the ith heavy metal, S_i is the required standard of the ith heavy metal, $AveP_i$ and MaxPi is the average and the maximum value of the pollution indices of all heavy metals respectively. Higher value for P_n indicates more serious pollution. The soil quality classification criterions based on Nemerow's Pollution index is indicated in Table 7.6.

Guan et al. (2014) modified the above traditional Nemerow's Pollution index by replacing the single factor index with I_{geo} as per following equation:

$$P_n = \sqrt{\left(I_{geomax}^2 + I_{geoave}^2\right)/2}$$

where, I_{geomax} is the maximum I_{geo} value and I_{geoave} is the arithmetic mean value of I_{geo}.

They proposed classification on soil quality based on the pollution index values as:

Class	Nemerow's pollution index	Interpretation
0	$0 < P_n \leq 0.5$	Uncontaminated
1	$0.5 < P_n \leq 1$	Uncontaminated to moderately contaminated
2	$1 < P_n \leq 2$	Moderately contaminated
3	$2 < P_n \leq 3$	Moderately to heavily contaminated
4	$3 < P_n \leq 4$	Heavily contaminated
5	$4 < P_n \leq 5$	Heavily to extremely contaminated
6	$P_n > 5$	Extremely contaminated

7.2.2.7 Contamination Factor

Hakanson (1980) proposed an overall indicator of contamination by integrating data for several heavy metals under study. This method is based on the calculation of a contamination factor (C_f^i) for each pollutant. The C_f^i is the ratio obtained by dividing the mean concentration of each metal in the soil (C_{0-1}^i) by the baseline or background value (concentration in unpolluted soil, C_n^i) as below.

$$C_f^i = \frac{C_{0-1}^i}{C_n^i}$$

where, C_{0-1}^i is the mean content of metals from at least five sampling sites and C_n^i is the pre-industrial concentration of individual metal. Based on C_f^i value for the specified element, four classes contamination levels are specified (Hakanson 1980). Major limitation of this method is that information on pre-industrial level of heavy metals concentration in soil is a prerequisite for determining contamination levels. This method was originally developed for contamination assessment for sediments, where bottom sediment was considered for assessing C_n^i. However, in absence of data on pre-industrial heavy metal concentration for the site, upper value of baseline concentration range can be taken as C_n^i. Researchers have also considered the concentration of elements in the Earth's crust as a reference value (Loska et al. 2004).

Hakanson (1980) categorized contaminated soils into four classes based on contamination factor values as below.

$C_f^i < 1$	Low contamination
$1 \leq C_f^i < 3$	Moderate contamination
$3 \leq C_f^i < 6$	High contamination
$6 \leq C_f^i$	Very high contamination

The calculated value of the contamination degree can be used for describing the role of an element in the global contamination of the reservoir by determining the proportion of a given element in the contamination (Hakanson 1980; Hakanson and Jansson 1983).

7.2.2.8 Contamination Degree

The C_f^i is the index of single element 'i' and the addition of contamination factors of all elements examined, describe the contamination degree (C_{deg}) of the environment.

$$C_{deg} = \sum_{i=0}^{n} C_f^i$$

Based on C_{deg} values, following four levels of contamination were recognized by Hakanson (1980). However, such categorization is dependent on number of contaminants considered in the study.

$C_{deg} < 8$	Low degree of contamination
$8 \leq C_{deg} < 16$	Moderate degree of contamination
$16 \leq C_{deg} < 32$	Considerable degree of contamination
$32 \leq C_{deg}$	Very high degree of contamination

7.2.2.9 Pollution Load Index

Each sampling site can be evaluated for the extent of the metal pollution by employing the method based on the pollution load index (PLI) developed by Tomlinson et al. (1980) as below:

$$PLI = \left(C_f^1 * C_f^2 * C_f^3 * \ldots \quad \ldots \quad \ldots * C_f^n\right)^{1/n}$$

where n is the number of metals studied and C_f^n is contamination factor for n^{th} element as mentioned above. The PLI provides simple but comparative means for assessing a site quality, where a value of $PLI < 1$ indicate uncontaminated site; $PLI = 1$ present that pollutants are present only at baseline level and $PLI > 1$ would indicate deterioration of site quality (Thomilson et al. 1980).

7.3 Assessment of Risks to Organisms Due to Heavy Metals Contamination in Soils

Soil analysis for heavy metals and their interpretation using above tools may indicate occurrence of contamination, its degree, spatial distribution and probable source. Such information has been extensively used in taking precautionary steps for the purpose protecting land resources. However, availability of heavy metals to organisms is not necessarily dependent on total contamination level. Chemical properties of metals as well as soil constituent and their interactions reflect on the transfer/availability to organisms. In order to have proper assessment of risks to organisms including plants and human health, several techniques have been developed for assessing the risk from contaminated sites as described below.

7.3.1 Soil–Plant Partition Coefficient or Transfer Factor

Heavy metal transfer from soil to plants can be calculated by soil–plant partition coefficients. This parameter has been used during screening crops suitable for growing in contaminated land as well as for screening accumulator plants during phytoremediation. The heavy metal transfer from soil to plant can be quantitatively estimated by soil–plant Partition Coefficient or Transfer Factor (TF) which is concentration Ratio as below (Hope 1995; Roca and Vallejo 1995; Dinelli and Lombini 1996; Rodriguez et al. 2002; Tome et al. 2003).

$$TF = \frac{C_{plant}}{C_{soil}}$$

where, C_{plant} and C_{soil} are the plant and soil concentrations of heavy metal, respectively. A prerequisite of the soil–plant transfer factor concept is the presence of a statistically significant relationship between the content of a given element in the soil and that in the plant (Bunzl et al. 2000). Such values are needed for many assessment models to predict the concentration of an element for a given plant species at an anticipated contamination level in the soil (Bunzl et al. 2000). Transfer factor (TF) describes the amount of an element expected to enter a plant from its substrate, under equilibrium conditions (Sheppard and Sheppard 1985; Davis et al. 1999). The index value above one indicates the plants as accumulators; the value around one indicates a rather undistinguished their behavior towards these metals and lower than one show the plants as excluders. This computation assumes a linear relationship between the concentrations of a certain element in the plant with that in the soil. This linearity does not hold for essential elements (the content of which is under strict metabolic control), but does apply to non-essential elements and pollutants within certain range.

7.3.2 *Bioaccumulation Factor (BAF)*

One of the major drawbacks of above-mentioned transfer factor is that it expresses metal transfer in a certain condition i.e. at one place, at the same time and under the same circumstances. But it does not incorporate the variation in such metal transfer behaviour in treated site compared to the control site. To overcome this, Baltrenaite et al. (2012) introduced following dynamic factors, where natural processes affecting metal transfer are integrated in the factor calculation.

7.3.2.1 Dynamic Factor of Metal Bioaccumulation (BA_{dyn})

Dynamic factor of metal bioaccumulation reflects the physiological sensitivity of plants to the general soil contamination degree and is calculated as:

$$\mathrm{BA_{dyn}} = \frac{C_{ContamPlant}}{C_{ContamSoil}} \times \frac{C_{ControlSoil}}{C_{ControlPlant}}$$

where,

$C_{ContamPlant}$= total concentration of a metal in plant (root + shoot) growing on contaminated agricultural soil;
$C_{ContamSoil}$= total concentration of the metal in contaminated agricultural soil;
$C_{ControlSoil}$= total concentration of metal in the control soil; and
$C_{ControlPlant}$= total concentration of the metal in the plant (root +shoot) growing on control soil.

7.3.2.2 Dynamic Factor of Metal Biophilicity

Dynamic factor of metal biophilicity reflects changes in metal participation in plant metabolism. Generally, metal biophilicity is the ratio of metal accumulation in living biomass and metal concentration in the Earth's crust. This indicates metal involvement in the metabolism in vegetation on the global basis. Dynamic factor of metal bilophicity ($\mathrm{BF_{dyn}}$) was calculated using following equation:

$$BF_{dyn} = \frac{BF_{contam}}{BF_{control}} = \frac{C_{ContamPlant}}{C_{EarthCrust}} \times \frac{C_{EarthCrust}}{C_{ControlPlant}} = \frac{C_{ContamPlant}}{C_{ControlPlant}}$$

where, $\mathrm{BF_{contam}}$ and $\mathrm{BF_{control}}$ are the biophilicities of a metal on contaminated and control sites, respectively; $C_{\mathrm{EarthCrust}}$ is the concentration of the metal in the Earth's crust and $C_{\mathrm{ContamPlant}}$ and $C_{\mathrm{ControlPlant}}$ are the concentrations of the metal in plant biomass on the contaminated and control sites, respectively.

7.3.2.3 Dynamic Factor of Metal Translocation (TR_{dyn})

Dynamic translocation factor (TR_{dyn}) of metals is calculated as:

$$TR_{dyn} = \frac{TR_{Contam}}{TR_{Control}} = \frac{C_{ContamShoot}}{C_{ContamRoot}} \times \frac{C_{ControlRoot}}{C_{ControlShoot}}$$

where,

TR_{Contam} is the translocation factor of a metal in plant on the contaminated site; $TR_{Control}$ is the translocation factor in plant on the control site;

$C_{ContamShoot}$ is total concentration of the metal in plant shoot growing on contaminated agricultural soil;

$C_{ContamRoot}$ is total concentration of metal in plant root growing on contaminated agricultural soil;

$C_{ControlRoot}$ is total concentration of metal in the plant root growing on control soil;

$C_{ControlShoot}$ is total concentration of metal in the plant shoot growing on control soil.

7.3.2.4 Assessment of Phytoremediation Potential of Plant

To evaluate the phytoextraction/ phytostabilization potential of the plant species, the following factors were calculated using the following formulae:

$$\text{Bioabsorption Coefficient [BAC]} = \frac{metal\ content\ in\ shoot}{metal\ content\ in\ soil}$$

$$\text{Bioconcentration Factor [BCF]} = \frac{metal\ content\ in\ root}{metal\ content\ in\ soil}$$

$$\text{Translocation Factor [TF]} = \frac{metal\ content\ in\ shoot}{metal\ content\ in\ root}$$

This criteria were used to classify plants as phytoextractors and phytostabilizers as in Table 7.7. Higher ratios of BAC and TF indicate better phytoextraction potential of plants.

Dynamic factor of phytoremediation (FR_{dyn}) reflects phytoremediation capability of a plant and its changes with time. Dynamic factor of phytoremediation (FR_{dyn}) is calculated using following equation:

$$FR_{dyn} = \frac{FR_{Contam}}{FR_{Control}}$$

where, FR_{Contam} and $FR_{Control}$ are the annual phytoremediation effectiveness factors of plant on the contaminated and control sites, respectively. In the case of a specific metal and a specific tree, they are calculated using equation:

Table 7.7 Criteria for assessment of phytoremediation potential of plants

Phytoremediation potential	BAC	BCF	TF
Good phytoextractor (Fitz and Wenzel, 2002)	>1	–	>1
Good phytostabiliser (Mendez and Maier, 2008)	–	>1	<1

$$FR_i = \frac{C_{i-plant} \times B}{1000 \times C_{i-soil} \times \rho \times d}$$

where FR_i is the annual metal phytoremediation factor (in kg ha^{-1}); $C_{i-plant}$ is the metal concentration in tree (in mg kg^{-1}); B is the annual tree increment (in kg ha^{-1}); C_{i-soil} is the metal concentration in root zone soil layer (in mg kg^{-1}); ρ is the soil bulk density (in g cm^{-3}); and d is the depth of root zone soil layer (cm).

7.4 Risk Assessment for Contaminated Soil on Human Health

Heavy metals in soil pose potential threats to the environment and can damage human health through various absorption pathways such as direct ingestion, dermal contact, and food. The human health risk models proposed by USEPA have been adopted worldwide. The model pertaining to health risk from polluted soil is summarized by Liu et al. (2013) as below:

Exposure pathway: Soil-ingestion (I)

$$\mathrm{CDI_I} = \frac{CS \times IRS \times EF \times ED}{BW \times AT} \times CF$$

Exposure pathway: Dermal contact (D)

$$\mathrm{CDI_D} = \frac{CS \times SA \times AF \times ABS \times EF \times ED}{BW \times AT} \times CF$$

Exposure pathway: Food (F)

$$\mathrm{CDI_F} = \frac{CF \times IRF \times EF \times ED}{BW \times AT}$$

CDI = chronic daily intake (mg kg^{-1} d^{-1}); CS = exposure-point concentration (mg/kg); EF = exposure frequency (day/year); ED = exposure duration (year); AT = averaging time (day); BW = body weight (kg); SA = exposed skin area (cm^2); AF = adherence factor (mg cm^{-2}); ABS = dermal absorption fraction; CF = unit's conversion factor (10^{-6} kg mg^{-1}); IRS = ingestion rate of soil (mg d^{-1}); IRF = ingestion rate of food (kg d^{-1}).

Cancer risk can be evaluated using the equation:

$$\text{Cancer risk} = \sum_{k=1}^{n} CDI_k \times SF_k$$

The non-carcinogenic risk from individual heavy metals can be expressed as the chronic hazard index (HI):

$$\text{Chronic hazard index (HI)} = \sum_{k=1}^{n} CDI_k \div RFD_k$$

where the chronic hazard index (HI) is the sum of more than one hazard quotient for multiple substances or multiple exposure pathways, CDI_k is the daily intake of the heavy metal 'k', SF is the slope factor of hazardous substances (mg kg^{-1} d^{-1}) and RFD_k is the chronic reference dose (mg kg^{-1} d^{-1}) for the heavy metal 'k'. HI values > 1 indicates that there is a chance of non-carcinogenic risk to occur, and when $HI < 1$ the reverse applies. Using above method of risk assessment, arable fields near industrial and waste mining sites in Zhejiang province of Southeast China were found unsuitable for growing leafy and root vegetables in view of the risk of elevated intake of heavy metals (Liu et al. 2013). Chromium and Pb were the primary heavy metals posing non-cancer risks while Cd caused the greatest cancer risk among local residents.

7.4.1 Human Health Risk Assessment for Heavy Metals from Contaminated Food

Risk to human health (Hazard Quotient, HQ) for intake of heavy metals through consumption of vegetables grown on sewage-irrigated soils is also calculated using the following relationship (Pierzynski et al. 2000):

$$HQ_{gv} = \frac{add}{RfD}$$

where, 'HQ_{gv}' is the hazard quotient to a human from consumption of green vegetables, '*add*' is the average daily dose (mg metal/kg body-weight/day) and '*RfD*' is the reference dose.

Average daily dose (*add*) is computed as below:

$$add = \frac{mc \times cf \times di}{dw}$$

where, '*mc*' is the metal concentrations in plant (mg kg^{-1}) on dry weight basis, '*cf*' the fresh to dry weight conversion factor, '*di*' the daily intake of green vegetable (kg) and '*bw*' the body weight (kg). A reference dose corresponds to the maximum amount of a metal that can safely be consumed per day, per kilogram of body weight. Integrated Risk Information System of United States Environmental Protection Agency (IRIS 2016) computed reference doses as 0.001 mg kg^{-1} d^{-1} for Cd; 0.035 mg kg^{-1} d^{-1} for Pb; 0.3 mg kg^{-1} d^{-1} for Zn; and 0.04 mg kg^{-1} d^{-1} for Cu. If HQ is greater than 1, then the *add* of that particular metal exceeds the RfD, indicating that there is a potential risk associated with consumption of food containing that metal.

7.5 Regulation in Respect of Heavy Metals Entry into the Agricultural Land

Life forms that have evolved through millions of years in metallic stress free environment are facing threat in the present times from heavy metal contaminated ecosystem. These metals in soil are known to affect environment in several ways: contamination of food through plants grown on it, decrease in crop productivity (Rooney et al. 2007) and depression of soil microbial activity (Brookes and McGrath, 1984; Chaudri et al. 1993; McGrath 1993, Akerblom et al. 2007). Oral intake of heavy metals through contaminated food for long duration is found to affect human and animal health adversely (ATSDR 2005; USEPA 2005a, b). Because of the potential threats from different pollution activities, many countries have set regulatory limits in respect of heavy metals in order to protect agricultural land; and such limits are based on (1) maximum concentrations in soils that can be allowed to receive further metal inputs, (2) annual pollutant loading limits into the soil from all the sources, (3) cumulative pollutant loading limits, (4) maximum allowed concentration of metals in amendment materials (like composts prepared from MSW or sewage-sludge) (McGrath et al. 1994). Different approaches of risk assessment are shown to result in widely differing numerical limits being set for the same constituent, which creates confusion among the regulatory authorities worldwide.

7.5.1 Heavy Metal Loading Limits in Soil

Different member countries of the European community, as well as the USA, have implemented laws to limit the load of pollutants via sewage sludge or compost application onto the soil. These include limit values for heavy metal concentrations in the soil to which sludge or compost is applied, in the sludge or compost material itself, and all maximum cumulative and annual quantities of heavy metals added to agricultural soils. The maximum permitted concentrations in agricultural soils varied from 1.5 to 20 mg kg^{-1} for Cd, 50 to 1500 mg kg^{-1} for Cr, 30 to 750 mg kg^{-1} for Cu,

Table 7.8 Maximum allowable concentration of trace elements in soils of various countries (mg kg^{-1})

Country	Cd	Cr	Cu	Ni	Pb	Zn	As	Hg
Austria	5	100	100	100	100	300	50	5
Poland	1–3	50–80	30–70	30–75	70–150	100–300	30	5
Germany	1.5	100	60	50	100	200	20	1
Russia	–	0.05 (Cr^{6+})	23 (soluble pool)	35	20	110	2	2.1
U.K	3–15	–	50	20	500–2000	130	10	–
U.S	20	1500	750	210	150	1400	–	8
European Commission	1–3	50–150	50–140	30–75	50–300	150–300	–	1–1.5

Adopted from (Kabata-Pendias 2000)

20 to 210 mg kg^{-1} for Ni, 20 to 2000 mg kg^{-1} for Pb, 100 to 1400 mg kg^{-1} for Zn, 2 to 50 mg kg^{-1} for As and 1 to 8 mg kg^{-1} for Hg (Table 7.8). Such wide variations probably result from different approaches adopted by the countries while setting the limit values, such as: (i) a comprehensive analysis of the pathways of pollutant transfer to selected target organisms and an assessment of the likely harmful effects that metals may have on their targets; (ii) setting limits consistent with the lowest observed effect concentrations reported in various toxicity studies, and (iii) the metal balance approach via the adjustment of allowable metal inputs against small losses of metals due to crop removal, soil erosion, and leaching (McGrath et al. 1994). In developing countries like India, sufficient information has not been compiled to set limit values based on the analysis of pollutant transfer pathways. On the other hand, setting limits based on the metal balance approach does not seem to be a viable strategy in India because of significant non-point source metal inputs as well as significant metal content in organic and inorganic fertilizers.

Due to such divergent approaches, wide differences in regulatory limiting values have been observed among different countries. For example, maximum concentrations of metal allowed in agricultural land treated with sewage-sludge by USA and EU countries differ by large factors; ×40 for Cd, ×19 for Cu, ×50 for Cr, ×14 for Ni, ×13 for Pb, ×14 for Zn, ×40 for Hg (McGrath et al. 1994).

7.5.2 USEPA Approach of Calculating Loading Limits of Heavy Metals

In order to appreciate the potential for widely varying estimates of an acceptable soil loading of heavy metals, it is necessary to look at the various factors that potentially affect this loading. USEPA calculated the estimated tolerable soil loading limit for heavy metals through the following steps.

Step 1
Total background intake (TBI, mg day^{-1}) of a heavy metal is determined by the equation:

$$TBI = DI + (AC \times IV) + (WC \times WV)$$

where,

DI= daily dietary intake of the metal (mg day^{-1}).
AC= concentration of the metal in air (mg m^{-3}).
IV= daily inhalation volume (m^3 day^{-1}).
WC= concentration of the metal in drinking water (mg L^{-1}).
WV= daily volume of water consumption (L day^{-1}).

Step 2
Adjusted daily reference intake (RIA, (μg day^{-1}) is determined by the equation:

$$RIA = \left\{ \left(RfD \times \frac{BW}{RE} \right) - TBI \right\} \times 10^3$$

Where,

RfD= reference dose of the heavy metal, i.e., the daily exposure which over a lifetime is likely to be without "appreciable" risk.
BW= body weight (kg).
RE= relative effectiveness of ingestion exposure to the metal (unitless).

Step 3
Reference cumulative soil application rate (RP, kg ha^{-1}) of the heavy metal is determined by following equation:

$$RP = \frac{RIA}{\sum (UC_i \times DC_i \times FC_i)}$$

Where,

UC_i= uptake response slope of crop group 'i' for the heavy metal M (μg M g^{-1} plant dry weight)/ (kg M applied ha^{-1}).
DC_i= daily dietary consumption of crop group 'i' (g dry weight day^{-1}).
FC_i= fraction of crop group 'i' grown on sludge-amended soil (unit-less).

Uncertainties, errors and biases in the estimates of crop uptake coefficients (UC), vegetable consumption (DC), and other variables in equation above compound the uncertainty inherent in the final heavy metal loading estimate (RP). Thus, USEPA's final estimate of acceptable metal loading (RP) has an undefined but presumably high degree of uncertainty.

Calculating Heavy Metals Update Coefficients for Leafy Vegetables: The USEPA 503 risk assessment for dietary exposure to heavy metal from leafy vegetables grown on sewage sludge-amended soil was based on a single uptake coefficient (UC) averaged from numerous field uptake experiments. The individual vegetable and soil loading data for Cd are tabulated in the technical support document (USEPA 1992). In addition, to determine a single UC for experimental crop uptake data with more than one Cd application rate, USEPA applied a linear best-fit function to the relationship between crop Cd concentration (y -axis) and Cd loading to the soil (x -axis).

7.5.3 Determination of Maximum Safe Concentration Limits of Heavy Metals Based on Sensitive Scenario

Risk based approach for regulatory guideline is normally adopted based on sufficient information to assess the risk adequately and establish thresholds below which it is considered to be negligible or acceptable. In India, requisite research to establish cumulative loading limit for toxic metals for soils is lacking. Due to variability in parent materials as well as climatic conditions, wide range of soil types is found in India and hence, wide variability in their heavy metal immobilizing capacity is expected owing to differences in their constituents and properties. Such wide variation in metal immobilizing capacity may lead to wide variation in protective limits in Indian soils. Saha et al. (2013) determined maximal protective concentration limits of heavy metals in a soil which was acidic in nature and contained lower levels of clay, organic matter, oxides of Fe and low CEC. All of these soil characteristics are associated with a lower metal immobilization capacity (McBride 1994). Hence, the maximal concentration limits determined for heavy metals in such a sensitive agro-ecosystem was assumed to protect agricultural land from heavy metal contamination in all other agro-ecological situations as well. Three different approaches, namely, 'phytotoxicity', 'food contamination,' and 'soil microbial activity diminution' were used for determination of these limits.

Phytotoxicity Threshold Concentration of Heavy Metals in Soil: These were measured in terms of its level in the soil at which a plant experiences 20% growth retardation (PT_{20}).

Threshold Concentration of Heavy Metals in Soil Based on Food Contamination: These were defined as the maximum amount of heavy metal concentrations in the soil, which do not increase their concentrations in the edible food crop above the background level (i.e., concentration range found in the crop grown in uncontaminated area). The upper limit of background metal concentration (C_{ul}) was calculated as:

$$C_{ul} = \bar{x} + 3\sigma$$

Where, $\bar{x}$ is the mean value of metal concentration in spinach leaf in control (no metal) treatment of the experiment and 3σ is standard deviation of the metal concentrations in spinach leaf from farmers' fields.

$$\text{Maximum safe concentration level of metal in soil} = \frac{C_{ul}}{UR}$$

Uptake response slope (UR) was calculated for the best fit regression line (passing through origin) obtained for soil-metal level ('x' axis) and leaf metal level ('y' axis).

Threshold Concentration of Heavy Metals in Soil Based on Impact on Microbial Activity: It is the heavy metal concentration in soil (ED_{20}), where the microbial mediated processes are inhibited by 20% (Doelman and Haanstra 1989; Kostov and Van 2001). All the biochemical reactions are catalyzed by enzymes (Tabatabai 1982), and depending on their nature and level of contamination, many heavy metals have been found to impede soil enzyme activities to different extents (Tyler 1981; Kuperman and Carreiro 1997; Yang et al. 2006). Assays of enzyme activities are considered to be important in assessing the impact of metal pollution on the soil environment (Tyler 1981; Kuperman and Carreiro 1997). From best-fit lines plotted for enzymatic activity *vs.* soil metal concentrations, soil heavy metal level with respect to 20% inhibition of enzyme activity is computed.

The levels of all the heavy metals in soil, corresponding to 20% decrease in biomass growth were found higher than the limiting levels calculated with the help of biochemical and food chain contamination approaches (Table 7.9). Thus, adverse effects on microbial activity and contamination of the food chain by heavy metals occurred much earlier than their adverse effects on plant growth. As the lowest among these maximal concentration limits determined by using the above three different approaches can protect all target organisms, 0.33 mg kg^{-1} Cd, 31 mg kg^{-1} Cr, 179 mg kg^{-1} Cu, 30.7 mg kg^{-1} Ni, 81 mg kg^{-1}Pb, and 392 mg kg^{-1} Zn

Table 7.9 Maximum safe concentration limits of heavy metals in soil determined by various methods

Metal	PT_{20}	Food contamination	ED_{20}
Cd	17.8	0.33	9.5
Cr	176.4	82.5	31.0
Cu	409	189.3	179
Ni	153	29.5	51.1
Pb	–	73.7	81.0
Zn	852	32.3	392

Saha et al. (2013)

was considered to be maximal safe concentration limits for these heavy metals in soil.

Maximum concentration limits determined in the above study for Zn and Cu in the soil were comparable to the limits set by France, Italy, and the UK, but were about four fold higher than the limits set by Sweden and Denmark. Limit values determined for Pb and Ni in this study were comparable to those set by many EU countries. The maximal concentration limit of Cr in the soil was almost similar to the limits set by Denmark and Sweden and was much lower than the limits set by other EU countries. In Sweden, the soil management policy of 'no accumulation of possibly hazardous elements in soil' has led to very low values of 'maximal allowable metal concentration' levels. Maximal soil concentration limits for Cd set by all the EU countries (0.5–3 mg kg^{-1} Cd) were higher than the value determined in this study. The limit values determined for all the heavy metals in this study were found to be considerably lower than the corresponding limit values set by the USA. Due to the variability in parent materials and climatic conditions, a wide range of soil types is found in India, and hence, wide variations in their heavy metal-immobilizing capacity are expected due to the differences in their constituents and properties. Although considerable information is available for the physical and chemical characteristics of different soil series/types (Bhattacharyya et al. 2013), information in respect to safe concentration limits of metals are absent in other soil types of India.

7.6 Soil Testing for Heavy Metal Contaminated Land

While soil analysis for total heavy metal content may indicate their accumulation or contamination, it does not indicate whether their levels can pose any risk to different components of the ecosystem. Toxicity of any metal in the soil essentially depends upon the readiness with which it is transferred to the targeted organisms. Such ease of transferability of metals normally depends upon the constituents and chemical environment of the soil, besides the total content. With increasing evidence of heavy metal pollution in soils worldwide, there is growing demand for methods to assess metal toxicity for the purpose of risk assessment before taking corrective measures if required. In most countries, risks to human health from the soil contamination are based on total concentrations of metals in soil and on the data obtained from exposure pathway (Chaney and Ryan 1994).

7.6.1 Factors Influencing Phytoavailability of Heavy Metals

While many researchers have tried to develop relationships between concentrations of metals in soils /plants and effects on plants or organisms or human, it is perhaps important to first characterize the major risk pathways for the metals of concern.

7.6.1.1 Absorption of Heavy Metals by Plant Roots

Like toxic heavy metals and trace elements, essential micronutrients are also present in very low concentration in soil. Using phytochelates, plant-induced pH changes and redox changes in rhizosphere, roots are able to solubilize and take up micronutrients from nearly insoluble precipitates. Plant absorbs these essential nutrients through ion channel or active transport systems. On the other hand, non-essential trace metals are either transported across the root membrane as substitute for nutrients through the same transport mechanisms or through breaks in the endodermal barrier at sites of lateral root initiation/ apices (McLaughlin et al. 1998). Roots mostly take up heavy metal from soil via soil solution. Further absorption of heavy metal by plant roots are highly related to their free ion activity in the soil solution (Parker and Pedler 1997). Also complexation of metals with inorganic ligands in soil can increase plant uptake of metals. For example, uptake of Cd is enhanced in soil by Cl salinity through the formation of $CdCl_n^{2-n}$ complexes (Smolders and Mc Laughlin 1996). Thus Cl concentration in soil solution (hence salinity) affects the interpretation of soil solution Cd concentration results for predicting its phytoavailability.

7.6.1.2 Soil and Plant Factors Influencing Phytoavailability of Heavy Metals

Soil constituents provide binding sites for heavy metals present in the soil which maintains equilibrium with soil solution. Solubility of different heavy metals depends on their affinity to:

(a) Complex with soluble humic substances.
(b) Chemisorb on soil minerals (e.g. Clays, oxides)
(c) Precipitate as insoluble sulfides, carbonates, phosphates or oxides.
(d) Co-precipitate in other minerals.

Owing to the variation in above chemicals properties among different soil types, concentrations of different heavy metals in soil solution vary widely at similar contamination levels.

Soil pH has tremendous influence on trace elements concentration in soil solution through quantity of variable charge sites available for sorption of metals. Generally metal sorption to soil particles is low at low pH and increases as soil pH increases due to increase in negatively charged sites. However, sorption of some metalloids for oxyanion (e.g. CrO_4^{-2} and AsO_4^{-3}) is high at low soil pH and therefore their concentration in soil solution decreases with reduction in pH (i.e. in acidic soil) (McLaughlin et al. 1998).

Expression of toxicity to plant and contamination of food-chain due to heavy metals depend on their nature, mobility, levels as well as properties of soil. While Cd, Cr(VI) and Cu are relatively more phytotoxic, Cd and Zn are mobile and have

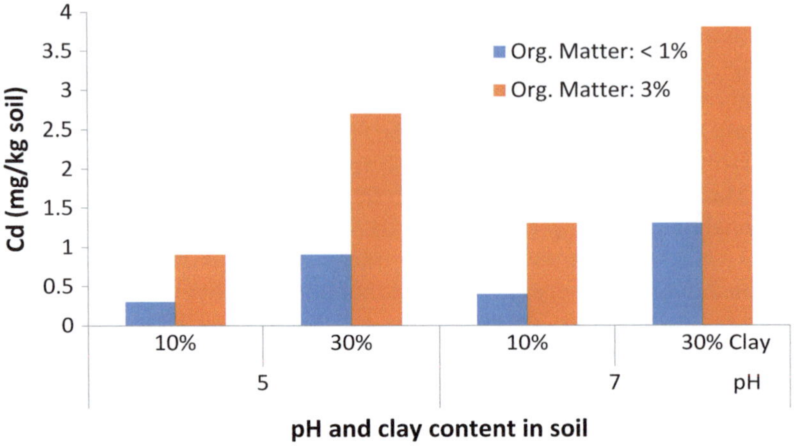

Fig. 7.3 Tolerable level of Cd as influenced by pH, clay and organic matter contents of soil (Adapted from Kloke et al. 1984)

potential to contaminate food chain due to higher bioaccumulation factor (Zn contamination is many a times preferred in underdeveloped and developing countries due to its essentiality in human/animal physiology). Soils have varying capacity to immobilize metals (through sorption, transformation, precipitation) due to variable types and contents of clay, organic matter, oxides, carbonates, phosphates, sulphides etc. along with chemical conditions (pH and Eh) prevailing in it and therefore, tolerable level for metals depends considerably on soil properties (Fig. 7.3).

Beside pH, following soil properties also influence sorption of metals and hence, their concentration in soil solution:

- Soluble organic and inorganic ligands: changes metal speciation and their charge.
- Ionic strength of soil solution: enhances competition for sorption sites.
- Redox potential: Controls reactivity of Fe and Mn oxides, chemical speciation of As, Cr and Se.

Growing plant can also influence heavy metal solubility (hence phytoavailablity). About 70% of C transferred from shoots to roots is estimated to be exerted as exudates in the soil. Such exudates influence soil solution composition by increasing solubility of minerals, modifying sorption sites (for metals), changing speciation of metals (Mc Laughlin et al. 1998), and changing solution pH.

7.6.2 *Soil Testing for Heavy Metals*

Soil testing methods have traditionally been developed and used for assessing the status for nutrients available to plants (i.e. soil fertility) for more than 40 years and for recommending fertilizer application rates for crops and cropping systems. However, there are several special consideration for assessment of hazards due to heavy metal contamination (McLaughlin et al. 2000), including the examination of effects on different components of the environment like humans, animals, plants and microorganisms. Although extensive researches have been conducted in India on soil testing for fertilizer recommendation under All India Coordinated Research Project on Soil Test Crop Response (STCR 2016), practically, no research effort has been carried out for developing soil testing methods for Indian soils.

7.6.2.1 Desirable Characteristics of Soil Test Methods

McLaughlin et al. (1999) suggested a number of criteria that should be applied to any soil test to diagnose or predict risks from soil contamination by heavy metals which are as below:

- Soil test should be simple, inexpensive and robust.
- It should have been developed and calibrated under field conditions across a wide range of soil types.
- It should have been validated independently at different soil types.
- It should account for the major environmental factors known to affect crop metal concentrations and toxic response to plants or organisms.
- It should be for prognosis (i.e. truly predictive).

7.6.2.2 Extractants for Soil Testing

There are a number of extractants and methods developed for predicting phytoavailability of heavy metals. These can be grouped as:

Complexing Extractants
Originally developed for predicting micronutrient deficiency, use of such extractants has been extended for polluted soils also. Several complexing reagents have been used to extract phytoavailable metals like diethylene triamine penta acetic acid (DTPA), ethylene diamine tetra acetic acid (EDTA), sodium pyrophosphate ($Na_4P_4O_7$) by various researchers (Latterell et al. 1978; He and Singh 1993; Mc Keague 1967). Such complexing extractants tend to extract metals from many other fractions (reserve pool), besides immediately available pool (soluble and

exchangeable). For this reason, several researchers have questioned their sustainability as extractants for soil test (McLaughlin et al. 2000; Cook 1998).

Neutral Salt Extractants
Unbuffered solution of neutral salts like $CaCl_2$, $NaNO_3$ and NH_4NO_3 have been indicated as best predictors of heavy metals uptake by plants in contaminated soils. This is because these reagents extract metals from immediately available pool (i.e. soil solution) as well as from the fraction which quickly replenishes the soil solution once depleted after absorption by plants roots (i.e. exchangeable fraction). Alloway and Morgan (1986) and Davies (1992) found that 1.0 *M* NH_4NO_3 was suitable predictor for heavy metals concentration in vegetable crops. Dilute $CaCl_2$ (0.01 *M*) has also been suggested as the best predictor of phytoavailable metals in soils by several researchers (Houba et al. 1990; Rayment 1994; Andrews et al. 1996). At equimolar concentrations, this solution may extract more Cd than NH_4NO_3 and $NaNO_3$ due to formation of chloro-complexes. Use of 0.1*M* $NaNO_3$ as extractants has also been recommended for predicting plant uptake of metals and their microbial toxicity (Gupta and Aten 1993). Major drawback of using unbuffered salt solutions is their very lower magnitude of extraction demanding costly and sensitive analytical instruments like Inductively Coupled Plasma - Optical Emission Spectrometer (ICP-OES), ICP-Mass Spectrometer (ICP-MS), High Performance Liquid Chromatography (HPLC)-ICP-MS etc.

Ion-exchange Resins
Although ion exchange resins are nowadays being advocated for assessing plant available macro-nutrients in agricultural soils, their use in soil testing of contaminated land is less common. Jing and Logan (1991) used Chelex™ resins beads (enclosed in dialysis bag to assess phyto-available Cd in sewage sludge amended soil; but failed to demonstrate any significant advantage of this over other extractants). However, Lee and Zhang (1993) found such resin a better predictor than DTPA or 0.01 *M* $CaCl_2$ for soil Cd phytoavailability.

7.6.2.3 Microbial Biosensor to Indicate Bioavailable Heavy Metal in Soil and Its Transfer to Plants

Bacterial sensor-reporters, which consist of living micro-organisms genetically engineered to produce specific output in response to target chemicals, offer an interesting alternative to monitoring approaches. Bacterial sensor-reporters detect bioavailable and/or bioaccessible compound fractions in samples. Currently, a variety of environmental pollutants can be targeted by specific biosensor-reporters. Although most of such strains are still confined to the lab, several recent reports have demonstrated utility of bacterial sensing-reporting in the field, with method detection limits in the nanomolar range (Tecon and van der Meer 2008). Method detection limits of MBS assays are often in the nanomolar range, thereby competing effectively with existing chemical analytics. Despite these aspects, MBS-assays are still rarely applied outside research laboratories (Harms et al. 2006). Convincing data have been produced which demonstrate field robustness, good measurement precision and accuracy of MBS-assays in comparison to chemical analytics, as in the case of arsenic in groundwater (Trang et al. 2005) or rice (Baumann and van der Meer 2007). However, due to GMO related issue, regulatory authorities in different countries do not accept MBS as realistic alternative for a variety of analytical procedures.

7.6.3 Soil Test Screening Levels of Heavy Metals for Assessment of Toxicity in an Agro-ecosystem

Unlike soil tests for nutrient elements, there is a dearth of good data relating soil tests to environmental endpoints for heavy metal toxicity. Moreover, soil testing methodology for the identification of heavy metal-contaminated soils is practically absent in India. A study was conducted India to determine the screening levels of different heavy metals by using a widely used soil test procedure (extraction with 0.01M $CaCl_2$ as recommended by Houba et al. 1990) for a sensitive agro-ecosystem in which heavy metals are likely to cause toxicity at an early stage and where sensitive agro-ecosystem was defined as an edible plant with a high metal uptake capacity, growing in a soil type having a low metal-fixing capacity (Saha and Panwar 2014).

The above-ground biomass of spinach was significantly reduced at $CaCl_2$-extractable levels of and higher than 0.080 mg kg^{-1} Cd, 0.582 mg kg^{-1} Cr, 1.581 mg kg^{-1} Cu, 1.63 mg kg^{-1} Ni, and 15.258 mg kg^{-1} Zn (single element contamination). Doses of Pb did not have any significant effect on the above-ground biomass of spinach. Soil test levels more than 0.100 mg kg^{-1} Cd, 0.465 mg kg^{-1} Cr, 1.357 mg kg^{-1} Cu, 1.350 mg kg^{-1} Ni, and 13.987 mg kg^{-1} Zn reduced biomass yield by more than 20% (Fig. 7.4).

Soil test screening levels of heavy metals for vegetable growing land were identified as 0.003 mg kg^{-1} Cd, 0.176 mg kg^{-1} Cr, 0.637 mg kg^{-1} Cu, 0.022 mg kg^{-1} Ni, 0.008 mg kg^{-1} Pb, and 0.068 mg kg^{-1} Zn; beyond which it might contaminate vegetables. Similarly, increasing levels $CaCl_2$ extractable heavy metals in soil had significant adverse effect on several soil enzymes which are considered indicator of soil microbial activities (Fig. 7.5). Soil test levels with more than 0.087 mg kg^{-1} Cd, 0.052 mg kg^{-1} Cr, 0.70 mg kg^{-1} Cu, 0.239 mg kg^{-1} Ni, 0.008 mg kg^{-1} Pb, and 3.80 mg kg^{-1} Zn were found to diminish any of the soil enzymes dehydrogenase, phosphatase, arylsulfatase, protease, lipase and esterase by more than 20%.

As the lowest among these maximal screening levels of soil test values determined by using the above three different approaches can protect all target organisms, 0.003 mg kg^{-1} Cd, 0.052 mg kg^{-1} Cr, 0.637 mg kg^{-1} Cu, 0.022 mg kg^{-1} Ni,

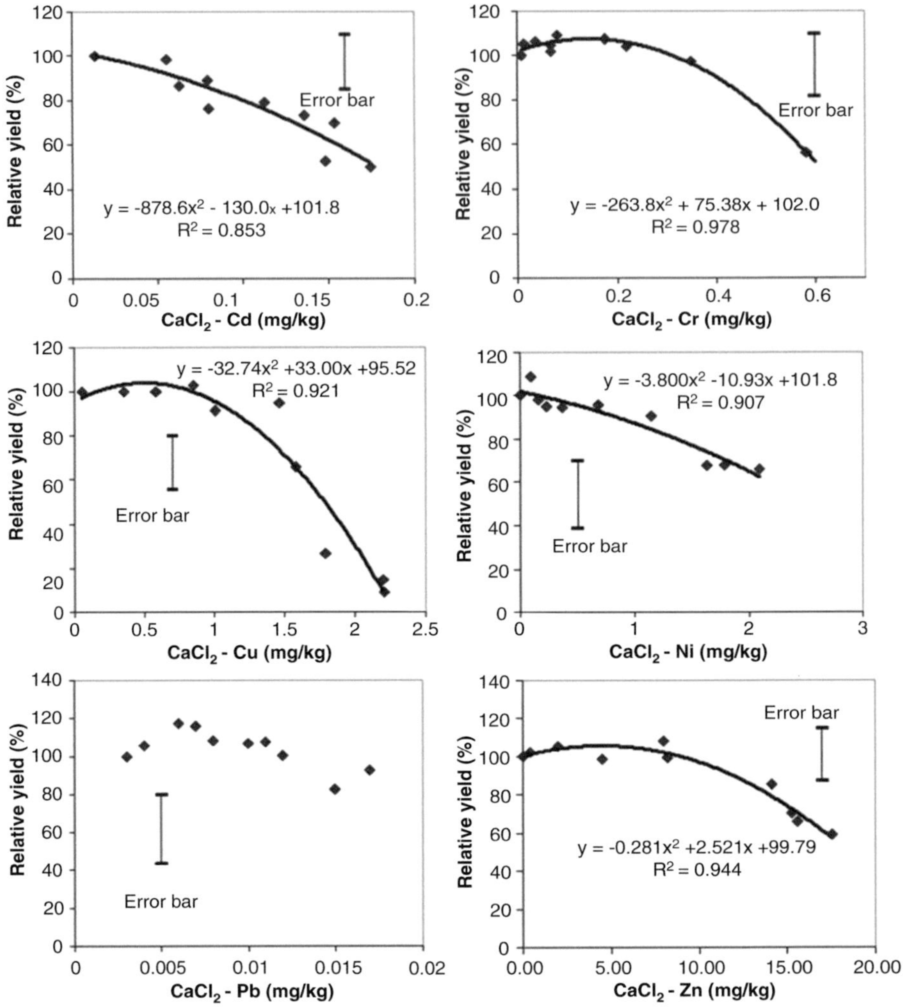

Fig. 7.4 Effect of $CaCl_2$-extractable metals in soil on above-ground biomass yield (Expressed as % yield after control treatment) of spinach. Error bar = Least significant difference between means at 5% probability level (Saha and Panwar 2014)

0.008 mg kg^{-1} Pb, and 3.800 mg kg^{-1} Zn were considered to be the screening levels for assessing heavy metals toxicity in municipal solid waste compost-amended soil for growing vegetables (Saha and Panwar 2014).

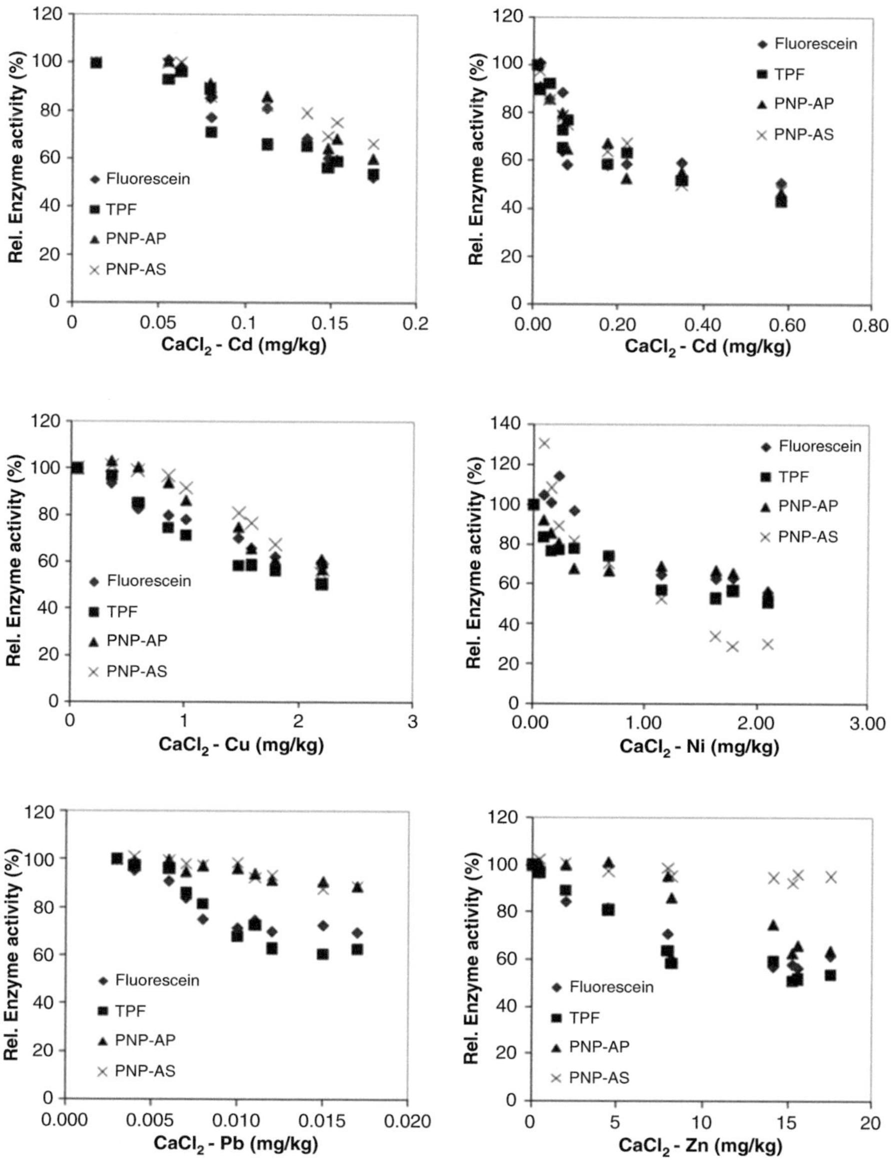

Fig. 7.5 Effects of $CaCl_2$-extractable metals in soil on relative soil enzyme activities (Fractions of control soil enzyme activities). Fluorescein = activities of esterase + lipase + protease; TPF = dehydrogenase activity; PNP-AP = acid phosphatase activity; PNP-AS = arylsulfatase activity (Saha and Panwar 2014)

References

Akerblom S, Baath E, Bringmark L et al (2007) Experimentally induced effects of heavy metals on microbial activity and community structure of forest mor layers. Biol Fert Soils 44:79–91

Alfaro MR, Montero A, Ugarte OM et al (2015) Background concentrations and reference values for heavy metals in soils of Cuba. Environ Monit Assess 187:4198–4207

Alloway BJ, Morgan H (1986) The behavior and availability of Cd, Ni and Pb in polluted soils. In: Assink JW, van den Brink WJ (eds) Contaminated soil. Martinus Nijhoff Publishers, Dordrecht, pp 101–113

Andersen S, Odegard S, Vogt RD, Seip HM (1994) Background levels of heavy-metals in Polish Forest soils. Ecol Engin 3:245–253

Andrews P, Town RM, Hedley MJ, Loganathan P (1996) Measurement of plant available cadmium in New Zealand soils. Australian J Soil Res 34:441–452

Angelone M, Bini C (1992) Trace Elements concentrations in soils and plants of Western Europe. In: Adriano DC (ed) Biogeochemistry of Trace Elements. Lewis Publishers, Boca Raton, pp 10–60

ATSDR (2005) Toxicological profile for nickel. Agency for Toxic Substances and Disease Registry, U.S. Department of Health And Human Services, Public Health Service, Division of Toxicology/Toxicology Information Branch, Atlanta

Bachman G, Freier R, Konietzka R (1995) Soil levels based on the German soil Protection Bill. In: Brink WJ, van der Bosman R, Arendt F (eds) Contaminated soil. Kluwer Academic Publishers, Dordrecht, pp 711–719

Baltrenaite E, Lietuvninkas A, Baltrenas P (2012) Use of dynamic factors to assess metal uptake and transfer in plants—example of trees. Water Air Soil Pollut 223:4297–4306

Baumann B, van der Meer JR (2007) Analysis of bioavailable arsenic in rice with whole cell living bioreporter bacteria. J Agr Food Chem 55:2115–2120

Bech J, Tume P, Longan L et al (2005) Baseline concentrations of trace elements in surface soils of the Torrelles and Sant Climent municipal districts (Catalonia, Spain). Environ Monit Assess 108:309–322

Bhagure GR, Mirgane SR (2011) Heavy metal concentrations in groundwaters and soils of Thane Region of Maharashtra, India. Environ Monit Assess 173:643–652

Bhattacharyya T, Pal DK, Mandal C et al (2013) Soils of India: historical perspective, classification and recent advances. Curr Sci India 104:1308–1323

Bini C, Dallaglio M, Ferreti O, Gragniani R (1988) Background levels of microelements in soils of Italy. Environ Chem Health 10:63–69

Blaser P, Zimmermann S, Luster J, Shotyk W (2000) Critical examination of trace element enrichments and depletions in soils: As, Cr, Cu, Ni, Pb, and Zn in Swiss forest soils. Sci Total Environ 249:257–280

Bradford GR, Chang AC, Page AL et al (1996) Background concentrations of trace and major elements in California soils. Karney Foundation Special Report of University of California, Riverside, Technical Report

Brookes PC, McGrath SP (1984) The effect of metal toxicity on the size of soil microbial biomass. J Soil Sci 35:341–346

Buat-Menerd P, Chesselt R (1979) Variable influence of the atmospheric flux on the trace metal chemistry of oceanic suspended matter. Earth Planetary Sci Let 42:398–411

Bunzl K, Albers BP, Schimmack W et al (2000) Examination of a relationship between ^{137}Cs concentrations in soils and plants from alpine pastures. J Environ Radioact 48:145–158

Campos ML, Pierangeli AP, Guilherme LRG et al (2003) Baseline concentration of heavy metals in Brazilian latosols. Commun Soil Sci Plant Anal 34:547–557

Chaney RL, Ryan JA (1994) Risk based standards for arsenic, lead and cadmium in urban soils. DECHEMA, Frankfurt, p 130

Chaudri AM, McGrath SP, Giller KE et al (1993) Enumeration of indigenous *Rhizobium leguminosarum* biovar trifolii in soils previously treated with metal –contaminated sewage-sludge. Soil Biol Biochem 25:301–309

Chen J, Wei F, Zheng C (1991) Background concentrations of elements in soils of China. Water Air Soil Pollut 57(58):699–712

Chen M, Ma LQ, Harris WG, Hornesby AG (1999) Background concentrations of trace metals in Florida surface soils: taxonomic and geographic distributions of total-total and total-recoverable concentrations of selected trace metals. Florida Center For Solid And Hazardous Waste Management, Florida

Cheng Q, Agterberg FP, Bonham-Carter GF (1996) A spatial analysis method for geochemical anomaly separation. J Geochem Explor 56:183–195

Cheng Q, Xu Y, Grunsky E (2000) Integrated spatial and spectrum method for geochemical anomaly separation. Nat Res Res 9:43–56

Chester R, Kudoja WM, Thomas A, Towner J (1985) Pollution reconnaissance in stream sediments using non-residual trace metals. Environ Pollut 23:213–238

Cicchella D, De Vivo B, Lima A (2005) Background and baseline concentration values of elements harmful to human health in the volcanic soil of the metropolitan and provincial area of Napoli (Italy). Geochem Explor Environ Anal 5:29–40

Cook N (1998) Bioavailability of trace metals in urban contaminated soils. Ph D thesis, McGill University, Quebec

da Silva FBV, do Nascimento CWA, Araújo PRM et al (2016) Assessing heavy metal sources in sugarcane Brazilian soils: an approach using multivariate analysis. Environ Monit Assess 188:457. doi:10.1007/s10661-016-5409-x

Dantu S (2009) Heavy metals concentration in soils of south-eastern part of Ranga Reddy district, Andhra Pradesh, India. Environ Monit Assess 149:213–222

Dantu S (2010) Geochemical patterns in soils in and around Siddipet, Medak District, Andhra Pradesh, India. Environt Monit Assess 170:681–701

Davies BE (1992) Trace metals in the environment: retrospect and prospect. In: Adriano DC (ed) Biogeochemistry of trace metals. Lewis Publishers, Boca Raton, pp 1–18

Davis PA, Avadhanula MR, Cancio D et al (1999) BIOMOVS II: an international test of the performance of environmental transfer models. J Environ Radioact 42:117–130

De Temmerman LO, Hoenig M, Scokart PO (1984) Determination of normal level and upper limit values of trace elements in soils. Z Pflanz Bodenkunde 147:687–694

Denneman CAJ, Robbersen JG (1993) Target values for soil and groundwater – the Netherlands integrated soil research programme. In: Eijsackers HJP, Hamers T (eds) Integrated soil and sediment research: a basis for proper protection: selected proceedings of the first European Conference on Integrated Research for Soil and Sediment Protection and Remediation (EUROSOL), Springer Dordrecht. pp 89–90

Dinelli E, Lombini A (1996) Metal distributions in plants growing on copper mine spoils in Northern Apennines, Italy: the evaluation of seasonal variations. Appl Geochem J 11:375–385

Doelman P, Haanstra L (1989) Short and long-term effects of heavy metals on phosphatase activity in soils: an ecological dose-response model approach. Biol Fert Soils 8:235–241

Dudka S (1992) Factor analysis of total element concentrations in surface soils in Poland. Sci Total Environ 121:39–52

Dudka S, Ponce-Hernandez R, Hutchinson TC (1995) Current levels of total element concentrations in the surface laver of Sudbury's soils. Sci Total Environ 162:161–172

Facchinelli A, Sacchi E, Mallen L (2001) Multivariate statistical and GIS-based approach to identify heavy metal sources in soils. Environ Pollut 114:313–324

Feller W (1971) An Introduction to Probability Theory and Its Applications, 3rd edn. Wiley, New York

Fitz WJ, Wenzel WW (2002) Arsenic transformations in the soil-rhizosphere-plant system: fundamentals and potential application to phytoremediation. J Biotechnol 99(3):259–278

Forstner U (1990) Contaminated sediments, Lecture Notes in Earth Science, vol 21. Springer, Berlin
Garcia R, Maize I, Millan E (1996) Heavy metal contamination analysis of roadsoils and grasses from Gipuzkoa (Spain). Environ Technol 17:763–770
Gough LP, Severson LC, Jackson LL (1994) Baseline element concentration in soils and plants, Bull Island, Cape Romain, National Wildlife Refuge, South Carolina, USA. Water Air Soil Pollut 74:1–17
Guan Y, Shao C, Ju M (2014) Heavy Metal Contamination Assessment and Partition for Industrial and Mining Gathering Areas. Int J Environ Res Public Health 11:7286–7303
Gupta SK, Aten C (1993) Comparison and evaluation of extraction media and their suitability in a simple model to predict the biological relevance of heavy metal concentrations in contaminated soils. Int J Environ Anal Chem 51:25–46
Hakanson L (1980) An ecological risk index for aquatic pollution control. A sedimentological approach. Water Res 14:975–1001
Hakanson L, Jansson M (1983) Principles of Lake Sedimentology. Springer, Berlin, pp 280–281
Harms H, Wells M, van der Meer JR (2006) Whole-cell living biosensors: are they ready for environmental application? Appl Microbiol Biotechnol 3:273–280
He QB, Singh BR (1993) Plant availability of cadmium in soils. I. Extractable cadmium in newly and long-term cultivated soils. Acta Agr Scand Sec B- Soil Plant Sci 43:134–141
Holmgren GGS, Meyer MW, Chaney RL (1993) Cadmiun, Pb, Zn, Cu and Ni in agricultural soils of the United State of America. J Environ Qual 22:335–348
Hope BK (1995) A review of models for estimating terrestrial ecological receptor exposure to chemical contaminants. Chemosphere 30:2267–2287
Houba VJG, Novozamsky I, Lexmond TM, Lee JJ (1990) Applicability of 0.01 *M* $CaCl_2$ as a single extraction solution for the assessment of the nutrient status of soils and other diagnostic purposes. Commun Soil Sci Plant Anal 21:2281–2290
Hu Y, Liu X, Bai J et al (2013) Assessing heavy metal pollution in the surface soils of a region that had undergone three decades of intense industrialization and urbanization. Environ Sci Pollut Res 20:6150–6159
Huang R (1987) Environmental pedology. Higher Education Press, Beijing
IRIS (2016) Integrated Risk Information System. US Environmental Protection Agency. https://www.epa.gov/iris. Accessed 16 June 2016
Jie-liang C, Zhou S, You-wei Z (2007) Assessment and mapping of environmental quality in agricultural soils of Zhejiang Province, China. J Environ Sci 19:50–54
Jing J, Logan TL (1991) Chelating resin method for estimation of sludge-cadmium bioavailability. Commun Soil Sci Plant Anal J 22:2029–2035
Kabata-Pendias A (2000) Trace Element in Soils and Plants, Third edn. CRC Press, Baton Raton
Kabata-Pendias A (2011) Trace elements in soils and plants. CRC Press/Taylor & Francis Group, Boca Raton
Kabata-Pendias A, Pendias H (1992) Trace elements in soils and plants. CRC ress, Baton Raton
Karageorgis AP, Katsanevakis S, Kaberi H (2009) Use of enrichment factors for the assessment of heavy metal contamination in the sediments of Koumoundourou Lake, Greece. Water Air Soil Pollut 204:243–258
Kloke A, Sauerbeck DR, Vetter H (1984) The contamination of plants and soils with heavy metals and the transport of metals in terrestrial food chain. In: Nriagu JO (ed) Changing metal cycles and human health. Springer, Berlin, pp 113–141
Kostov O, Van O (2001) Nitrogen transformation in copper- contaminated soils and effects of line and compost application on soil resiliency. Biol Fert Soils 33:10–16
Krishna AK, Govil PK (2007) Soil contamination due to heavy metals from an industrial area of Surat, Gujarat, Western India. Environ Monit Assess 124:263–275
Krishna AK, Govil PK (2008) Assessment of heavy metal contamination in soils around Manali industrial area, Chennai, Southern India. Environ Geol 54:1465–1472

Krzysztof L, Wiechula D, Korns I (2004) Metal contamination of farming soils affected by industry. Environ Int 30:159–165

Kuhad MS, Malik RS, Singh A (1989) Background levels of heavy metals in agricultural soils of Indo-Gangetic Plains of Haryana. J Indian Soc Soil Sci 37:700–705

Kuperman RG, Carreiro MM (1997) Soil heavy metal concentrations, microbial biomass and enzyme activities in a contaminated grassland ecosystem. Soil Biol Biochem 29:179–190

Latterel JJ, Dowdy RH, Larso WE (1978) Correlation of extractable metals and metal uptake of snap beans grown on soil amended with sewage sludge. J Environ Qual 7:435–440

Lee DY, Zheng HC (1993) Chelating resin membrane method for estimation of soil cadmium phytoavailability. Commun Soil Sci Plant Anal 24:685–700

Lide DR (2005) CRC Handbook of Chemistry and Physics. CRC Press, Boca Raton

Lima A, De Vivo B, Cicchella D et al (2003) Multifractal IDW interpolation and fractal filtering method in environmental studies: an application on regional stream sediments of Campania region (Italy). Appl Geochem 18:1853–1865

Liu Q, Diamond ML, Gingrich SE et al (2003) Accumulation of metals, trace elements and semivolatile organic compounds on exterior window surfaces in Baltimore. Environ Pollut 122(1):51–61

Liu X, Song Q, Tang Y et al (2013) Human health risk assessment of heavy metals in soil-vegetable system: a multi-medium analysis. Sci Total Environ 463-464:530–540

Loska K, Wiechula D, Korus I (2004) Metal contamination of farming soils affected by industry. Environ Int 30:159–165

Ma LQ, Tan F, Harris W (1997) Concentrations and distributions of eleven metals in Florida soils. J Environ Qual 26:769–775

Marandi A, Karro E (2008) Natural background levels and threshold values of monitored parameters in the Cambrian–Vendian groundwater body, Estonia. Environ Geol 54:1217–1225

McBride MB (1994) Environmental chemistry of soils. Oxford University Press Inc, New York

McGrath SP (1993) Soil quality in relation to agricultural use. In: Eijsackers HJP, Hamers T (eds) Integrated soil and sediment research: a basis for proper protection. Springer, Dordrecht, pp 187–200

McGrath SP, Chang AC, Page AL, Witter E (1994) Land application of sewage sludge: scientific perspectives of heavy metal loading limits in Europe and the United States. Environ Rev 2:108–118

McKeague JA (1967) An evaluation of 0.1 *M* pyrophosphate and pyrophosphatedithionite in comparison with oxalate as extractants of the accumulation products in podsols and some other soils. Canadian J Soil Sci 47:95–99

McLaughlin MJ, Smolders E, Merckx R (1998) Soil root interface: physiochemical processes. In: Huang PM (ed) Soil Chemistry and Ecosystem Health. Soil Science Society of America, Wisconsin, pp 233–277

McLaughlin MJ, Parker DR, Clarke JM (1999) Metals and micronutrients – food safety issues. Field Crops Res 60:143–163

McLaughlin MJ, Zarcinas BA, Stevens DP, Cook N (2000) Soil testing for heavy metals. Commun Soil Sci Plant Anal 31:1661–1700

Mendez MO, Maier RM (2008) Phytostabilization of Mine Tailings in Arid and Semiarid Environments—An Emerging Remediation Technology. Environ Health Perspect 116 (3):278–283

Mico C, Peris M, Recatala L, Sánchez J (2007) Baseline values for heavy metals in agricultural soils in an European Mediterranean region. Sci Total Environ 378:13–17

Mico C, Recatala L, Peris M, Sánchez J (2006) Assessing heavy metal sources in agricultural soils of an European Mediterranean area by multivariate analysis. Chemosphere 65:863–872

Murthy NN (2008) A case study of soil contamination at Katedan Industrial area, Hyderabad. NGRI Proc: Contaminated groundwater monitoring and soil assessment, Hyderabad, India, p 1–12

Namaghi HH, Karami GH, Saadat S (2011) A study on chemical properties of groundwater and soil in ophiolitic rocks in Firuzabad, east of Shahrood, Iran: with emphasis to heavy metal contamination. Environ Monit Assess 174(1-4):573–583

Panwar NR, Saha JK, Adhikari T et al (2010) Soil and water pollution in India: some case studies, IISS technical bulletin. Indian Institute of Soil Science, Bhopal, pp 1–40

Parker DR, Pedler JF (1997) Reevaluating the free-ion activity model of trace metal availability to higher plants. Plant Soil 196:223–238

Parth V, Murthy NN, Saxena PR (2011) Assessment of heavy metal contamination in soil around hazardous waste disposal sites in Hyderabad city (India): natural and anthropogenic implications. J Environ Res Manage 2:27–34

Pierzynski GM, Sims JT, Vance GF (2000) Soils and environmental quality. CRC Press/LLC, NW Corporate Blvd, Boca Raton

Purushotham D, Lone MA, Rashid M et al (2012) Deciphering heavy metal contamination zones in soils of a granitic terrain of southern, India using factor analysis and GIS. J Earth Sys Sci 121 (4):1059–1070

Rao AS, Saha JK (2012) Sustaining agricultural productivity through integrated soil management. Policy paper, National Academy of Agricultural Sciences, New Delhi, pp 24

Ratha DS, Sahu BK (1993) Sources and distribution of metals in urban soils of Bombay, India using multivariate statistical techniques. Environ Geol 22:276–285

Rayment GE (1994) Cadmium in Queensland vegetables and associated soils. MSc Thesis, School of Agriculture, University of Queensland, St Lucia, Australia

Reimann C, de Caritat P (2000) Intrinsic flaws of element enrichment factors (Efs) in environmental geochemistry. Environ Sci Technol 34:5084–5091

Roca MC, Vallejo VR (1995) Effect of soil potassium and calcium on cesium and strontium uptake by plant roots. J Environ Radioact 28:141–159

Rodriguez PB, Tome FV, Lozano JC (2002) About the assumption of linearity in soil-to plant transfer factors for uranium and thorium isotopes and 226Ra. Sci Total Environ 284:167–175

Rooney CP, Zhao F, McGrath SP (2007) Phytotoxicity of nickel in a range of European soils: influence of soil properties, Ni solubility and speciation. Environ Pollution 145:596–605

Saha JK, Panwar N (2014) Screening of soil for assessment of toxicity of heavy metals to organisms. Curr Sci India 106(2):300–304

Saha JK, Panwar N et al (2013) Risk assessment of heavy metals in soil of a susceptible agro-ecological system amended with municipal solid waste compost. J Indian Soc Soil Sci 61 (1):15–22

Saur E, Juste C (1994) Enrichment of trace elements from long-range aerosol transport in sandy podozolic soils of southwest France. Water Air Soil Pollut 73:235–246

SEPAC (1995) State Environmental Protection Administration of China, Beijing. Chinese Environmental Quality Standard for Soils (GB15618-1995)

Sheppard MI, Sheppard SC (1985) The plant concentration concept as applied to natural uranium. Health Physics 48:494–500

Smolders E, McLaughlin MJ (1996) Chloride increases cadmium uptake in Swiss chard in resin-buffered nutrient solution. Soil Sci Soc Am J 60:1443–1447

STCR (2016) Online fertilizer recommendations. All India Coordinated Research Project on Soil Test Crop Response. http://www.stcr.gov.in/. Accessed 26 July 2016

Sutherland RA (2000) Bed sediment-associated trace metals in an urban stream, Oahu, Hawaii. Environ Geol 39:611–627

Tabatabai MA (1982) Soil enzymes. In: Page AL, Miller RH, Keeney DR (eds) Methods of soil analysis, part 2, chemical and microbiological properties, Monograph no, vol 9. ASA-SSSA, Madison, Wisconsin, pp 903–947

Taylor SR, Mclennan SM (1995) The geochemical evolution of the continental crust. Rev Geophys 33:241–265

Tecon R, Van der Meer JR (2008) Bacterial biosensors for measuring availability of environmental pollutants. Sensors 8:4062–4080

Thomilson DC, Wilson DJ, Harris CR, Jeffrey DW (1980) Problem in heavy metals in estuaries and the formation of pollution index. Helgolander Meeresunters 33:566–575

Thornton I (1982) Implications of geochemical background data in relation to hazardous waste management. In: Hemphill DD (ed) Trace substances in environmental health. University of Missouri, Columbia, pp 57–67

Tome FV, Rodriguez MB, Lozano JC (2003) Soil to plant transfer factors for natural radionuclides and stable elements in a Mediterranean area. J Environ Radioact 65:161–175

Trang PTK, Berg M, Viet PH, Mui NV, van der Meer JR (2005) Bacterial bioassay for rapid and accurate analysis of arsenic in highly variable groundwater samples. Environ Sci Technol 39:7625–7630

Tyler G (1981) Heavy metals in soil biology and biochemistry. In: Paul EA, Ladd JN (eds) Soil biochemistry. Marcel Dekker Inc., New York, pp 371–414

USEPA (1992) United States Environmental Protection Agency. Pesticide industry sales and usage, 1990 and 1991 market estimates. United States Environmental Protection Agency Publishing, Washington, DC

USEPA (2005a) Ecological soil screening levels for cadmium. U.S. EPA, Office of Solid Waste and Emergency Response, Washington, DC

USEPA (2005b) Ecological soil screening levels for chromium. U.S. EPA, Office of Solid Waste and Emergency Response, Washington, DC

Vencelides Z, Hrkal Z, Prchalova H (2010) Determination of the natural background content of metals in ground waters of the Czech Republic. Appl Geochem 25:755–762

Yang Z, Liu S, Zheng D, Feng S (2006) Effects of cadmium, zinc, and lead on soil enzyme activities. J Environ Sci 18:1135–1141

Zhang HH, Li FB, Wu ZF et al (2008) Baseline concentrations and spatial distribution of trace metals in surface soils of Guangdong province, China. J Environ Qual 37:1752–1760

Zhao FJ, McGrath SP, Merrington G (2007) Estimates of ambient background concentrations of trace metals in soils for risk assessment. Environ Pollut 148:221–229

Chapter 8
Urban Activities in India Leading to Soil Pollution

Abstract Due to rapid growth in industry and service sectors, urbanization has seen phenomenon increase in many countries across the world with consequent rise in solid and liquid wastes. Per capita generation and nature of such urban wastes has been found highly related to level of economic activities of a country. Due to insufficient treatment facilities for these wastes, urbanization has been found polluting agricultural land considerably, particularly in lower and middle income countries. Unscientific processing of solid wastes as well as release of sewage effluents in environment and their use for benefit to crop production have been found an important route for contamination of agricultural land with toxic trace metals and organic pollutants. Safe and beneficial recycling of municipal solid wastes in agricultural land after appropriate processing and treatment is facing obstacles due to lack of appropriate quality and hazardousness of the end products. Widely prevalent marginal quality sewage water use for irrigation in agriculture in urban and peri-urban areas in developing countries like India has also been discussed in details through different perspectives like carbon sequestration, crop productivity, plant nutrient supply as well as risks from various pollutants and pathogens.

Keywords Urban solid waste • Wastewater • Compost • Sewage sludge • Heavy metals • Crop • Organic pollutants

Availability of better living amenities and job opportunities in urban areas than in rural areas attracts more people to settle in urban areas across the world. For instances, worldwide people living in the urban areas are more than 54% and this figure may rise to 66% by 2050. Urbanization and its associated processes alter several natural ecosystem processes in following ways:

- Shift in land utilization towards roadways, housing, drainage systems etc.
- Sealing of land surface leading to cessation of interaction between atmosphere and lithosphere
- Alteration of rainwater flow pathways and its partitioning between groundwater and surface water bodies

J.K. Saha et al., *Soil Pollution - An Emerging Threat to Agriculture*, Environmental Chemistry for a Sustainable World 10, DOI 10.1007/978-981-10-4274-4_8

- Enhanced disposal/emission of gaseous, liquid and solid wastes containing toxicants/xenobiotics and their type & dynamism of interaction with natural processes

All of these ecosystem alterations in the urban areas have tremendous adverse impact on biodiversity and quality of life processes. Different anthropogenic activities cause incremental pollution leading to the accumulation of various toxic chemicals in the environment. The major urban processes contributing to soil pollution in India are indiscriminate use of xenobiotic chemicals as well as disposal of liquid and solid wastes in unsafe manner. This chapter describes treatment and disposal of urban wastes generated with India in focus.

8.1 Urban Solid Wastes and Their Use in Agriculture

8.1.1 Municipal or Urban Solid Wastes Generation

Worldwide annual municipal or urban solid waste (MSW) generation in 2010 was about 1.3 billion metric tonnes. It is expected to reach ≈2.2 billion metric tonnes by 2025 (Daniel and Perinaz 2012). Such increase may not only be due to expansion of cities and their population, perhaps the per capita waste generation has also increased from 1.2 to 1.42 kg day^{-1}. Further the high-income countries generate more total daily per capita MSW when compared to the low-income countries (Fig. 8.1). Countries under OECD generate on an average maximum per capita MSW (2.2 kg day^{-1}) which is about five times the amount generated in south Asian countries (Fig. 8.2). Higher per capita waste generation in high-income countries has often been linked to increased consumerism and better in lifestyle.

Human population in India is more than 1.2 billion and this constitutes approximately 18% of human population worldwide and ranks second in the world with respect to total human population. Urban population grew with the pace of 31.8% and during the last decade, and exceeded 377 million which is accounted as 31% or more of the India's total population with population density of more than 400 km^{-2}. The total urban solid waste (MSW) generated in India is 68.8 million metric tonnes per year (TPY) or 1, 88, 500 metric tonnes per day (TPD). The MSW generation in an area is directly linked to the total population residing. A survey reflected that the metropolitan cities and bigger towns of the country generate more MSW than smaller cities. For example, among the metropolitan cities in India, Kolkata generated maximum MSW of 11,520 TPD or 4.2 million TPY (Annepu 2012). Among different Indian geographical regions, Northern India generated the highest amount of urban solid wastes (40,500 TPD or 14.8 million TPY) which accounts 30% of total MSW generated in the country; and Eastern India (23,500 TPD or 8.6 million TPY) generated the least, i.e. only 17% of MSW generated. Among states, Maharashtra, West Bengal, Uttar Pradesh, Tamil Nadu and Andhra Pradesh generated higher scale of MSW i.e. 22,200; 15,500; 13,000;

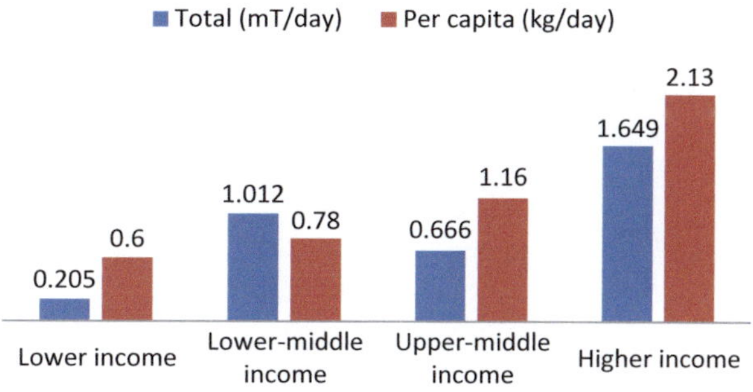

Fig. 8.1 Average total and per capita MSW generation in countries under different group

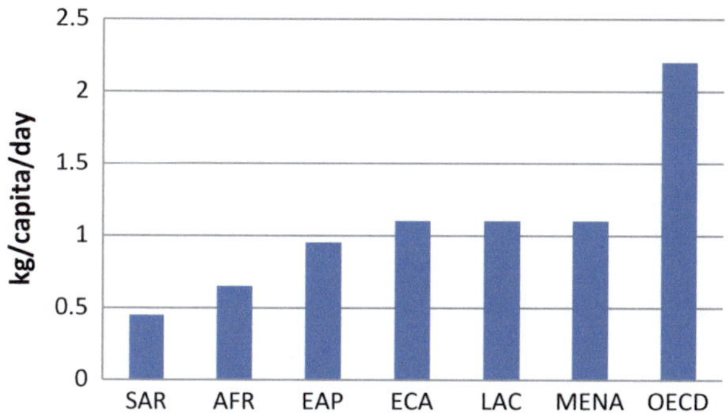

Fig. 8.2 Solid waste generation rate of various regions (Adapted from Daniel and Perinaz 2012). *SAR* South Asia, *AFR* Africa, *EAP* East Asia and Pacific, *ECA* Europe and Central Asia, *LAC* Latin America and the Caribbean region, *MENA* Middle East and North Africa, *OECD* Organisation for Economic Co-operation and Development

12,000; and 11,500 TPD or 8.1, 5.7, 4.75, 4.3, and 4.15 million TPY, respectively. The MSW generation in India is increasing at significant rate and may reach approximately to 270 million TPY by the year of 2047 (Singhal and Pandey 2001).

In India, waste generation per capita per day varies according to total population of cities and their lifestyle; but in general it ranges between 200–870 g day^{-1}. The annual per capita waste generation is increasing at the rate of 1.3% (CPCB 2012). Moreover per capita waste generation in cities located in the western part of India was lower at 440 g day^{-1}, followed by 500, 520 and 560 g day^{-1} for cities located in Eastern, Northern, and Southern India, respectively. Among the smaller Indian states, Goa recorded high per capita waste generation (620 g day^{-1}) due to predominance of tourism industry. Tamil Nadu, Jammu & Kashmir and Andhra Pradesh stand for larger per capita waste generators among states with 630, 600 and 570 g MSW day^{-1}, respectively. Among Union Territories, Andaman and Nicobar Islands generated the highest per capita waste of 870 g day^{-1}, probably due to high tourist influx.

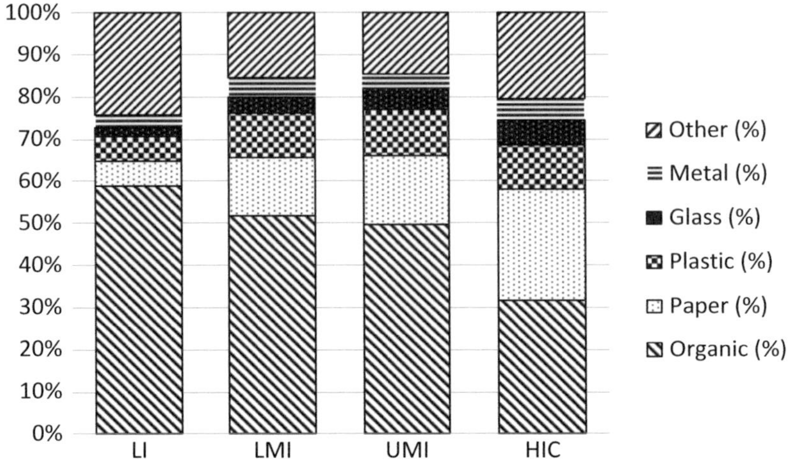

Fig. 8.3 Average composition of MSW generated by countries under different income groups (Daniel and Perinaz 2012)

8.1.2 Composition of MSW

The solid wastes generated in cities contain biodegradable organic matter, non-biodegradable organic materials and inorganic substances. Specifically, MSW originates from home (including kitchen), fruits and vegetable market, hotels and restaurants, animal carcasses, garden pruning and clippings, street sweepings, construction debris etc. and consists of decomposable organic matter, papers, plastics, cloth wastes, rubbers, glass, metals, stones, concretes, soil etc. Average composition of MSW varies widely in different countries; such as content of organic matter 4–88%, paper 2–68%, plastics 1–35%, glass 1–13%, metals 1–20%, and other materials 1–63%. Organic fraction of MSW can be beneficially recycled in agriculture for enhancing soil productivity. In general, MSW generated from lower income countries contains higher organic fraction as compared middle income and higher income countries (Fig. 8.3). Paper content is generally maximum in MSW generated from high income countries (4–68%) and is minimum in that from lower income countries (2–21%).

Solid wastes from Indian cities are found to contain 30–40% organic fractions, followed by 30–40% ash and fine earth, 3–6% paper along with less than 1% each of plastics, glasses and metals. The C/N ratio and the calorific value of these wastes range between 20–30 and 800–1000 kcal kg^{-1}, respectively (NEERI 1995).

8.1.3 Existing Methods of MSW Management in India

Urban solid wastes are managed and disposed off in various ways/means. Each method has its own advantages, disadvantages and its implications on environment.

Major methods of urban solid wastes management are discusses in the following Table 8.1.

Methods of treatment and disposal of MSW varies widely with countries. The most commonly practiced waste disposal methods in high-income countries are landfilling and waste incineration (Daniel and Perinaz 2012). Whereas low- and lower middle-income countries generally adopt open dumping (Fig. 8.4). In case of middle-income countries (including India), landfills are poorly managed; and therefore such kind of disposal is likely to be classified as controlled dumping.

For appropriate processing and disposal, the Municipal Solid Waste (Management and Handling) Rules, 2000 in India recommend collection and transportation of source-specific wastes. However current MSW disposal and management system in India is poorly developed, particularly in respect of collection, transportation and storage of MSW, segregation, processing and disposal of wastes (Sharholy et al. 2008; Kumar et al. 2009). The wastes generated in India are largely (more than 90%) disposed of by unscientific land filling or uncontrolled dumping on outer reaches of towns and cities. These faulty practices have serious environmental implications of global warming (act as a source of greenhouse gases emission) (Sharholy et al. 2008; Narayana 2009). A survey in 59 cities conducted few years back indicated that almost all the municipalities (58) follow uncontrolled dumping of MSW, while 34 of them practice composting (Kumar et al. 2009). In Class-l cities, about 94% of waste is thrown in land-dumping area and 5% of waste is converted to compost. Incineration and biomethanation methods to generate energy are followed only in one and three cities, respectively. Composting is considered as most recommended method of disposal of municipal solid waste (MSW) by Ministry of Environment and .Forest, Government of India; as it causes minimum impact on environment and generates useful product for nutrient deficit agriculture.

8.1.3.1 Solid Waste Collection and Disposal

As per the ideal solid waste management practice, MSW needs to be properly collected; and biodegradable fraction needs to be properly segregated, composted and disposed in the agricultural land after decomposition. If left uncollected and decomposed, they are a cause of several problems such as blockage of drains, barrier to movement of water, foul smell, generation of methane gas (a green house gas, GHG), release of heavy metals and many chemicals to pollute the soil and water flowing on its surface and propagate diseases particularly if contaminated with hospitals wastes. Majority of Indian cities collect wastes through community bins and dump in landfill areas; and a small fraction of uncollectable MSW lie on road-side land areas (Kumar et al. 2009). Although composting remains most preferred option, this form of waste processing is still very small (about 10% of the total MSW collected).

In several parts of the world, landfilling is commonly practiced and is considered as one of the cheapest waste management option. It requires large land space for disposal and soil quality in the landfilling site is also degraded. Due to unscientific

Table 8.1 Major methods of urban solid management and associated risks

Methods	Advantages	Disadvantages
Reuse	Resources can be saved due to reuse of original materials and space for landfill is also saved.	Adverse public perception; proper cleaning & decontamination is required for safe reusability; may increase health risk to public
Recycling	New usable products are produced by recycling of wastes; saves energy and resources for disposal; and generates revenue by selling the recyclable material	May require pretreatment to ensure safe usability; perception of public to use these products is unfavorable
On-site burial	Saves transportation cost; lowers the chance of pathogens spread; cost effective; easily adopted; and rapid method	Continuous environmental risk assessment and monitoring is required; major concern on future land use, requires lot of space, boundaries on approved sites; adverse public response; requires control of vermin and other vectors
Scientific landfill disposal	Highly suitable and properly organized with adequate control measures	Needs extensive transportation; enhances chance of pathogens spread from biological wastes; unhygienic environment and odor problem; limited capacity of each landfill
Open burning	Saves on-road waste transportation; more suitable in remote areas and in cold climatic areas lacking better disposal options	Prohibited in certain sites as well as for specific types of wastes; severe atmospheric pollution and associated health issue; periodic air monitoring is required; many municipalities prohibit burning; not effective in poor weather conditions; poor public feedback; chances of contamination to other areas
Incineration	Toxicity of waste, disease causing agents and volume of waste is reduced; electricity is generated from conversion of heat energy	Requires extensive transportation and associated costs; residues still require safe treatment and disposal; may produce undesirable end products.
Anaerobic digestion (biomethanation)	Produces manure for agriculture and methane gas for energy generation	Transportation concerns/costs; limited waste handling capacity; suitable only for segregated biodegradable wastes; residues require proper handling and technical support.
Composting	Most environment friendly, economically efficient; potentially useful end product; saves space; destroys human pathogens and minimizes their spread; largely accepted by municipalities and industry	Needs large separate space; requires continuous care and periodic monitoring during composting; emits foul smell; requires run-off and pest control; requires normal ambient temperature for effective composting; poor public response regarding usability of compost

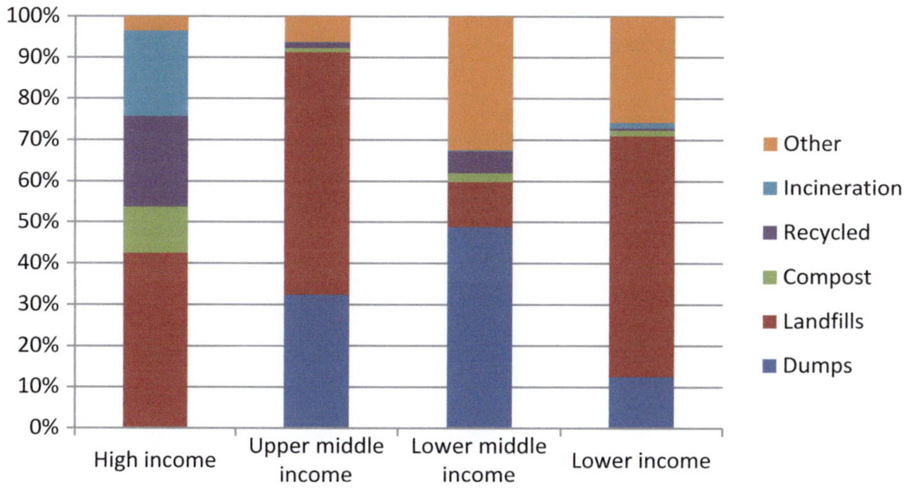

Fig. 8.4 Method of MSW disposal by countries under different income group

method of waste dumping in most of Indian cities, leachates from landfill sites may contaminate nearby land area as well as groundwater with heavy metals (Kanmani and Gandhimathi 2013). According to an estimation by The Energy and Resources Institute (TERI) report, the storage of solid waste generated in India during post-Independence until 1997 requires nearly 71,000 football fields stacked with 9 m high. In another scenario; it was reported that it has occupied 237.4 km^2 by 2001 and may increased to 590.1 km^2 by 2021 which is greater than the area of Hyderabad (583 km^2), the largest city in India area-wise. Moreover, if municipal solid wastes are not properly handled, more than 1400 km^2 of land would be required for solid waste disposal by the end of 2047. Majority of Indian cities are having community bin method of waste collection; and composting is the primary method of waste processing wherever followed (Fig. 8.5) (Kumar et al. 2009).

Impact of Uncontrolled Dumping of Wastes in Landfill Area: In most of the big cities, mixed MSW is being dumped over decades on the nearly fixed landfill area. Due to methane generation in deeper layer of MSW dump and its oxidation in presence of air, low level burning of wastes is a frequent phenomenon during dry seasons (Fig. 8.6). Such MSW burning releases huge amount of dioxins in the environment causing severe impact on human health and contamination of soil in surrounding area (Minh et al. 2003; Devanathan et al. 2012). High intake of POPs through contaminated air for respiration is suspected to be one of several causes of early slow down among men in the metro cities (Subramanian and Tanabe 2007). Also leaching from the unscientific landfill area causes contamination of surface and groundwater bodies with toxic heavy metals and organic pollutants, resulting in further spread of impact zone.

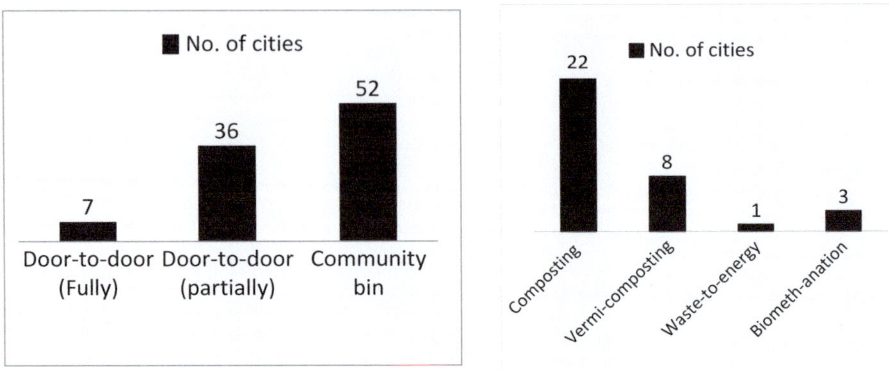

Fig. 8.5 Waste collection and processing methods in Indian cities

Fig. 8.6 Toxic smoke generation due to burning of MSW at landfill area dumpsite

8.1.4 Composting Methods Followed in Indian Cities

Composting is considered the cheapest method of MSW disposal that diverts the organic waste materials from landfills to useful manure for agricultural production. The composting of urban solid waste had been focused, promoted and encouraged by Government of India (GOI) in the early initiatives of MSW management. For this activity, Ministry of Food and Agriculture provided several promotional offers like soft loans to urban local bodies in the 1960s. In continuation of this programme, loans and grants were also provided to state governments during 1969–1974 for establishing MSW composting plants. In the later stages (in 1974), the MSW composting was revived particularly in the cities with large number of population (over 0.3 million) through modified scheme introduced by GOI. Further, during 1975–1980, highly sophisticated and mechanized large capacity composting plants (150–300 T day^{-1}) were set up in the metropolitan and big cities under the central government scheme of MSW disposal.

In pursuance of the directions of honorable Supreme Court of India, an Inter-Ministerial Task Force was constituted by Government of India to look into the possibility of safe use of city compost for integrated plant nutrient management (IPNM). The report submitted by the Task Force in 2005 suggested that providing sufficient technical and financial support for setting up of compost plant is a viable option for promoting IPNM. However, majority of different cities don't have any facilities for composting even if, Municipal Solid Wastes (Management and Handling) rules 2000 has been implemented to a certain extent. According to report by Central Pollution Control Board (CPCB) regarding implementation status of Schedule-IV, composting facilities existed in 63 cities of fourteen states and were proposed to start in another 56 cities (CPCB 2015).

Annual composting potential from MSW in India has been estimated to vary, depending on different composting methods, between 5 and 14 million tonnes of compost which can provide about 1.2 to 2.5 lakh tonnes of N, P_2O_5 and K_2O for use in agriculture. Windrow method of composting is followed by majority of the manufacturers, where mostly non-segregated or partially segregated wastes are allowed to decompose for several months with or without turning. Composting is performed in concrete pits in few cities. For proper and speedy decomposition of biodegradable segment of the waste, microbial inoculants are sprayed in majority of the composting plants. In 23% manufacturing sites surveyed, vermicomposting is being adopted for producing good quality composts in shorter time (Saha et al. 2010a).

8.1.5 *Quality of MSW Composts Prepared in India*

Samples of MSW composts from 34 manufacturers in 29 cities of the country were characterized for their physico-chemical properties as well as fertilizing potential and heavy metal contamination potential (Saha et al. 2009). Different types of preprocessing were adopted during MSW compost preparation. They are (i) Mixed waste compost (MWC), (ii) Partially segregated waste compost (PSWC), and (iii) Biowaste compost (BWC). In MWC method, non-segregated wastes are dumped as such in heaps for several months for decomposition with or without turning. Pre-processing is either followed poorly or not followed at all. This type of composting process was predominantly adopted in many cities. In case of PSWC method, large sized plastics, rubber, metals, etc. particularly non-degradable materials were manually removed prior to composting at composting yard. This is adopted in several composting plants in the country. In few cities and towns, biowaste compost (BWC) is prepared, where the segregation of biodegradable wastes is done at generator level in two ways. The wastes are collected from either biodegradable waste generating source like hotels and restaurant, vegetable market, slaughterhouse etc. or individual households deliver segregated biodegradable wastes separately to municipal organization during 'door-to-door' collection. The later method had been followed in few cities/towns such as Namakkal, Suryapet and Vijaywada. Thereafter these biodegradable wastes are composted in pits or heaps using earthworms. Compost prepared from biodegradable waste is termed as 'biowaste compost (BWC)'.

A survey indicated that compost prepared in Indian cities had low organic matter and major nutrients (N and P); and heavy metal contents were quite high (Saha et al. 2010a). Composts prepared from biowastes (as in BWC) were of better quality with higher amount of plant nutrients and lower degree of heavy metals contamination when compared to those prepared from mixed wastes. On an average, the former had 57%, 77% and 78% more organic matter, total N and total P, respectively than the later. On the other hand, heavy metals content like Zn, Cu, Cd, Pb, Ni and Cr was lesser in the former to the extent of 63%, 78%, 64%, 84%, 50%, and 63%, respectively (Saha et al. 2010a). Even the partial manual segregation followed at composting site did not significantly improve quality of compost. It had low fertilizing parameters (organic matter, N, P) and high heavy metals. The quality of compost produced from majority of MSW in Indian cities was below the standards of guidelines of 'The Fertilizer (Control) Order 1985' in respect of total organic C, total P, total K as well as heavy metals Cu, Pb and Cr. A compost grading scheme was developed based on fertilizing value (index based on plant nutrients, organic matter contents and relative importance of these constituents on crop productivity) and contamination potential (index based on contents and relative toxicity of different heavy metals). As per the scheme, composts are classified into 'marketable' grades (A, B, C and D with decreasing quality in sequence), 'restricted use' grades (RU-1, RU-2 and RU-3; not marketable, but can be used for specific purposes) and 'unsuitable' grade (may be incinerated or landfilled) (Saha et al. 2010a). A user-friendly software 'Compost Quality Evaluator' can be downloaded from www.iiss.nic.in/Software.html and used for evaluating the quality of MSW composts based on this scheme. Quality of 'biowaste compost' prepared from source separated biodegradable wastes was always better than 'mixed waste compost' and 'partially segregated waste compost' due to higher organic matter and plant nutrient contents, and lower concentrations of heavy metals (Fig. 8.7). About 50% of the composts produced from mixed wastes were not suitable for agricultural soil application because of very high metals content, 41% belonged to restricted use (RU) classes (8% in RU-1 and 33% in RU-3) and remaining were classified under marketable class D. Variability in feedstock mixture and degree of segregation has caused such a large variation in grades of composts prepared from partially segregated MSW.

In order to safeguard the land resources from pollution and to ensure minimum manurial value, Government of India enforced quality control limits for MSW compost through Fertilizer Control Order (FCO). In order to be eligible for

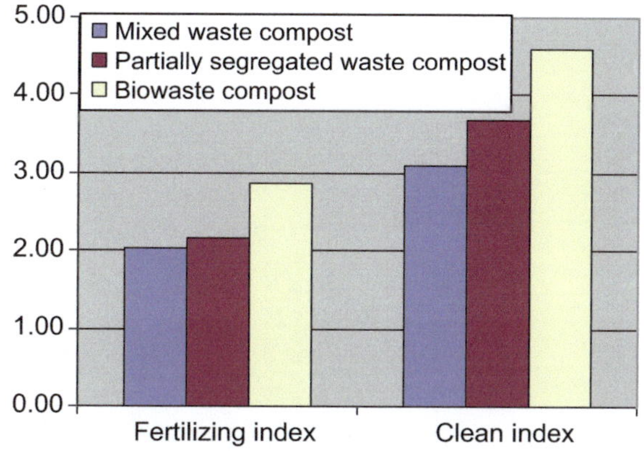

Fig. 8.7 Impact of segregation of wastes on fertilizing potential and heavy metal pollution potential of MSW composts. Higher 'Fertilizing index' indicate better manurial value and higher 'Clean index' indicates lower toxicity due to heavy metals

marketing, MSW compost sample has to comply with all the 21 quality parameters specified in the FCO. However, none of the compost samples from 29 cities complied fully with the FCO limits, despite sincere efforts by several municipalities.

Although information on the total heavy metals concentration in compost is necessary for quality control purpose, this is insufficient for evaluating exposure risks and potential toxic effect to environment. Toxic effect of heavy metals present in MSW composts generally depend on their distribution and extractability in different fractions (Petruzzelli et al. 1989). Distribution in different size fractions and their extractability by dilute acid and alkali were investigated on MSW compost samples prepared in different Indian cities (Saha et al. 2013). Results indicated that bigger size particles (>0.5 mm) constituted the major size fraction in all the composts samples and contained relatively less organic carbon and heavy metals as compared to finer size fractions. Concentration of the heavy metals increased with decreasing sizes of compost particles and the effect was more pronounced in mixed wastes and partially segregated wastes composts (Fig. 8.8). Acid extracted Cd, Zn, Ni and Pb to a greater extent than these extracted by alkali; and this difference in extraction was wider in high mineral matter containing MWC. On the contrary, dilute alkali extracted Cu and Cr more, particularly in composts having relatively higher amount of organic matter (i.e. BWC). Contents of water-soluble metals (except Cr) were comparatively lower in biowaste compost than either in mixed waste compost or compost prepared from partially segregated wastes or both. The major factors that influenced the water-soluble metals in the composts were total and water-soluble C and pH, rather than the total HM contents itself. The heavy metals occluded by fine particles present in MWC and PSWC has the potential to move to the deeper soil profile layers and, thus, can pollute greater volumes of soil in the rhizosphere.

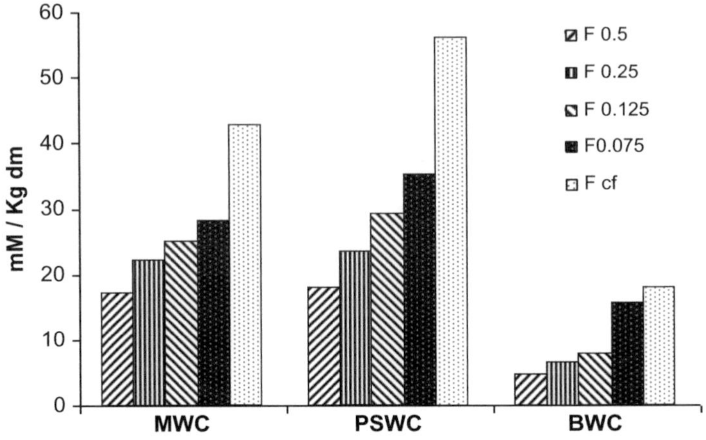

Fig. 8.8 Effect of segregation of waste feedstock on concentration of total heavy metals (Cd, Cr, Cu, Ni, Pb and Zn) in different size fractions of composts

8.1.6 Impact of MSW Compost on Soil and Crop Quality

MSW compost may increase soil heavy metal concentrations after its repeated application in agricultural land as organic fertilizer or as soil amendment. It is reported that different field and horticultural crops can take up heavy metals from MSW compost amended soils (Pinamonti et al. 1999). The availability of such trace heavy metals varies with compost maturity. As compost matures, the availability of metals decreases as they bind to the humic materials in the compost and binding strength tend to increase during the maturity of compost. The food chain contamination due to application of MSW compost in soil depends by and large on metal concentration in the compost and the source of raw-material used for composting. When compost is prepared from mixed MSW, heavy metals concentration tends to increase as compared with feedstock. On the other hand, the biowaste compost i.e. MSW compost prepared from segregation of waste at source contains minimum metal concentrations. Moreover, the lower heavy metal content in the finished product is observed when the sorting of waste is done at the earliest possible stage, such as while during collection of wastes or before initiation of the composting process. Total heavy metals content in the composts generally is not thought to be crucial for mobility and bioavailability. The extraction or absorption experiments can be adopted to assess their contents in mobile and bioavailable fractions. Bioavailable fractions of heavy metals can be estimated by extracting composts with water, neutral salt solutions, and chelating reagents like Ethylenediaminetetraacetic acid (EDTA) and diethylenetriaminepentaacetic acid (DTPA).

The occurrence of organic toxins in MSW compost has been reported by many workers and these are highly chlorinated and may persist long time in the environment. The sources of organic pollutants in MSW are primarily household wastes such as solvents and oils, printed paper products, and pesticides. It was confirmed from the survey made on various municipal and commercial wastes that phthalate esters were more prevalent among the xenobiotic compounds present in MSW composts (Düring and Gath 2002). Further, the mixed MSW composts had higher concentrations of polychlorinated biphenyls (PCB), dioxin/furans (Logan et al. 1999). On the other hand, risks to ecosystem due to PCBs were estimated to be small with use of source-segregated biowaste compost in agricultural land. Application of waste derived composts contaminated with persistent organic pollutants to pasture land may pose health risks to grazing animals and subsequently to human through milk/meat products (Smith et al. 2001).

8.2 Use of Urban Wastewater in Agriculture

Agricultural production and productivity in India largely depends on the availability of irrigation water due to unpredictability of monsoon rainfall in large part of the country. Demand for irrigation water is mostly met from groundwater surface water bodies like rivers and lakes. Although irrigation with surface water bodies is cheaper as compared groundwater sources, its availability is major constraints in most areas. Even if inefficient, groundwater irrigation covers more than half of the total irrigated area and contributes for 70% of production. However, groundwater reserve in most of the areas is depleting fast due to its higher withdrawal rate as against its rate of replenishment. In view of shortage of irrigation water vis-à-vis its requirement, farmers try to exploit whatever water sources available to them at lower cost. Due to its availability at almost free of cost, irrigation of agricultural lands with urban wastewater (UWW) is common practices in peri-urban areas (Sharma and Kansal 1986; Kansal 1992; Ghimire 1994; Dotaniya et al. 2015). Farmers in peri-urban areas usually grow vegetable crops or other perishable commodity like fodder throughout all seasons of the year with high cropping intensity (300–400%) and generate higher (four times more) income from a unit land than the freshwater irrigated farmers (Minhas and Samra 2004).

Worldwide, more than 800 million farmers are involved in urban agriculture. Out of which about 200 million farmers use marginal quality water for crop production due to scarcity and unavailability of good-quality water (Qadir et al. 2010). However, statistics on the extent of irrigation with UWW is lacking and no clear estimates are available at present on this regard. It is estimated that effluents generated from large cities have the potential to irrigate 21,000 ha to a depth of 7.5 cm on daily basis or 7.8 million hectares on annual basis (Minhas and Samra 2004). WHO (2006) has estimated that about 1.0–1.5 million hectares of land area is irrigated annually with sewage waters in India.

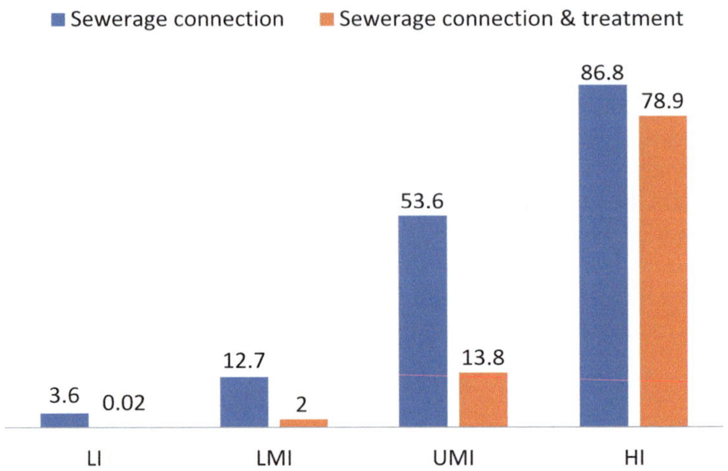

Fig. 8.9 The population (in %) under different income group countries having access to sewerage connection in 2010 (Adapted from Baum et al. 2013)

8.2.1 Wastewater Generation and Treatment Capacity: A Meta-analysis

Urban wastewater from cities is the highest quantity of waste generated from human activities. Input into this form of waste occurs from three main sources: domestic, commercial and runoff/storm water during certain climatic events. Discharge of UWW into surface water bodies, particularly in the middle and low income countries, has been ascribed as major factors responsible for deterioration of lakes and rivers water quality due to contamination with above pollutants. Lack of sewerage connection and sewage treatment facilities has been ascribed as principal causes for this (Fig. 8.9). According to the 4th World Water Development Report, only small part (20%) of UWW generated worldwide is properly treated to remove or reduce pollutants load to an acceptable level (UNESCO 2012). Wastewater treatment capacity in general depends on the income level of the country with the figure being 70% of the generated wastewater in high-income countries as compared to only 8% in low-income countries (Sato et al. 2013).

Major fraction (75–85%) of the water used for domestic purpose comes out as wastewater which requires appropriate treatment before disposal (Qadir et al. 2010). According to CPCB report, about 38,254 million liter per day (MLD) sewage water was being generated from Class-I cities and Class-II towns, out of which only 11,554 MLD (32%) were subjected to treatment before throwing away (CPCB 2009). The projected UWW generation for the country has been estimated as 122,000 MLD by 2050 (Bhardwaj 2005). UWW generated by domestic, municipal and industrial activities finds its way in large numbers of open and covered channels and finally mixes up with surface water bodies. Cities around Ganga basin were generating 2637.7 MLD of UWW with treatment capacity of 44.2% only

(CPCB 2009). Therefore, a large fraction of untreated UWW pollutes the river Ganga. Due to operation and maintenance related problems, existing sewage treatment plants (STPs) are not being effectively utilized.

8.2.2 Impact of Urban Waste Water on Soil Quality

Urban wastewater is an obvious by-product of urban society and originates mainly from households (domestic wastewater) and also due to municipal activities. Such domestic wastewater may often be contaminated with industrial effluent (depending on the industrial profile in particular area/region) and occasionally with run-off from rainwater. A number pollutants and contaminants are found in UWW which can be classified into:

- Nutrient elements for plants and microbes (N, P, K, S, Ca, Mg and other micronutrients)
- Heavy metals and toxic elements (Cd, Pb, Ni, Cr, Hg, Cu, As etc.)
- Pathogens (viruses, salmonella, coliforms, helminthes and many other bacteria and protozoa)
- Organic pollutants (pesticides, PCBs, PAHs, detergents, cosmetics, antibiotics and other pharmaceuticals)

Composition of urban wastewater varies widely depending upon treatment methods (if any), quality of water supplied in the cities, contamination of industrial effluents (if any). Typical composition of untreated UWW is given as below in Table 8.2. In general, untreated UWW contains higher organic substances, N and P as compared to wastewater from treated textile industry and contaminated groundwater (Table 8.3).

8.2.3 Carbonaceous Components in UWW and Its Impact on Soil Processes

Urban domestic wastewater (sewage) is mainly composed of carbonaceous organic materials. High content of carbonaceous organic materials in sewage also indicates higher concentrations of contaminants. High organic materials leads to development anaerobic processes in the water body and consequently, emission of offensive odour. Domestic and/or agro-industrial activities (e.g., food processing) are the main source of the organic substances detected in the UWW. The estimated contribution of organic C by domestic use (g^{-1} $capita^{-1}$ day^{-1}) is faeces 17, urine 5, dishwashing and food preparation 8, and laundry 7. The rate of generation of organic N through faeces, urine, dishwashing, and food preparation is normally 1.5, 1.7, 0.2 and 0.2 g^{-1} $capita^{-1}$ day^{-1} respectively. The organic matter found in sewage effluent is present in different physical phases. The organic C in the effluent is present as soluble, fine

Table 8.2 Typical composition of untreated UWW of domestic origin in developing countries

Parameter	Concentration Range
Total solids (ppm)	700–1350
Suspended	200–450
Dissolved	500–900
Organic load (mg L^{-1})	
Total organic C	80–290
BOD_5	250–400
COD	450–800
Total Nitrogen (mg L^{-1})	35–60
Organic N	15–25
NH_4–N	20–35
NO_3–N	Trace – 2
Total phosphorus (mg L^{-1})	4–15
Organic P	1–6
Inorganic P	3–9
pH	6.7–8.0
EC (mS cm^{-1})	0.7–1.2
Heavy metals (μg L^{-1})	
Cd	Trace – 4
Cr	10–40
Cu	30–100
Pb	25–80
Hg	Trace – 3
Ni	10–40
Ag	3–10
Zn	100–300
Pathogen (cfu mL^{-1})	
Total coliforms	10^7–10^{10}
E. coli	10^6–10^9
Faecal streptococci	10^4–10^7
Helminth eggs	1–1000

Sperling (2007) and Henze and Comeau (2008)

suspended and coarse suspended forms. The contents of C (as % of total) in soluble, fine suspended particle and coarse suspended particle were found at 52%, 9% and 39% respectively in the effluent obtained from activated sludge process, and at 69%, 6% and 25% in the effluent obtained from trickling filter plants respectively (Hunter and Kotalic 1974). Manka et al. (1974) identified different organic constituents of sewage water as below in Table 8.4.

Soil organic matter plays important role in influencing soil structure and soil stability as well as in supplying nutrient elements essential for plant growth. The C cycle in soil consists of various reactions in which microorganisms play significant roles. Due to its content of considerable carbonaceous material, UWW also adds

Table 8.3 Average chemical characteristics of contaminated irrigation water from different cities of Madhya Pradesh

Parameters	Contaminated groundwater near Ratlam industrial area	Wastewater from Nagda industrial area	Urban wastewater from Bhopal city
	mg L^{-1}		
NH_4–N	Tr.	9.13	19.11
NO_3–N	7.34	2.15	5.5
K	3.76	35.9	23.36
P	Tr.	1.34	21.6
SO_4–S	64.64	534.4	5.59
Organic C	76	105	214

Saha et al. (2014)

Table 8.4 Typical organic constituents of sewage water

Constituents	Stabilization pond	Domestic sewage water (extended aeration activated sludge)
COD range (mg L^{-1})	120–153	106–167
Anionic detergents (% of COD)	12.2	16.0
Carbohydrates (% of COD)	7.8	4.6
Tannins (% of COD)	2.1	1.0
Proteins (% of COD)	21.1	23.1
Fulvic acid (% of COD)	26.6	24.0
Humic acid (% of COD)	14.7	6.1
Hymatomelanic (% of COD)	6.7	4.8

Adapted from Manka et al. (1974)

significant quantity of organic matter to the soil. However during standard treatment processes, most of the organic matters get removed from raw UWW. Consequently, the characteristic chemical oxygen demand (COD), biological oxygen demand (BOD) and total organic carbon (TOC) levels found in treated effluents are much lower than in the original UWW. Irrigation with UWW adds considerable but variable amount of organic matter in agricultural land. The quantities of organic C added to the soils through treated effluent irrigation are relatively lower compared to organic C added through manure. For instance, an annual addition of 1000 mm of typical sewage effluent containing a TOC level of 50 mg L^{-1} adds only 500 kg ha^{-1} of organic carbon. Ciaccio (1971) estimated that sewage water contains about 34 mg L^{-1} non-volatile (15.2% of total soluble C) and 25 mg L^{-1} volatile (11.3% of total soluble C) organic acids, which have significant metal complexing capacity besides providing C (@ 90 mg L^{-1}) to the soil. The organic acids in UWW include formic acid, acetic acid, butyric acid, propionic acid, valeric acid, caproic acid, besides humic acid and fulvic acid. These organic acids have significant influence on several soil processes due their high metal complexing ability.

Friedel et al. (2000) studied the effect of long-term sewage water irrigation (upto 80 years) on soil organic matter, soil microbial biomass and biological activities in agricultural soils of Mexico. Total organic C content increased 2.5 fold from initial level along with significant increase in potential mineralizable C and dissolved organic C in Ap horizons of Vertisols. Soil microbial biomass C and their activities increased with years of irrigation (25–80) and TOC accumulation which consequently increased potential mineralizable C. However, ratio of potential mineralizable C and total organic C (C_{minp}: TOC) didn't change appreciably with TOC accumulation indicating that organic matter from UWW is no different from native organic matter in the surface horizon (Ap). Aapplication of untreated UWW for 5 years leads to significant and maximum increase in organic C (SOC) in 15–30 cm depth of soil profile (Saha et al. 2010b). Carbon added through sewage water might have percolated down through cracks and macropores and accumulated in the subsoil layers. The uppermost soil layers showed little SOC build-up. This was because of frequent tilling of upper layers resulting in considerable oxidation and mineralization of the accumulated C. On the contrary, C entered in the deeper layer are sequestered for longer period since being less exposed to microbial attack. High metal build-up due to application of contaminated organic matter in soil has been found to increase accumulation of organic matter in soil and to decrease soil microbial biomass (Chander and Brooks 1991). While surveying contaminated sites in Italy, Valsechhi et al. (1995) observed close positive relationship for organic matter with heavy metals content in soil. Negative relationships were observed for the heavy metals contents with both soil respiration and specific respiratory activity (ratio between evolved CO_2-C and microbial biomass C per unit time). Therefore sequestration of C applied through UWW in soil depends on level of heavy metals carried along with it. Friedel et al. (2000) did not find C sequestration potential of UWW applied C due to lower heavy metals concentration. Amorphous iron and aluminum oxides usually found in sewage-sludge also play an important role in soil organic C accumulation.

Sewage irrigation for several decades also increased denitrification capacities, but decreased adenylate energy charge ratio (AEC) in soils (Friedel et al. 2000). Measurement of the AEC ratio indicates the growth state of natural microbial communities and the effect of environmental changes on them. These changes were ascribed due to the addition of surfactants, particularly alkyl benzene sulfonates (effect on denitrification capacity) and addition of Na and salts (effect on AEC) through sewage. Denitrification loss of soil/sewage N in the form of N_2O has tremendous global warming potential.

8.2.4 Nutrient Potential of Sewage Water

Untreated UWW has a good nutrient potential in crop husbandry. Typical nutrients concentrations in sewage water are 20–85 mg N L^{-1}, 4–36 mg P L^{-1}, 7–20 mg K L^{-1}, 10–50 mg S L^{-1} besides other nutrients. Since the amount of N in sewage effluents

Table 8.5 Influence of different types of wastewaters on soil chemical properties

	Unpolluted area		Polluted area	
	Mean	Range	Mean	Range
Industrial waste water				
Organic C (%)	*0.5[a]	0.3–0.8	0.7[b]	0.3–1.0
Available-P (μg g^{-1})	11.3[a]	5.9–22.3	18.5[b]	14.9–23.0
Available-K (μg g^{-1})	177[a]	110–300	270[b]	145–454
Available S (μg g^{-1})	16[a]	8–19	659[b]	372–1104
Conductivity (mS/m)	0.74[a]	0.32–1.24	4.03[b]	0.35–12.9
ESP	1.51[a]	1.1–1.9	31.18[b]	2.2–46.1
Polluted groundwater				
Organic C (%)	0.5[a]	0.1–0.9	0.7[b]	0.4–1.0
Available-P (μg g^{-1})	16.6[a]	11.5–43.7	18.7[a]	16.0–22.4
Available-K (μg g^{-1})	189[a]	128–293	230[b]	171–332
Available-S (μg g^{-1})	15.1[a]	3.2–22.1	28.2[b]	16.8–57.3
Conductivity (mS/m)	0.32[a]	0.22–0.99	1.83[b]	0.49–5.01
ESP	2.71[a]	1.2–4.3	6.96[b]	1.3–24.2
Untreated urban wastewater				
Organic C (%)	0.50[a]	0.30–0.77	1.13[b]	0.45–1.8
Available-P (μg g^{-1})	9.75[a]	4.6–19.3	26.5[b]	5.1–140.9
Available-K (μg g^{-1})	144[a]	73–219	669[b]	375–821
Available-S (μg g^{-1})	21[a]	7–40	31[b]	5–59
Conductivity (mS/m)	0.17[a]	0.12–0.21	0.31[a]	0.13–0.52
ESP	1.3[a]	0.90–1.9	1.6[a]	1.2–2.6

*The same letter in a row indicated on the mean values are not statistically different among them at 5% probability level

is high, the C/N ratio is low (typical values are 5), and a release of mineral N from sewage borne organic substances is favoured. Such sewage water, mainly in its raw (untreated) form in India, is used extensively as source of irrigation water in the nearby area of sewage creeks mainly for fodder and vegetable production. In general, yield of most crops were significantly higher in UWW irrigated fields as compared to those in groundwater irrigated fields; and crops in sewage effluent treated area respond less to NPK application compared to those in groundwater treated area. Changes in fertility parameters of Vertisol due to irrigation with different types of polluted water (contaminated groundwater, treated industrial wastewater and untreated UWW) was compared (Saha et al. 2014). Long-term use of different types of polluted irrigation water increased soil organic C (SOC) contents considerably; the maximum average magnitude of increase was recorded with UWW (125%). Irrigation with UWW also increased available P and K content in soil to a maximum level, followed by industrial wastewater (Table 8.5). Although concentrations of K in both UWW and industrial wastewater (IWW) were almost similar, the high available of K in UWW irrigated soil may be due to its exhaustive use in crop production.

Several on-farm experiments have shown considerable enhancement of crop yields due to use of treated or untreated UWW vis-à-vis use of groundwater, indicating considerable nutrient supplying potential of the former. In a field experiment at Bhopal, untreated UWW irrigation (containing on an average, 214, 27, 22 and 23 mg L^{-1} of total organic C, N, P and K respectively) of about 150 cm over a period of 5 years on pristine agricultural land (having no history of sewage irrigation) didn't affect any major soil chemical parameters adversely, except for a slight increase in salinity (Saha et al. 2010b). Sewage water irrigation for 5 years to wheat crop added about 321 kg N, 262 kg P and 282 kg K ha^{-1} to the soil. This increased the available N, P and K contents by 11.4%, 44% and 17% respectively, in UWW irrigated root zone soils as compared to ground water irrigated soils. While fertilizer P application didn't enhance its content in sub-surface layer, continuous irrigation with UWW showed significant increase in P content in the below root zone depth, due to its (mainly in organic P form) percolation down the profile through cracks in soil. Irrigation with UWW also significantly improved availability of micronutrients like Zn, Fe and Mn in the soil. However, sole application of sewage water was not sufficient to supply nutrients necessary to maintain the productivity of wheat that obtained through the recommended level of fertilizers. An additional N application (50% of recommended level) through fertilizers was required to obtain higher productivity of wheat under sewage farming. Irrigation with UWW in wheat crop had significant residual effect on grain (11% more) and straw (33% more) yields of following soybean crop. Recoveries of N, P and K by the aboveground biomass of the cropping system were considerably more from the nutrients supplied by UWW than those supplied by fertilizers and manures (Saha et al. 2016). In another study, UWW increased population of nitrogen fixing bacteria, P solubilising organisms and pH in rhizosphere of peas in Hill soil, West Bengal (Saha and Tarafdar 1996). The long-term effects of waste water irrigation on soil fertility and nutrient supply to lucerne (*Medicago sativa*) was investigated near Mexico City (Siebe 1998). Wastewater irrigation increased total N and available P but diminished the Ca saturation of the soil by displacing Ca by Na. In consequence, Ca concentration decreased, while P and Na increased in lucerne tissue after 80 years of WW irrigation. Similar report of considerable increase in total and available P content has been observed in farmers' field (Vertisol) at Bhopal due to long-term untreated sewage water irrigation (Saha et al. 2014). Potential impacts of long-term sewage irrigation on soil health were investigated at peri-urban area of Delhi, where crop fields were being irrigated with UWW for more than 20 years and results were compared with an adjacent rainfed *Leucaena leucocephala* plantation system (Masto et al. 2009). The study revealed that irrigation with sewage caused substantial build-up of DTPA extractable Zn (by 314%), Cu (by 102%), Fe (by 715%), Mn (by 197.2%), Cd (by 203%), Ni (by 1358%) and Pb (by 15.2%). Sewage irrigated soil showed a significant decrease in microbial biomass carbon (by −78.2%), soil respiration (by −82.3%), phosphatase activity (by −59.12%) and dehydrogenase activity (by −59.4%). In the study, microbial biomass carbon was chosen as the most sensitive indicator for assessing the effect of UWW irrigation. On the contrary, irrigation with about 150 cm of

Table 8.6 Changes in microbial properties of soil due to sewage irrigation

	Parameters	Groundwater irrigated plots	UWW irrigated plots
1	Dehydrogenase activity (mg [$]TPF/g soil/h)	[#]0.133[b]	0.144[a]
2	Alkaline phosphatase activity (mg [*]PNP/ g soil/h)	9.44[a]	7.84[a]
3	Urease activity (μg NH_4–N/g soil/2 h)	51.9[a]	44.9[a]
4	Microbial biomass C (μg/g)	37.0[b]	53.9[a]
5	Basal respiration (g CO_2–C /m^2/ day)	2.47[b]	3.01[a]
6	Fungi population (10^3 cfu/g)	6.6[b]	14.8[a]
7	Bacterial population (10^{10} cfu/g)	15.9[a]	23.7[a]
8	Actinomycetes population (10^3 cfu/g)	57.2[b]	100.2[a]
9	Total coliform (cfu/g)	0.7[b]	16.0[a]

Saha et al. (2010b)
[#]The same letter in a row indicates means are not significantly different (@ $P \leq 0.05$)
[$]*TPF* triphenyl formazan
[*]*PNP* p-nitrophenol

untreated UWW over five years caused significant increases in several enzyme activities as well as population of fungal, and actinomycetes in vertisol (Table 8.6).

8.2.5 *Plant Nutrient Recovery from UWW*

Recycling of nutrients removed by crops is advocated as a strategy under integrated nutrient management to mitigate fertilizer demand in agriculture. It has been estimated that the daily nutrient supplying potential of waste water generated in Indian cities is about 750, 250 and 760 tonnes of N, P and K, respectively based on the huge volume of UWW generated and its average nutrient content (CPCB, 2009). This value was corresponding to about 3.6% of the total fertilizer nutrient (N + P + K) consumption during 2007–2008. Therefore recycling and recovering of nutrients from wastewater can be very useful and additionally, it delivers the dual benefits of purification of polluted water and reduction in the amount of fertilizers requirements. Saha et al. (2016) had studied the nutrient recovery from untreated UWW through a field experiment where, nutrient removal by wheat-soybean cropping system was estimated. The highest nutrient recoveries from UWW were observed in crops grown with no additional nutrient application. During the six years of experiment, about 80%, 40% and 87% of N, P and K supplied through UWW has been recovered by above-ground biomass of the crops. The highest N had been recovered by wheat grain, whereas the highest P and K were taken up by soybean straw. Moreover, the recovery of 31% N, 22% P and 69% K supplied through UWW was found in the straw portion of the crops, which can be further recycled back into the agricultural land through different residue recycling technologies. In addition, available status of these nutrients in soil has increased with continuous untreated UWW irrigation, extent of their recovery is expected to increase with the progress of sewage farming. The study concluded that sewage farming can be promoted as a way for nutrient recovery through non-vegetable and non-fodder crops like wheat-soybean cropping system, which is predominantly followed in central India. Recovered nutrients in the plant biomass can be converted into organic manure for application in groundwater irrigated agricultural land. In this way, costly fertilizer inputs in agriculture can be reduced and nutrient enrichment of surface water bodies (eutrophication) due to mixing/accumulation of polluted sewage effluent in the water bodies can be minimized.

8.2.6 Environmental Concern due to Use of UWW for Irrigation to Crop

8.2.6.1 Environmental Impact due to High N Content in UWW

Loss of N from agricultural land has severe environmental implications. Nitrous oxide escaped into atmosphere through denitrification process may contribute to global warming. Significantly higher loss of N (about 40–55%) has been observed in UWW irrigated fields (either fertilized or non-fertilized) through denitrification as compared to fresh water irrigated plots where N loss was in the range of 10–40% (Dag et al. 1984). Further downward movement of considerable amount of NO_3^- in the profile irrigated with sewage water during 7 years of study has been documented (Feigin et al. 1991).

In agricultural areas near Santiago, about 780 kg N per ha was supplied each year in vegetables growing fields (principally spinach beet, lettuce and celery) through irrigation with untreated contaminated sewage water/effluent generated from the cities (Schalscha and Vergara 1982). Long-term (>25 years) use of such untreated sewage water exclusively for irrigation led to leaching of about 350 kg nitrate-N ha $^{-1}$ per year, which contaminated groundwater through downward movement of percolating water (Schalscha et al. 1979).

The oxygen level in the soil profile is reduced due to use of sewage water/effluent with a high BOD value, which in the consequence may enhance denitrification process in the soil. Denitrification is microbially mediated loss of NO_3^- as N_2O and N_2 gases under anoxic conditions and has considerable agricultural and environmental implications (N_2O is an important greenhouse gas). In a green house experiment, significant increase in denitrification rate was observed with sewage effluent irrigation of clay soil (Feigin et al. 1981).

8.2.6.2 Crop Contamination with Pathogens

Choice of crops grown with UWW is very important as there is several human health issues involved with sewage farming. As the UWW contains various kinds of pathogenic organisms like bacteria, viruses, protozoa, helminthes, etc., use of such waters for irrigation may expose the public to the peril of infections. Therefore microbiological quality of the sewage water governs its suitability for use in agricultural activities. According to WHO guidelines, no crops which is eaten raw, should be grown with sewage water if it contains more than 100 coliform organisms in 100 mL. Transmission of diseases to human being by sewage farming may occur, directly when the farmers have physical contact with the contaminated waste water and/or indirectly by consumption of contaminated food materials/ produce grown. The pathogenic microorganisms that reach the soil and the crop plants through these route may survive for days, weeks and at times for months. For instances, the presence of significant amount of total coliform population in contaminated wastewater treated plots were observed even after harvest of the wheat crop (Table 8.6) (Saha et al. 2010b). These are a group of pathogenic organisms that cause acute gastroenteritis in mammals. Analysis of 344 samples of edible portions of various crops indicated that vegetables and fodder crops cultivated with UWW over a longer period were severely contaminated with coliforms while the contamination of grains was insignificant (Minhas et al. 2006). Further it has been pointed out that the plant parts that have direct contact with untreated waste water are severely loaded with these pathogenic microbes. Sun drying of such produce (e.g. grain) in the field reduced the contamination level. The microbial loads in vegetables enhanced considerably due to the storage of vegetables harvested from sewage irrigated area. Contaminated waste water use for irrigation or its accidental entry in pasture lands by overflow increases the chances of animal sickness due to *Salmonella aberdeen* (Bicknell 1972).

8.2.6.3 Soil Pollution with Heavy Metals due to Urban Waste Water

Several reports have also indicated that heavy metal contents in sewage water are within the permissible limit (Brar et al. 2000; Rattan et al. 2002; Yadav et al. 2002). A comparison among polluted irrigation water from different sources indicated that heavy metals contaminations were generally lower in municipal waste water than the industrial wastewater and contaminated groundwater near industrial area (Table 8.7). Though sewage water/effluents contain low concentration of heavy metals, their accumulation in soils may take place through long term intensive application of the contaminated waste waters in land for agricultural purposes. The degree and amount of metal build-up in the soils irrigated with contaminated waste water is mainly related to the frequency and period of its application (Bansal et al. 1992; Palaniswami and Sree Ramulu 1994). Number of studies conducted worldwide demonstrated different level of heavy metal accumulation in waste water irrigated soils (Nyamanagara and Mzezwa 1999; Cao and Hu 2000; Singh et al. 2004; Mapanda et al. 2005). In India too, several studies have reported buildup of trace metals in soils at variable degree due to long term application UWW for irrigation (Brar et al. 2002; Rattan et al. 2005). Several researchers reported increases in DTPA-extractable trace metal elements in the root zone soil layer to the range of 30–400% in various cities of India due to several decades of UWW

Table 8.7 Average chemical characteristics of water used for irrigation purposes in Bhopal, Nagda and Ratlam

	Ratlam		Nagda		Bhopal	
	Unpolluted groundwater	Contaminated groundwater	Unpolluted groundwater	Industrial wastewater	Unpolluted groundwater	UWW
	Concentrations in mg L^{-1}					
Zn	0.02[a1]	2.66[b]	0.025[a]	0.260[b]	0.015[a]	0.070[a]
Cu	0.010[a]	0.068[b]	0.014[a]	0.046[b]	0.011[a]	0.035[a]
Pb	0.003[a]	0.032[b]	0.004[a]	0.007[a]	0.004[a]	0.021[b]
Cd	Trace[a]	0.001[a]	Trace[a]	Trace[a]	Trace[a]	0.001[a]
Cr	Trace[a]	Trace[a]	Trace[a]	0.008[a]	0.005[a]	0.027[b]
Ni	0.002[a]	0.026[b]	0.002[a]	0.014[a]	0.013[a]	0.015[a]

Saha et al. (2014)
[1]Means followed by the same letter in a row for a particular location are not statistically different among each other at $p < 0.05$ significance level

irrigation (Rattan et al. 2005; Simmons et al. 2006; Dheri et al. 2007; Kharche et al. 2011). Bouwer and Chaney (1974) estimated that at normal application rate, greater amounts of metal are likely to contaminate the soil in one year through biosolid (sludge) than the magnitude of contamination expected through a century of effluent irrigation. There are several reports of heavy metal accretion in soil due to sewage effluent irrigation over a long period. Friedel et al. (2000) found significantly high amount of Cd, Cu, Pb and Zn in soils irrigated with wastewater for 25–80 years as compared to nearby soils under rainfed agriculture in Mexico. In chile, continuous application of sewage water for long period resulted significant migration of Cu, Cr, Cd, Co, Zn and Ni downward from upper soil (ploughed) layer (Schirado et al. 1986); which was supposed to be due to increased mobility through complexation by dissolved organic C.

8.2.6.4 Irrigation with Municipal Sewage Water and Heavy Metals Contamination in Crops

Crops grown on the heavy metal contaminated soils can cause health problems to animals and human beings when the food materials from such area accumulate these toxic metals in excessive quantities (Tiller 1986; Rattan et al. 2001; Gupta et al. 2010). Vegetables raised on UWW irrigated soils in various parts of India have shown significant contamination with many heavy metals (Brar et al. 2000; Khurana et al. 2004; Ghafoor et al. 2004; Sharma et al. 2007; Gupta et al. 2008; Gupta et al. 2010; Saha et al. 2015). Accumulation of metals in soil due to continuous application wastewater in and around Varanasi, India for irrigating vegetable crops has been studied (Singh and Agarwal 2010). The enrichment factor (indicates accumulation in reference to pre-contamination era) in soil for Cd, Cr, Zn, Cu and Pb was 1.9, 2.9, 4.8, 6.5 and 15.5, respectively and the degree of contamination (indicates accumulation in reference to background level) of these metals in soil was also found to be 2.9, 2.0, 2.3, 2.7 and 2.2, respectively. High enrichment factor values (>2) indicates considerable accumulation of heavy metals (except Cd) in soil due to continuous sewage irrigation (refer Chap. 7 for details on interpretation). The maximum concentration of Zn, Cu, Cr, Pb and Cd was observed in *Brassica oleracea* and the corresponding metal levels in plants were 63.80, 12.25, 10.65, 3.63, and 0.56 mg kg^{-1}, respectively. However leafy vegetable spinach, grown with untreated sewage water, had high heavy metals uptake and translocation ratio. Among the metals, maximum uptake was observed for Mn, followed by Zn, Cu, Pb, Ni Cr and Cd, respectively. In another study from untreated sewage water irrigated areas of Kolkata city, concentrations of Cd, Pb and Ni in the most of the commonly grown vegetables and tubers were above the permissible limit (Saha et al. 2015). Cauliflower had the highest mean transfer coefficients (TCs) of 0.59 for Zn, Colocasia for Cu (0.67) and Cd (1.02), Amaranthus for Pb (0.93) and Ni (1.09), respectively. Enrichment factors computed for the area indicated that soils were highly contaminated with different toxic metals such as Cd (452.04, indicating extremely high contamination), Pb (17.32, indicating high contamination), Zn (5.99, indicating significant contamination), Cu (3.14, indicating significant contamination) and Ni (3.07, indicating moderate contamination level). This indicates, use of UWW for irrigation for a long period not only contaminate the soil by significant heavy metals loading, but also resulted in heavy metal enrichment of crop plants beyond the acceptable critical values of WHO, which therefore, may not be appropriate for human consumption.

8.2.6.5 Soil Contamination with Organic Pollutants due UWW Irrigation

The sewage water generated from different domestic and industrial activities often contains numerous synthetic organic compounds. However, very few studies have been carried out on characterization of organic pollutants in wastewater. Advancement in analytical science and facilities of sophisticated laboratory equipment facilitated the quantitative determination extremely low amounts of organic pollutants present in soil and water in recent years. Although domestic and industrial wastewaters generally undergo various treatments before discharge into sewage canal, treated effluents however, may still contain organic contaminants several of which are extremely toxic to human and animals (Tables 8.8 and 8.9). Over 100 such chemicals were identified in sludge and wastewater (Overcash 1983). Several organic priority pollutants (OPPs) and pesticides were however not decreased during treatment. Analysis of UWW from New York City during 1989–1993 indicated that commonly identified OPPs included five volatile compounds, one semi-volatile compound, and one pesticide (Gamma BHC). With few exceptions, volatile and semi-volatile concentrations were <100 ppb, and pesticide concentrations were <0.2 ppb (Table 8.9). Comparisons of effluent concentrations to most stringent standards for receiving water showed five compounds to occasionally exceed the limit, with only a single volatile compound, tetrachloroethene,

Table 8.8 Major organic pollutants in the secondary effluents

Chemical group	Major compounds with highest concentration	Value (μg L^{-1})
Halogenated aliphatics	Methylene chloride	0.16–48
	Methyl chloride	
Substituted aromatics	Toluene	0.21–18
	Pentachlorophenol	
Phthalate acid esters	Bis(2-ethylhexyl) phthalate	0.21–21
	Di-n-butil phthalate	
Polynuclear aromatics	Phenanthrene	0.16–0.68
	Anthracene	

Adapted from Overcash (1983)

Table 8.9 Concentrations for the most commonly occurring organic priority pollutants and pesticides in municipal wastewater from 1996

Compounds	Influent	Effluent
Organic priority pollutants		
	Highest recorded concentrations (in ppb)	Highest recorded concentrations (in ppb)
Chloroform	17 (99)[a]	45 (86)
Tetrachloroethene	78 (94)	69 (74)
Toluene	47 (87)	110 (24)
Bis(2-ethylhexyl) phthalat	180 (48)	50 (12)
Methylene chloride	160 (39)	29 (31)
Gamma BHC	0.09 (37)	0.12 (33)
1,1,1 -trichloroethane	72 (33)	57 (17)
Pesticides		
	Concentration (in ppb)	Concentration (in ppb)
Alpha BHC	0.037–0.054 (2)	0.007–0.09 (4)
Gamma BHC (Lindane	0.012–0.09 (37)	0.013–0.12 (33)
Aldrin	0.024–1.1 (26)	0.008–0.44 (23)
4,4-DDT	0.012–0.053 (5)	0 (0)
Endosulfan sulfate	0 (0)	0.26 (1)
Heptachlor	0.021–0.35 (7)	0.02–0.447 (10)
Endosulfan II	0.22 (1)	0 (0)
Dieldrin	0 (0)	0.012–0.028 (2)
Endrin aldehyde	0.032 (1)	0 (0)

Stubin et al. (1996)
[a]Percent detection in influent/effluent during the current study shown in parentheses

exceeding them with any regularity (Stubin et al. 1996). Paxéus (1996) detected wide range of organic compounds in the wastewater discharged from treatment plants in Sweden, which included aromatic hydrocarbons, solvents, antioxidants, plasticisers, preservatives, food and household related compounds, flame retardants and washing and cleaning related compounds. The concentrations of individual compounds in effluents were varying from 0.5 to 50 ppb. Moreover, concern in this aspect arises when domestic sewage water gets contaminated with industrial water (Shi 2009). A range of toxic compounds, such as phenols, polynuclear aromatic hydrocarbons and organic acids and bases were found in industrial wastewater.

But the risks related with use of treated sewage water for irrigation in respect to trace organics are normally not very high, due to either complete removal or substantial reduction in their concentration during effluent treatment. Furthermore, degrading actions by various kinds of soil microorganisms on these compounds and various physic-chemical reactions with soil particles (similar to the fate of pesticides) can reduce the concentration and impact of organic pollutants in the soil-water system.

Trace organic pollutants derived from sewage effluent experience adsorption, volatilization and biodegradation reactions within soil matrix. As a result, the level of these original contaminants present in the waste water/sewage effluent may reduce significantly upon entry in the land mass. The isothermal adsorption equations, such as Freundlich and Langmuir equations describe the adsorption pattern of the trace organics on soil matrix and explain their fate and bioavailability in rhizosphere. Soil organic matter content has been considered as best indicator of degree of adsorption for non-ionic and polar organic species; whereas, pH of soil solution generally governs the sorption process for ionic or ionizable substances (Bronner and Goss 2011). Partitioning coefficient (K_w) is adopted as indicative parameter for volatilization of trace organics from dilute aqueous solutions. K_w is expressed as the ratio of the equilibrium concentration of volatile organic compounds in the water and in the air at 298°K. Biodegradation is the process of involvement microbes and organisms present in the soil that reduce the level of organic pollutants in the soil. The rate of decomposition the organic pollutants in soils was determined by type of microorganisms and soil properties as well as the chemical properties of the organic pollutant itself. Degradation of organic pollutants generally follows a first-order reaction at low substrate concentrations (Alexander 1985). The available organic C fraction in the soil from other sources can enhance the biodegradation of trace pollutants, while their adsorption in soil matrix reduces its decomposition rate. Bioaccumulation of trace pollutants, (i.e., uptake through plant roots) has been detected by few researchers (Chang and Page 1985); however, the significance of this process in soil-plant systems has not been mentioned.

8.3 Use of Sewage Sludge in Agriculture

The sewage sludge is generated after treatment of effluents from cities and towns using different technologies. Sewage treatment and consequent production of sewage sludge compost provides a win-win situation for both urban society and farming community as it allows beneficial recycling of wastes. Although estimates of volume of sewage being treated are available, quantitative information on sludge generation from sewage treatment plants (STPs) in India is not available. The CPCB evaluated performance of 152 sewage treatment plants located in 15 states across the country. The total treatment capacity of these plants was reported to be 4716 MLD (CPCB 2013). However actual performance of these plants was only 66% of the total capacity (3126 MLD) as revealed from the study. Moreover BOD and COD of treated effluents from 49 and 7 STPs respectively exceeded permissible levels for discharge into the water bodies, indicating inefficiency of several STPs in the country. Average sewage effluent treatment capacity of different states followed the order: Tamil Nadu with 16.9% rank top position followed by Uttar Pradesh (16.4%), Andhra Pradesh (15%), Punjab (14%), West Bengal (10%), Haryana (7%), Maharashtra (6%), Gujarat (4.9%), Madhya Pradesh (3.6%), Bihar (3.4%), Uttarakhand (1.1%), Karnataka (0.9%), Delhi (0.4%) and Goa (0.2%) in the order. The maximum efficiency of BOD removal (90%) was observed in the STPs designed on Trickling filter and Sequential Batch Reactor (SBR) technologies.

Composition and characteristics of sewage sludge varies widely depending on UWW quality and its treatment processes (Table 8.10). Organic and inorganic pollutants (non-essential trace metals, toxic metals), salt ions, microorganisms and pathogens, beneficial plant macro- and micro-nutrients and some organic compounds are widely found in sewage sludge during chemical analysis. Composts prepared from municipal sewage sludge have proved useful in agriculture because of presence of higher amount of organic carbon and plant nutrients.

8.3.1 Nutrient Potential of Sewage Sludge

Sewage sludge contains considerable amount of organic carbon and plant nutrients that are essential for crop production. Therefore, application of sewage-sludge in agricultural land has shown to enhance crop yield considerably (Singh and Agrawal 2010; Özyazıcı 2013). A recent review has documented the advantages of sludge application for improving several soil physico-chemical parameters like organic carbon, bulk density, aggregate stability, infiltration capacity, air recirculation, porosity, cation exchange capacity and available nutrients (Mtshali et al. 2014). Urban sewage sludge compost has the potential to correct soil acidity due their high contents of bases and organic matter. Sludges generally contain higher amount of salts as compared to soil and therefore, their application may result in development/ enhancement of soil salinity particularly in low rainfall area.

Table 8.10 Typical characteristics and composition of municipal sewage sludge

Parameters	Range
Bulk density (g mL^{-1})	0.33–0.8
Organic matter (%)	7–80
N (%)	0.48–3.23
P (%)	0.08–2.22
K (%)	0067–1.44
S (%)	0.4–0.88
Ca (%)	0.37–4.9
Mg (%)	0.07–1.1
As (μg g^{-1})	1.1–87
Pb (μg g^{-1})	22–335
Cd (μg g^{-1})	0.5–16
Cr (μg g^{-1})	14.7–1290
Cu (μg g^{-1})	150–484
Mn (μg g^{-1})	32–605
Ni (μg g^{-1})	5–96
Hg (μg g^{-1})	Tr.–6.5
Zn (μg g^{-1})	261–1490

Ogilvie (1998)

8.3.2 Heavy Metals in Sewage Sludge

Although several benefits can be harvested from use of sewage sludge compost in crop production, this byproduct from sewage treatment plants are screened and monitored by government organization before permitting for its safe use in sustainable agriculture production. During treatment process, toxic substances in waste water are concentrated into solid phase, making it riskier to be applied on agricultural lands. The risks due to the presence of undesirable substances such as heavy metals/trace elements, potentially toxic synthetic organic compounds, pathogenic organisms; and possible element imbalance in composts have become major apprehension for potential compost users including farmers. The heavy metals concentrations in sewage sludges are vary widely (Table 8.10). It usually contains 30–50% higher metals concentrations as compared to mechanically-segregated composts (Smith 2009). The heavy metal contents in sludge are especially higher if UWW is contaminated with industrial effluent. Therefore their level depends largely on industrialization profile of a city. Continuous application of sludge compost in fields may buildup heavy metals in soils beyond the permissible levels for its safe use in agriculture. Composting of sewage sludge tends to diminish heavy metals availability and therefore threat of metals transfer to plants with composted sludge is generally lower as compared to other forms of stabilized sludge (Smith 2009). Most of the researchers observed 'plateau response' type of heavy metal uptake with increasing rates of sludge application (hence, increasing contamination) due to consequent increase in metal fixation capacity (because of high contents of hydroxides of Fe, Mn and Al in the sludge) (Mahler and Ryan 1988). However, few

Table 8.11 Maximum safe allowable concentrations of heavy metals in sewage sludge and compost to be used soil application (μg g^{-1} on dry weight basis)

Country	Cd	Cr	Cu	Hg	Ni	Pb	Zn
European union	20–40	–	1000–1750	16–25	300–400	750–1200	2500–4000
Germany-sewage sludge	10	900	800	8	200	900	2500
Germany-compost[a]	1.5 (1)	100 (70)	100 (70)	1 (0.7)	50 (35)	150 (100)	400 (300)
France	20	1000	1000	10	200	800	3000
USA ceiling concentrations	85	3000	4300	57	420	840	7500
USA pollutant concentrations	39	1200	1500	17	420	300	2800
Russia	30	1200	1500	15	400	1000	4000
India	5	50	150	2.5	50	300	500

During and Gath (2002) and MOEF (2000)
[a]Quantity of manure –20 tonnes ha^{-1}, and in brackets for 30 tonnes ha^{-1} in 3 years

instances of increasing metal uptake response with repeated additions of compost were also observed due to mineralization of metal-bound organic matter, thereby supporting 'time-bomb model' proposed by Beckett et al. (1979).

Contamination of soils with potentially toxic elements due to amendment with sewage sludge and MSW compost is subjected to strict control in several countries in reference to total permissible metal concentration, soil properties and intended use. Different countries have set limits for maximum metal concentration in sewage sludge and compost as well as in agricultural soils for the purpose of minimizing risks associated with use of such hazardous wastes (Tables 8.11 and 8.12).

8.3.3 *Organic Pollutants in Sewage Sludge*

Numerous organic pollutants (>500) with different physico-chemical properties also find their way into waste water drainage systems and subsequently accumulate in the sludge during wastewater treatment. Several pollutants have the potential to contaminate the food chain when contaminated sludges are used in agricultural land as soil amendment. These pollutants can be classified into one or more of the following groups: aliphatic, chlorobenzene, nitrosamines, organotin, monocyclic hydrocarbons, POPs, flame retardants, pesticides, phenols, sterols and stanols, phthalate, triaryl/alkyl phosphate esters polychlorinated biphenyls (PCBs), polycyclic aromatic hydrocarbons (PAHs), surfactants, etc. (Harrison et al. 2006). These pollutants differ widely in expression of toxicity and degradability. Significantly high concentrations of several organic contaminants like dichloromethane, toluol and surfactants, as well as their degradation products (linear alkylbenzene sulfonate) had been reported in sewage sludges of the north-eastern Germany

Table 8.12 Maximum concentrations of metals (in mg kg^{-1}) allowed in agricultural soils through application of sewage sludge

Country	Cd	Cu	Cr	Hg	Ni	Pb	Zn
European Countries	1–3	50–140	100–150	1–1.5	30–75	50–300	150–300
France	2.00	100	150	1.00	50.0	100	300
Germany[a]	1.50	60.0	100	1.00	50.0	100	300
Italy	3.00	100	150	–	50.0	100	300
Spain	1.00	50.0	100	1.00	30.0	50.0	150
The Netherlands	0.80	36.0	100	0.30	35.0	85.0	140
United Kingdom[b]	3.00	135	400	1.00	75.0	300	300
Denmark	0.50	40.0	30.0	0.50	15.0	40.0	100
Finland	0.50	100	200	0.20	60.0	60.0	150
Norway	1.00	50.0	100	1.00	30.0	50.0	150
Sweden	0.50	40.0	30.0	0.50	15.0	40.0	100
United States	20.0	750	1500	8.00	210	150	1400

Adapted from McGrath et al. (1994)
[a]Data shown for soil pH more than 6 (at pH 5–6, Cd 1.0 ppm and Zn 150 ppm); [b]Data for soil pH 6–7

(Schnaak et al. 1997). Alcock et al. (1999) identified several hazardous chemicals like polychlorinated naphthalenes, chlorinated paraffins, brominated diphenyl ethers, quintozene, chloronitrobenzenes, polydimethylsiloxanes, and a range of pharmaceutical compounds in sewage sludge. Most these compounds are extremely resistant to microbial degradation and therefore persistent in soil. Toxicity potential of the organic pollutants in soil is, however, considered to be low. Because of hydrophobic nature of toxic organic chemicals, these are adsorbed rapidly with sludge-soil matrix and as a result availability of these compounds to plants are reduced (O'Çonnor 1996). However, these toxic chemicals may enter through direct ingestion (of contaminated soil) into the body of animals during grazing or into the body of children while playing on sludge amended land. Application of sludge in pasture land led to the intake and accumulation of Dibenzofurans (PCDD/Fs), phenols, some chlorinated benzenes, PCBs and Polychlorinated Dibenzo-p-Dioxins in livestock (Duarte-Davidson and Jones 1996). For site specific risk assessment and management at Superfund sites in USA, USEPA determined soil test screening limits (SSLs) for 16 inorganic and 93 organic pollutants. These limits are not standards used for regulatory purpose, but can be used as guidelines for the purpose of determining whether the site is clean after remediation. Harrison et al. (2006) used these SSL values to evaluate quality of sludge in respect of organic pollutants concentration through extensive survey of reports and research articles. Presence of 516 organic compounds in sludge had been reported, of which 90 were priority pollutants (requiring remediation technology) and 101 were target compounds (requiring monitoring). About 15% of the organic pollutants detected in sludge had SSL values. On the other hand 86% of the compounds exceeded at least one SSL. Therefore, repeated application of sewage sludge may raise their concentrations in soil.

References

Alcock RE, Sweetman A, Jones KC (1999) Assessment of organic contaminant fate in waste water treatment plants. I: Selected compounds and physicochemical properties. Chemosphere 38:2247–2262

Alexander M (1985) Biodegradation of organic chemicals. Environ Sci Technol 19(2):106–111

Annepu RK (2012) Sustainable solid waste management in India. M.Sc. thesis. Columbia University, New York

Bansal RL, Nayyar VK, Takkar PN (1992) Accumulation and bioavailability of Zn, Cu, Mn and Fe in soils polluted with industrial waste water. J Indian Soc Soil Sci 40:796–799

Baum R, Luh J, Bartram J (2013) Sanitation: a global estimate of sewerage connections without treatment and the resulting impact on MDG progress. Environ Sci Technol 47:1994–2000

Beckett PHT, Davies RD, Brindley P (1979) The disposal of sewage sludge onto farmland- the scope of the problems of toxic elements. Water Pollut Control 78:419–445

Bhardwaj RM (2005) Status of wastewater generation and treatment in India. IWG-Env, International Work Session on Water Statistics, Vienna, 20–22 June 2005, p 9. http://unstats.un.org/unsd/environment/envpdf/pap_wasess3b6india.pdf. Accessed 12 Sept 2013

Bicknell SR (1972) *Salmonella aberdeen* in cattle associated with human sewage. J Hyg 70:121–126

Bouwer H, Chaney RL (1974) Land treatment of waste water. Adv Agron 26:133–176

Brar MS, Mahli SS, Singh AP et al (2000) Sewer water irrigation effects on some potentially toxic trace elements in soil and potato plants in north western India. Can J Soil Sci 80:465–471

Brar MS, Khurana MPS, Kansal BD (2002) Effect of irrigation by untreated sewage effluents on the micro and potentially toxic elements in soils and plants. In: Proceeding of 17th world congress of soil science held at Bangkok, Thailand from 14–21 August, Symposium no.-24 198, pp 1–6

Bronner G, Goss KU (2011) Sorption of organic chemicals to soil organic matter: influence of soil variability and pH dependence. Environ Sci Technol 45(4):1307–1312

Cao ZH, Hu ZY (2000) Copper contamination in paddy soils irrigated with wastewater. Chemosphere 41:3–6

Chander K, Brookes PC (1991) Effects of heavy metals from past applications of sewage sludge on microbial biomass and organic matter accumulation in a sandy loam and silty loam UK soil. Soil Biol Biochem 23:927–932

Chang AC, Page AL (1985) Fate of wastewater constituents in soil and groundwater: trace organics. In: Pettygrove GS, Asano T (eds) Irrigation with reclaimed municipal wastewaster – a guidance manual. Lewis, Chelsea, pp 15.1–15.20

Chang AC, Granato TC, Page AL (1992) A methodology for establishing phytotoxicity criteria for chromium, copper, nickel, and zinc in agricultural land application of municipal sewage sludges. J Environ Qual 21:521–536

Ciaccio LL (1971) Water and water pollution handbook. Marcel Dekker, New York

CPCB (2009) Sewage treatment in India. Central pollution control board, New Delhi

CPCB (2012) Management of municipal solid waste. Central Pollution Control Board, Delhi. http://www.cpcb.nic.in/divisionsofheadoffice/pcp/management_solidwaste.pdf. Accessed 25 Nov 2012

CPCB (2013) Performance evaluation of sewage treatment plants under NRDC. Central Pollution Control Board, Delhi. http://cpcb.nic.in/upload/NewItems/NewItem195STPREPORT.pdf. Accessed 12 Dec 2013

CPCB (2015) Annual review report 2013–14: implementation of municipal solid wastes (management and handling) rules, 2000, Central Pollution Control Board, Delhi. http://cpcb.nic.in/upload/Latest/Latest_98_Anual_Rew_Rep_MSW_2013_14.pdf. Accessed 23 Aug 2016

Dag J, Feigin A, Giskin M et al (1984) Response of cotton to effluent irrigation in a grumusol soil in Yizre'el Valley; 1976–1982 experiments 1: Cotton yield. Agriculture Research Organization, Israel. Pamphlogy 228:47–54

Daniel H, Perinaz BT (2012) What a waste: a global review of solid waste management. Urban development series knowledge papers. World Bank 15:1–98

Devanathan G, Subramanian A, Sudaryanto A et al (2012) Brominated flame retardants and polychlorinated biphenyls in human breast milk from several locations in India: potential contaminant sources in a municipal dumping site. Environ Int 39:87–95

Dheri GS, Brar MS, Malhi SS (2007) Heavy metal concentration of sewage contaminated water and its impact on underground water, soil, and crop plants in alluvial soils of north-western India. Commun Soil Sci Plant Anal 38:1353–1370

Dotaniya ML, Saha JK, Meena VD (2015) Sewage water irrigation boon or bane for crop production. Indian Farm 65:24–27

Duarte-Davidson R, Jones KC (1996) Screening the environmental fate of organic contaminants in sewage sludge applied to agricultural soils: II. The potential for transfers to plants and grazing animal. Sci Total Environ 185:59–70

Düring R, Gath S (2002) Utilization of municipal organic wastes in agriculture: where do we stand, where will we go? J Plant Nutr Soil Sci 165:544–556

Feigin A, Feigenbaum S, Limoni H (1981) Utilization efficiency of nitrogen from sewage effluent and fertilizer applied to corn plants growing in a clayey soil. J Environ Qual 10:284–287

Feigin A, Ravina I, Shalhevet J (1991) Irrigation with treated sewage effluent management for environmental protection. Adv Ser Agric Sci 17:224

Friedel JK, Langer T, Siebe C, Stahr K (2000) Effects of long-term waste water irrigation on soil organic matter, soil microbial biomass and its activities in central Mexico. Biol Fertil Soils 31:414–421

Ghafoor A, Qadir M, Sadiq M et al (2004) Lead, copper, zinc and iron contaminations in soils and vegetables irrigated with city effluent on urban agricultural lands. J Indian Soc Soil Sci 52:114–117

Ghimire SK (1994) Evaluation of industrial effluents toxicity in seed germination and seedling growth of some vegetables. M.Sc. dissertation, Central Department of Botany, Tribhuvan University, Kirtipur, Kathmandu, Nepal

Gupta N, Khan DK, Santra SC (2008) An assessment of heavy metal contamination in vegetables grown in wastewater-irrigated areas of Titagarh, West Bengal, India. Bull Environ Contam Toxicol 80:115–118

Gupta S, Satpati S, Nayek S, Garai D (2010) Effect of wastewater irrigation on vegetables in relation to bioaccumulation of heavy metals and biochemical changes. Environ Monit Assess 165:169–177

Harrison EZ, Oakes SR, Hysell M et al (2006) Organic chemicals in sewage sludges. Sci Total Environ 367:481–497

Henze M, Comeau Y (2008) Wastewater characterization. In: Henze M, van Loosdrecht MCM et al (eds) Biological wastewater treatment: principles modelling and design. IWA Publishing, London, pp 33–52

Hunter JV, Kotalik TA (1974) Chemical and biological quality of treated effluents. In: Sopper WE, Kardos LT (eds) Conference on recycling treated municipal wastewater through forest and cropland. US Environmental Protection Agency EPA-660/2–74-003, pp 6–27

Kanmani S, Gandhimathi R (2013) Assessment of heavy metal contamination in soil due to leachate migration from an open dumping site. Appl Water Sci 3:193–205

Kansal BD (1992) Effect of waste disposal on soils and plants. In: Dhaliwal GS, Hansra BS, Jerath N (eds) Changing scenario of our environment. Punjab Agricultural University, Ludhiana, pp 197–206

Kharche VK, Desai VN, Pharande AL (2011) Effect of sewage irrigation on soil properties, essential nutrients and pollutant element status of soils and plants in a vegetable growing area around Ahmednagar city in Maharashtra. J Indian Soc Soil Sci 59:177–184

Khurana MPS, Singh J, Nayar VK (2004) Assessment of heavy metal contamination in soils and plants irrigated with sewage waters containing industrial effluents in district Amritsar, Punjab. Indian J Environ Ecoplan 8:221–280

Kumar S, Bhattacharyya JK, Vaidya AN et al (2009) Assessment of the status of municipal solid waste management in metro cities, state capitals, class I cities, and class II towns in India: an insight. Waste Manag 29:883–895

Logan TJ, Henry CL, Schnoor JL et al (1999) An assessment of health and environmental risks of trace elements and toxic organics in land applied municipal solid waste compost. Compost Sci Util 7:38–53

Mahler RJ, Ryan JA (1988) Cadmium sulfate application to sludge-amended soils: III. Relationship between treatment and plant available cadmium, zinc, and manganese. Commun Soil Sci Plant Anal 19:1771–1794

Manka J, Rebhun M, Mandelbaum A, Bortinger A (1974) Characterization of organics in secondary effluents. Environ Sci Technol 8:1017–1020

Mapanda F, Mangwayana EN, Nyamangara J, Giller KE (2005) The effect of long-term irrigation using wastewater on heavy metal contents of soils under vegetables in Harare, Zimbabwe. Agric Ecosyst Environ 107:151–165

Masto RE, Chhonkar PK, Singh D, Patra AK (2009) Changes in soil quality indicators under long-term sewage irrigation in a sub-tropical environment. Environ Geol 56:1237–1243

McGrath SP, Chang AC, Page AL et al (1994) Land application of sewage sludge: scientific perspectives of heavy metal loading limits in Europe and the United States. Environ Rev 2:108–118

Minh NH, Minh TB, Watanabe M et al (2003) Open dumping site in Asian developing countries: a potential source of polychlorinated dibenzo-p-dioxins and polychlorinated dibenzofurans. Environ Sci Technol 37:1493–1502

Minhas PS, Samra JS (2004) Wastewater use in peri-urban agriculture impacts and opportunities. Technical Bulletin 02/2004. Central Soil Salinity Research Institute (CSSRI), Karnal, Haryana, India

Minhas PS, Sharma N, Yadav RK, Joshi PK (2006) Prevalence and control of pathogenic contamination in some sewage irrigated vegetable, forage and cereal grain crops. Bioresour Technol 97:1174–1178

MOEF (2000) Municipal solid wastes (management and handling) rules, 2000. Notification, dated 25th September, 2000. Ministry of Environment and Forests, New Delhi. Available via http://www.moef.nic.in/legis/hsm/mswmhr.html. Accessed 14 Sept 2016

Mtshali JS, Tiruneh AT, Fadiran AO (2014) Characterization of sewage sludge generated from wastewater treatment plants in Swaziland in relation to agricultural uses. Resour Environ 4:190–199

Narayana T (2009) Municipal solid waste management in India: from waste disposal to recovery of resources? Waste Manag 29:1163–1166

NEERI (1995) Strategy paper on SWM in India. National Environmental Engineering Research Institute, Nagpur

Nyamangara J, Mzezewa J (1999) The effect of long-term sewage sludge application on Zn, Cu, Ni and Pb levels in a clay loam soil under pasture grass in Zimbabwe. Agric Ecosyst Environ 73:199–204

O'Connor GA (1996) Organic compounds in sludge-amended soils and their potential uptake by crop plants. Sci Total Environ 185:71–81

Ogilvie D (1998) National study of the composition of sewage sludge. New Zealand Water and Wastes Association. PO Box 13–880, Onehunga, Auckland

Overcash MR (1983) Land treatment of municipal effluent and sludge: specific organic compounds. In: Page AL et al (eds) Utilization of municipal wastewater and sludge on land. University of California Press, Riverside, pp 199–231

Ozyazıcı MA (2013) Effects of sewage sludge on the yield of plants in the rotation system of wheat-white head cabbage-tomato. Eur J Soil Sci 2:35–44

Palaniswami C, Sree Ramulu US (1994) Effects of continuous irrigation with paper factory effluent on soil properties. J Indian Soc Soil Sci 42:139–140

Paxéus N (1996) Organic pollutants in the effluents of large wastewater treatment plants in Sweden. Water Res 30:1115–1122

Petruzzelli G, Szymura I, Lubrano L, Pezzarossa B (1989) Chemical speciation of heavy metals in different size fractions of compost from solid urban wastes. Environ Technol Lett 10:521–526

Pinamonti F, Nicolini G, Dalpiaz A, Stringari G, Zorzi G (1999) Compost use in viticulture: effects on heavy metal levels in soil and plants. Commun Soil Sci Plant Anal 30:1531–1549

Qadir M, Wichelns D, Raschid-Sally L et al (2010) The challenges of wastewater irrigation in developing countries. Agric Water Manag 97:561–568

Rattan RK, Datta SP, Chhonkar PK et al (2001) Effect of long term application of sewage effluents on available water status in soils under Keshopure Effluent Irrigation Scheme in Delhi. Agric Water Manag 9:21–26

Rattan RK, Dutta SP, Chandra S, Saharaan N (2002) Heavy metals in environments-Indian scenario. Fertil News 47:21–40

Rattan RK, Datta SP, Chhonkar PK, Suribabu K, Singh AK (2005) Long-term impact of irrigation with sewage effluents on heavy metal content in soils, crops and groundwater – a case study. Agric Ecosyst Environ 109:310–322

Saha N, Tarafdar J (1996) Quality of sewage as irrigation water and its effect on beneficial microbes around pea (*Pisum sativum*) rhizosphere in hill soil. J Hill Res 9:69–72

Saha JK, Panwar N, Singh MV (2009) Determination of lead and cadmium concentration limits in agricultural soil and municipal solid waste compost through an approach of zero tolerance to food contamination. Environ Monit Assess. doi:10.1007/s10661-009-1122-3

Saha JK, Panwar N et al (2010a) An assessment of municipal solid waste compost quality produced in different cities of India in the perspective of developing quality control indices. Waste Manag 30:192–201

Saha JK, Panwar N, Srivastava A et al (2010b) Chemical, biochemical, and biological impact of untreated domestic sewage water use on Vertisol and its consequences on wheat (*Triticum aestivum*) productivity. Environ Monit Assess 161:403–412

Saha JK, Panwar N, Coumar MV (2013) Effect of methods of preparation on distribution of heavy metals in different size fractions of municipal solid waste composts. Environ Monit Assess 185:8815–8821

Saha JK, Sharma AK, Srivastava A (2014) Impact of different types of polluted irrigation water on soil fertility and wheat grain yield in clayey black soils of central India. Environ Monit Assess 186:2349–2356

Saha S, Hazra GC, Saha B et al (2015) Assessment of heavy metals contamination in different crops grown in long-term sewage-irrigated areas of Kolkata, West Bengal. India Environ Monit Assess. doi:10.1007/s10661-014-4087-9

Saha JK, Srivastava A, Kundu S, Rao AS (2016) Comparison between fertilizers and untreated sewage effluent as sources for nutrient recovery by wheat (*Triticum aestivum*)-soybean (*Glycine max*) cropping system. J Plant Nutr 39:470–478

Sato T, Qadir M, Yamamoto S, Endo T, Zahoor A (2013) Global, regional, a country level need for data on wastewater generation, treatment, and use. Agric Water Manag 130:1–13

Schalscha BE, Vergara FI (1982) The nitrogen balance of vegetable crops irrigated with untreated effluent. Plant Soil 67:221–226

Schalscha EB, Vergara I, Schirado T, Morales M (1979) Nitrate movement in a Chilean agricultural area irrigated with untreated sewage water. J Environ Qual 8:27–30

Schirado T, Vegara I, Schalscha EB, Pratt PF (1986) Evidence for movement of heavy metals in a soil irrigated with untreated wastewater. J Environ Qual 15:9–12

Schnaak W, Kulcher T, Kujawa M et al (1997) Organic contaminants in sewage sludge and their ecotoxicological significance in the agricultural utilization of sewage sludge. Chemosphere 35:5–11

Sharholy M, Ahmad K, Mahmood G, Trivedi RC (2008) Municipal solid waste management in Indian cities – a review. Waste Manag 28:459–467

Sharma VK, Kansal BD (1986) Heavy metals contamination of soils and plants with sewage irrigation. Policy Res 4:86–91

Sharma RK, Agarwal M, Marshall F (2007) Heavy metal contamination of soil and vegetables in suburban areas of Varanasi, India. Ecotoxicol Environ Saf 66:258–266

Shi H (2009) Point sources of pollution: local effects and control – volume 1. Industrial wastewater-types, amounts and effects, pp 191–203. ISBN:978–1–84826-167-9 (eBook)

Siebe C (1998) Nutrient inputs to soils and their uptake by alfalfa through long-term irrigation with untreated sewage effluent in Mexico. Soil Use Manag 14:119–122

Simmons RW, Blummel M, Reddy R, Khan AA (2006) Impact of wastewater irrigation on Cd and Pb concentrations in rice straw and paragrass: implications for food safety. A report of International Water Management Institute (IWMI), Regional Office, South Asia

Singh A, Agrawal M (2010) Effects of municipal waste water irrigation on availability of heavy metals and morpho-physiological characteristics of *Beta vulgaris* L. J Environ Biol 31:727–736

Singh KP, Dinesh M, Sinha S, Dalwani R (2004) Impact assessment of treated/untreated wastewater toxicants discharged by sewage treatment plants on health, agricultural, and environmental quality in the wastewater disposal area. Chemosphere 55:227–255

Singhal S, Pandey S (2001) Solid waste management in India: status and future directions. TERI Inf Monit Environ Sci 6:1–4

Smith SR (2009) A critical review of the bioavailability and impacts of heavy metals in municipal solid waste composts compared to sewage sludge. Environ Int 35:142–156

Smith KEC, Green M, Thomas GO, Jones KC (2001) Behavior of sewage sludge-derived PAHs on pasture. Environ Sci Technol 35:2141–2150

Sperling MV (2007) Wastewater characteristics, treatment and disposal. IWA Publishing, London

Stubin AI, Brosnan TM, Porter KD, Jimenez L, Lochan H (1996) Organic priority pollutants in New York City municipal wastewaters: 1989–1993. Water Environ Res 68:1037–1044

Subramanian A, Tanabe S (2007) Persistent toxic substances in India. In: Li A, Tanabe S, Jiang G et al (eds) Developments in environmental science, vol 7. Elsevier Ltd. doi:10.1016/S1474-8177(07)07009-X

Tiller KG (1986) Essential and toxic heavy metals in soils and their ecological relevance. Trans XIII Congress. Int Soc Soil Sci 1:29–44

UNESCO (2012) Managing water under uncertainty and risk. The United Nations World Water Development Report 4. United Nations Educational, Scientific and Cultural Organization, Paris

Valsecchi G, Gigliotti C, Farini A (1995) Microbial biomass, activity, and organic matter accumulation in soils contaminated with heavy metals. Biol Fertil Soils 20:253–259

WHO (2006) Guidelines for the safe use of wastewater, excreta and greywater. Policy and regulatory aspects, vol 1. World Health Organization, 20 Avenue Appia, 1211 Geneva 27, Switzerland

Yadav RK, Goyal B, Sharma RK, Dubey SK, Minhas PS (2002) Post-irrigation impact of domestic sewage effluent on composition of soils, crops and ground water – a case study. Environ Int 28:481–486

Chapter 9
Industrial Activities in India and Their Impact on Agroecosystem

Abstract Industrial profile of a city indicate nature of pollution that surrounding environment is facing or is likely to encounter. Though investigation on land resources for changes in soil quality due their activity has not been made adequately in India, an assessment of mining & industrial activities may indicate the potential threat to surrounding agroecosystem. The country has considerable environmentally destructive mining activities in about 57 million ha land area in several states altering the land morphology and existing land use and releasing heavy metals in the environment. About 595 metallic mines along with associated smelters, 145 thermal power plants, quite a large number of large & small scale metallurgical, chlor-alkali industry and tannery industry are likely to have adverse impact on land and water bodies through release of metals. More than 70,000 chemical and petrochemical industries in the country have the potential to contaminate land and water resources with toxic organic pollutants affecting various life forms. Considerable number of high volume effluent generating industries like textile, distillery, tannery, paper mill etc. may increase salinity of soil and groundwater. This chapter describes extent, distribution and possible impacts of polluting industries in India indicating potential threat faced by the environment including agroecosystem.

Keywords Industry • Mining • Waste • Impact • Soil • Environment • India

Industrial activities are mostly responsible for entry of pollutants into agricultural land in India. Post 1980s, the country saw economic liberalization and achieved a further impetus since mid 1991. The industrial sector is estimated to account for around 29.5% of the India's gross domestic product (GDP) and it employs over 17% of the total workforce in the year 2015–2016. Their growth rate reached to 7.6% in 2015–2016. In that year, the mining and quarrying sector contributed 2.4%, the manufacturing sector contributed 17.2%, and the water supply, gas, and electricity sector contributed 2.4%. This shows that Industry Growth Rate in India GDP has been on the rise over the last few years. This rise in growth in the resource intensive manufacturing sector is enabled and facilitated by an ever-increasing rate of material use leading to manifold impacts to the environment. According to the assessment done by the authors of the report, the annual cost of environmental degradation in India amounts to about Rs. 3.75 trillion ($80 billion) equivalent to

J.K. Saha et al., *Soil Pollution - An Emerging Threat to Agriculture*, Environmental Chemistry for a Sustainable World 10, DOI 10.1007/978-981-10-4274-4_9

5.7% of GDP (World Bank 2013). Of this total, cost due to croplands degradation was about 1.07% of GDP. While estimating cost due to croplands degradation, only salinity, erosion and water-logging were used as degradation criteria. However degradation of soil quality (hence crop productivity) from build-up of pollutants and associated damage to the soil ecosystem due to industrial activity was not taken into account in the above computation. The contribution of the manufacturing sector to environmental degradation primarily occurs during all the stages: procurement and use of natural resources, industrial processes and activities, product use and disposal. The Central Pollution Control Board (CPCB) has identified 17 highly polluting industries, the majority of which are manufacturing industries and the identified industries are: aluminum, cement, chlor-alkali, copper, distillery, dyes and Intermediates, fertilizers, iron and steel, oil, refineries, pesticides, petrochemicals, pharmaceuticals, pulp and paper, sugar, tannery, power plant, zinc. There were about 36,165 numbers of HW Generating Industries in India. Total quantity of HW generated is about 6.23 Mt. out of which landfillable HW is about 2.73 Mt. (44%), incinerable HW is about 0.41 Mt. (7%) and recyclable HW is about 3.09 Mt. (49%) (CPCB 2009). State wise hazardous waste generation from industries is indicated in Table 9.1. The highest hazardous waste generation has been recorded in Maharashtra state followed by Gujarat and Andhra Pradesh.

Major industrial sectors in the country are not having its appropriate waste disposal system, effluent treatment plant (ETP) or having ETP with insufficient capacity. This chapter attempts to assess the status of industrial activities in the country and associated environmental hazards.

9.1 Mining Industries in India

Mining & quarrying in India is one of the core industrial activities and a main sector for driving economic growth of the country contributing about 2.39% to our GDP in 2014–2015. Besides its direct contribution to GDP, it also plays a crucial role in growth of other core industries like power, steel, cement, etc., which, in turn, are vital for the overall development of the economy. The mineral production in India is practiced since the ancient times as the mining activities can be traced as far as 6000 years ago or so. These are grouped into fuel and non-fuel minerals. The fuel minerals are coal, lignite, petroleum and natural gas. Atomic minerals (for nuclear energy) are also clubbed under this category. Further, the non-fuel minerals are again divided into metallic and non-metallic minerals including minor minerals. Metallic minerals are, melted to make new products, iron ore, copper, gold, lead, zinc, silver, tin, etc. Non-metallic minerals are usually sedimentary rocks like limestone, mica, gypsum, dolomite, asbestos, etc. Owing to meet the fuel need of many core industries such as steel, railways and power plants, coal have received the maximum attention. The public sector is involved in the entire production of lignite, petroleum and natural gas, copper, lead, zinc ores, gold, silver, diamond, tungsten concentrates, pyrites, rock phosphate, etc. About 5,70,715 km^2 land area in

Table 9.1 State-wise hazardous waste (HW) generating industries and hazardous waste generation

State	HW generating Industries (No's) as per HWM Rules, 2000/2003	Total HW generation (t a^{-1})
Andhra Pradesh	1532	507,046
Assam	23	4000
Bihar	31	3182
Chandigarh	271	8425
Delhi	1777	17,000
Goa	49	9932
Gujarat	6052	1,207,000
Haryana	889	14,972
Himachal Pradesh	575	–
Karnataka	1589	92,013
Kerala	423	83,530
Maharashtra	4355	1,407,480
Madhya Pradesh	818	182,590
Orissa	257	74,918
Pondicherry	66	30,320
Punjab	1448	15,769
Rajasthan	512	183,737
Tamil Nadu	2177	181,624
Uttar Pradesh	1633	82,375
West Bengal	568	236,449
Chattisgarh	149	40,216
Meghalaya	39	37,412
Daman, Diu & DNH	598	30,862
Jharkhand	169	–
Uttaranchal	39	7514
Tripura	187	–
Total	26,226	4,458,366

Adopted from CPCB (2009)

India is under mining belt/terrain. Out of 3318 reporting mines during 2014–2015, 498 were located in Rajasthan followed by Andhra Pradesh (444), Gujarat (362), Madhya Pradesh (326), Tamil Nadu (272), Jharkhand (250), Chhattisgarh (201), Karnataka (178), Odisha (173), Maharashtra (150), Telangana (145) and West Bengal (120) (Annual Report 2014-15, Ministry of Mines, Govt. of India). These 12 States together accounted for 94% of total number of mines in the country. It is reported that India produces 89 minerals in total which includes 4 fuel minerals, 48 non-metallic minerals, 10 metallic minerals, 3 atomic minerals, and 24 minor minerals. Among the mines reported in 2014–2015, 575 involved in coal & lignite mining, 595 involved in metallic minerals mining and 2148 involved in non-metallic mineral mining. The major metals produced in the country are iron-ore, copper-ore, manganese ore, chromite, zinc concentrates, lead concentrates, gold, bauxite, and silver. Among the mineral mines, 560 (19% of total) were coal mines, 553 (19% of total) were limestone mines and 316 (11% of total) were iron ore mines. Besides these, significant number of mines are engaged in extraction of bauxite (189), manganese (141), dolomite (116) and steatite (113) (FICCI 2013). The statewise distribution of mining area is given in Table 9.2. Maximum area (as proportion to total geographical area) under mining is in Goa followed by Andhra Pradesh, Karnataka, Odisha, Chhatisgarh, Rajasthan, Jharkhand-Bihar (combined) in the decreasing order.

Environmental Threats from Mining Industry The extraction of mineral from nature often creates imbalances which adversely affect the environment. The key environmental impacts of mining are on wild life and fishery habitats, the water balance, local climates, sedimentation, depletion of forests and disruption of ecology. Therefore, management of country's minerals resources must be closely associated with overall economic development and environmental protection and preservation strategy.

9.2 Thermal Power Plants

India's power requirement is largely met through thermal power production. It has quite a significant number (145) of thermal power plants operating all over the country and more than 62% of electricity demand is met through the country's vast coal reserves. Installed Capacity of Power Station in India is about 200,000 MW.

Fly ash is a by-product material being generated by thermal power plants across the globe from combustion of pulverized coal. Ash content in Indian coal is found to be in range of 30–50%. The quantum of fly ash produced depends on the quality of coal used and the operating conditions of thermal power plants. When pulverized coal is burnt to generate heat, the residue contains 80% Fly Ash and 20% bottom ash. Fly ash particles ranging in size from 0.5 to 300 micron in equivalent diameter, being light weight, have potential to get airborne easily and pollute the environment. If not managed properly fly ash disposal in sea/rivers/ponds can cause damage to aquatic life also. Slurry disposal in lagoons/settling tanks can become breeding grounds for mosquitoes and bacteria. It can also contaminate the underground water resources with traces of toxic metals which may be present in fly ash. Huge investments/expenditures are made just to get fly ash out from the thermal power plants and dump it in the ponds. If understood and managed properly, it can prove to be a valuable resource material. Annual production of fly ash in India was about 184 Mt. in the year 2014–2015 and is expected to cross 225 Mt. by the year 2017 with more than 90,000 acre of land requirement for ash ponds. Such a huge quantity of fly ash does pose challenging problems, in the form of land usage, health

Table 9.2 State-wise and mineral-wise details within the obvious geological potential areas

	State OGP	Gold	Diamond & precious stones	Base metal	Pt-group	Fe-ore	Mn-ore	Chro-mite	Coal & lignite	Bauxite
Andhra Pradesh	1,31,500	3000	1,17,000	33,000	300	400	500		11,000	6000
Rajasthan	1,02000	25,000		85,350					16,000	
Karnataka	80,000	35,000	62,000	2000	4000	2130	1360	360		300
Chhattisgarh	57,250	2800	45,000			205			7450	350
Odisha	47,025	8680	29,000	4800	1400	700	1110	1400	1725	19,000
Madhya Pradesh	31,300	5650	18,400	9000					5600	350
Maharashtra	28,100	5500	18,000	5500	1000		430		3100	750
Gujarat	25,100	5500		18,300					5800	1000
Jharkhand & Bihar	23,550	11,180		12,120	430	300	600	430	3350	250
Tamil Nadu	17,300	1000	3000	1500	1000	800		500	3300	200
Uttar Pradesh	9100	4500	5600	4500						
Kerala	6000	1000	2000							3000
West Bengal	5240	2580		3330					1940	
Meghalaya	2510								760	
Goa	1500	1500				600	600			1050
Haryana	1300									
Sikkim	1000			1000						
Assam	940			750					190	
Total	**570,715**	**112,890**	**300,000**	**181,150**	**8130**	**5135**	**4600**	**2690**	**60,215**	**32,250**

MoM (2015)

hazards and other environmental dangers. About 56% of the fly ash generated was utilized in the year 2014–2015 by different purposes like fly ash based building products, manufacturing of Portland pozzolana cement, construction of roads/ highways/flyovers, reclamation of low lying areas, back filling and stowing of mines, waste land development, construction of Roller Compacted concrete dams etc.

Soil fly-ash mixture tend to have lower bulk density, higher water-holding capacity and lower hydraulic conductivity than soil alone due to its textural manipulation through fly-ash mixing. Most of the fly-ash produced in India is alkaline in nature; hence, its application to agricultural soils could increase the soil pH and thereby neutralize acidic soils. Under Fly Ash Mission, application of flyash has been found to increase in yields of different crops on experimental fields as well as in farmers' fields at several locations, the magnitude of increase however depended on crop species (Kumar and Jha 2014). Presently about 1.9% of the totals ash generated is being utilized in agriculture.

Environmental Threat from Use of Fly Ash In general, the major problems associated with the application of flyash for crop production and dumping includes high electrical conductivity, unfavourable physical properties associated with pozzolonic nature of the material, boron toxicity and accumulation heavy metal particularly chromium and arsenic in soil (Yao et al. 2014; Pandey and Singh 2010; Pandey et al. 2011; Ram et al. 2015). Few investigations involving use of ash in agriculture showed undesirable effects on crop yield and on development of plants. The most frequently cited cause of these effects is heavy metals and boron toxicity (Ferraiolo et al. 1990). In some cases, ash is shown to induce P deficiency, salt injury, pozzolonic effects and heavy metal toxicity to crops (Shukla and Mishra 1986). Fly ash amendments significantly decreased the root, shoot and panicle length and augmented sterility in rice. Interestingly, ICP-MS analysis for metal accumulation revealed that the total accumulation of toxic metals, particularly Cd, Cr, Pb and As were 14–15 fold higher in roots and shoots and 4–20 fold higher in grains for the plants grown on fly ash amended soil (Singh et al. 2016).

9.3 Chemical Industries

Chemical industry products include petroleum products, polymers, fertilizers, dyes and dye intermediates, drugs and pharmaceuticals, pesticides, edible and industrials oils, synthetic paints, inks and numerous specialty organic and inorganic chemicals. At present there are about 70,000 chemical manufacturing units located in India (Draft National Chemical Policy-December 2013 of Department of Chemicals and Petrochemicals, Government of India), comprising of both small and large-scale units. Most of the chemical industries are located in seven states viz., Andhra Pradesh, Gujarat, Kerala, Maharashtra, Orissa, Tamil Nadu and West Bengal. Maharashtra and Gujarat account for over 84% of the chemical production units.

Major fraction of the chemical industries is covered under medium & small scale sector, which are generally responsible for release untreated effluent into environment. Chemical industries sector has witnessed growth rate of about 14% in the last 5 years while petrochemicals industries have registered 24% growth during this period (Table 9.3). The reason behind India's chemical industry growth could be wide range of products in conjunction with their high domestic demand, and export potential (Table 9.4).

Environmental Threat from Chemical Industries One of the major problems associated with the chemical manufacturing is generation of significant quantities of hazardous waste. The major generators of hazardous wastes are dyes and dye intermediates, bulk drugs, pesticides and petrochemicals industries. Electrical and electrochemical industries generate effluents containing significant quantities of toxic heavy metals. Treatment and disposal of hazardous waste have gained momentum in the last decade. However, the issue of hazardous waste treatment and disposal has been addressed to a reasonable extent, only in the states of Maharashtra, Gujarat and Andhra Pradesh. Toxic pollutants are released into the environment via atmosphere, water bodies and land mass as gaseous, liquid and solid waste form and causes highly adverse effect on environment, magnitude of which depends on nature of chemicals, their toxicity and quantum of release.

The two parameters that govern the hazardousness of effluents from sector specific chemical industries are volume and risk of chemicals released as effluent (Table 9.5). Generally the risk of a chemical is indicated by its persistency, toxicity, bioaccumulation/biomagnifications, carcinogenicity/mutagenicity and dispersibility. Accordingly, oil refinery and petrochemical industries are identified as the sectors requiring maximum attention due to their threat to the environment. Although, industries related to production of bulk drugs, synthetic dyes and dye intermediates generate highly toxic chemicals, the volume of waste generation is considerably low.

9.4 Chlor-Alkali Industries

Chlor-alkali industry is associated with the production of caustic soda, soda ash and liquid chlorine. Out of a total of 35 caustic soda manufacturing units in India, 21 units employ membrane cell process, 11 units employ mercury based process and 2 units employ both the processes of manufacturing. Gujarat has the maximum number of caustic soda manufacturing units followed by Tamil Nadu. Chlor–alkali production sector is the largest intentional user of Hg in the world. Significant fraction of Hg used in chlor–alkali production is released into the environment via atmospheric emission, water discharges, and sodium hydroxide and hydrogen products.

Environmental Impacts of Chlor-Alkali Industries The direct impacts are due to discharge of effluent from the Effluent Treatment Plant (ETP) into a river water

Table 9.3 Production of major chemicals and petrochemicals (group-wise) during 2007–2008 to 2014–2015 (Figures in 000'MT)

	2007–2008	2008–2009	2009–2010	2010–2011	2011–2012	2012–2013	2013–2014	2014–2015
A. Major chemical industries								
Alkali Chemicals	5501	5545	5830	6271	6478	6487	6481	6619
Inorganic Chemicals	889	782	819	898	881	873	892	922
Organic Chemicals	1692	1414	1448	1550	1640	1686	1792	1619
Pesticides (Tech.)	124	130	135	144	156	155	179	187
Dyes & Pigments	173	168	220	245	241	240	284	285
Total (A)	**8380**	**8039**	**8452**	**9108**	**9396**	**9441**	**9628**	**9632**
B. Basic major petrochemical industries								
Synthetic Fibers	2660	2479	2835	3123	3105	3124	3144	3527
Polymers	5304	5060	4791	5292	6211	6424	6784	6533
Elastomers (S.Rubber)	105	96	106	105	100	96	105	172
Synth. Detergent Intermediates	585	552	618	638	623	627	597	596
Performance Plastics	733	804	990	976	969	945	783	766
Total (B)	**9386**	**8991**	**9340**	**10,135**	**11,008**	**11,216**	**11,412**	**11,594**

Source: Chemicals & Petrochemicals Statistics at a Glance: 2015. Ministry Of Chemicals & Fertilizers, Government of India http://www.chemicals.gov.in

Table 9.4 Major chemicals produced in India during 2014–2015

Chemicals category	Production (thousand t)
Pesticides: mancozab, acephate, 2, 4-D, aluminium phosphide, glyphosate, profenofos, chlorpyriphos, phorate, D.D.V.P, D.D.T, cypermethrin, zinc phosphide, atrazin, isoproturon, pendimethalin, thiamethoxam, captan & captafol, permethrin, phenthoate, pretilachlor, triazophos, ethion, quinalphos, dimethoate, malathion	186.8
Alkali Chemicals: Soda ash, caustic soda, liquid chlorine	6619.5
Inorganic chemicals: carbon black, calcium carbonate, hydrogen peroxide, calcium carbide, titanium dioxide, aluminium fluoride, red phosphorus, potassium chlorate	921.6
Organic Chemicals: ethyl acetate, formaldehyde, chloro methanes, methanol, acetic acid, acetic anhydride, nitrobenzene, acetaldehyde, phenol, aniline, PNCB, acetone, menthol, ONCB, pentaerythritol, ethanolamines, ortho nitro toluene, isobutylbenzene, maleic anhydride, MEK	1619.1
Dyes and pigments: reactive dyes, organic pigment, disperse dyes, optical whitening agents, acid direct dyes, inorganic pigments, azo dyes, pigment emulsion, sulphur dyes, solvent dyes, vat dyes, food colours, ingrain dyes	285.2
Petrochemicals: Synthetic fibres / yarn, polymers, synthetic rubber, synthetic detergent intermediates, performance plastics, fibre intermediates, olefins, aromatics, vinyl chloride monomer, C4-Raffinate, ethylene dichloride, phthalic anhydride, isopropanol, di-ethylene glycol	29,347

Note: Chemicals listed in each category are arranged in decreasing order of production

Table 9.5 Priority matrix with respect to risk and volume

	Volume		
Risk	High	Medium	Low
High	Oil refinery, Petrochemicals	Pesticides	Bulk-drugs, dyes & dye intermediates
Medium	Fertilizers, Chlor-alkali	–	–
Low	Soda ash	Paints, inks & printing inks	–

Adopted from www.cpcb.nic.in

causing decline in water quality in terms of high level of mercury, TDS, free chlorine, alkalinity/acidity and affecting biota in the river. The loss of mercury is 100% during the production of caustic soda, and the magnitude of loss is about 394 g t^{-1} of caustic soda produced. The indirect impacts are ambient air SO_2 rise due to stack emissions which may deposit on soil and contaminate crops; and decline in water quality due to rise in temperature of water bodies receiving cooling water discharge from the nearby industry. In view of high adverse impact on environment, Government of India has banned the commissioning of new mercury cell based chlor-alkali plants since 1991 and has made mandatory for new chlor-alkali plants to install the modern membrane cell based technology.

9.5 Tannery Industries

In India, leather and leather products are processed from about 2334 major production centers. The main step process with adverse environmental consequences in these industries is tanning of raw hides by using chromium salts. These industries are concentrated in a few pockets of Tamil Nadu, West Bengal and Uttar Pradesh (having 88.5% of the total tanneries in the country) due to the existence of tanning operations since long and proximity to the port for exporting the products.

Environmental Impacts of Tannery Industries The environmental impacts of tanning industry wastes are contamination of surface water bodies with effluent having high oxygen demand, and toxic chemical constituents and discoloration of water bodies (Song et al. 2000). Tannery waste contains both organic (chlorinated phenols) and inorganic (chromium) pollutants (Mwinyihija et al. 2006). Other pollutants of concern with the tanning industry include azodyes, cadmium compounds, cobalt, copper, antimony, barium, lead, selenium, mercury, zinc, arsenic, polychlorinated biphyenls (PCB), nickel, formaldehyde resins and pesticides residues. All these pollutants that released into the environment due to the tanning process lead to deterioration of air, water and soil quality. According to WHO report, more than 8000 workers in the tanneries of Hazaribagh, India suffer from gastrointestinal, dermatological, and other diseases, and they die mostly before the age of 50 (Maurice 2001). In other separate studies in Kanpur, India, Rastogi et al. (2008) has also indicated that there is a significantly higher prevalence of morbidity among tannery industry workers, mostly from respiratory diseases because of chromium exposure.

9.6 Pesticide Industries

The agro-chemical industries in India contribute significantly to the country's economy. The agrochemicals market in the country grew at a rate of 11% from USD 1.22 billion in 8th five year plan to USD 1.36 billion in 9th five year plan. India is the fourth largest producer of agro-chemicals globally after US, Japan and China. However the per hectare consumption of agrochemicals in India is very low (only 0.58 kg) when compared to US (4.5 kg ha^{-1}) and Japan (11 kg ha^{-1}). Among the crops, paddy consumes maximum share of pesticide (around 28%), followed by cotton (20%) in India. There are about 150 industrial units technically involved in the manufacturing of pesticides and more than 500 units engaged in preparation of formulations in India. India is the largest producer of pesticides in Asia and ranks 12th in the world in terms of pesticide use. Pesticides are not only used for agricultural purposes, but also in industries (for wood and carpet, wood preservation, etc.), public health (malaria, dengue fever and cholera) and domestic purposes such as fly killer, ant killer, repellants. Rodenticides, fungicides, etc.

Table 9.6 Persistence of pesticides in soils

Pesticide and their group	Persistence
Organophosphates	7–84 days
DDT	10 years
Carbamates	2–8 weeks
Aliphatic acids	3–10 weeks
Diuron	16 months
BHC	11 years
Toxaphene	6 years

Environmental Impacts of Pesticide Industries The wastewater generated from pesticide industries consists of reaction water from chemical process, process solvent water, product wash water, stream water. The waste water contains pollutants like halomethanes, cyanides, haloethers, phenols, heavy metals, pesticides, etc. However, the BOD, COD and TSS contents are very low. Many of the chemicals used are persistent in nature (Table 9.6) and severely affects soil ecology and find its way in food chain through plant food materials and affects the human health.

9.7 Phosphatic Fertilizer Industry

There are about 93 fertilizer plants producing single super phosphate in India with an installed capacity of 7.7 Mt. per year, which are located in AP (9), Assam (1), Bihar (1), Chhatisgarh (2), Gujrat (7), Haryana (1), Karnataka (2), MP (16), Maharashtra (17), Rajasthan (22), TN (3), UP (7), WB (5). Besides these 12 fertilizer plants are manufacturing di-ammonium phosphates with an installed capacity of 8.3 Mt. per year. Phosphoric acid is mainly used for manufacturing DAP and complex fertilizers. Presently, there are 12 phosphoric acid plants having a combined installed capacity of about 2.0 Mt. per annum as P_2O_5.

The phosphate fertilizer plant produces large amount of waste in the form of liquid and solid effluents. Phosphogypsum is generated as a by-product of the phosphoric acid based fertilizer industry and is well recognized as a source of sulphur (16%) and calcium (21%), chemically present in the form of $CaSO_4.2H_2O$. It also contains some amount of phosphorus (0.2–1.2% P_2O_5), almost half of which is present in water soluble form. Different fertilizer industries generated around 11 Mt. of phosphogypsum in 2012–2013 (based on the assumption that 5 t of phosphogypsum generated per tonne of phosphoric acid production). Besides, old stocks of over 12 Mt. (nearly 2.2 Mt. of sulphur and 3.0 Mt. Ca) are lying indisposed at different plant sites as huge waste piles called "stacks". The phosphogypsum, unlike other sulphur sources, offers all the desirable fertilizers characteristics (besides supplying calcium), i.e., readily available to the growing plant. The product as soil amendments has already been notified by Bureau of Indian Standards (BIS) for reclamation of sodic soils.

Environmental Threat from Phosphogymsum One of the main concerns on the use of phosphogypsum is the higher amounts of the radionuclide. During acidifying process on rock phosphate to produce phosphoric acid, about 80–90% of ^{226}Ra (due to similarity with Ca in chemical properties) is concentrated in phosphogypsum while nearly 86% of U and 70% of Th end up in the phosphoric acid (Poole et al. 1995). Radium^{226} in phosphogypsum further decays to produce Radon gas (Rn^{222}) emitting carcinogenic alpha particles. USEPA has classified phosphogypsum and rock phosphate as "Technologically Enhanced Naturally Occurring Radioactive Materials" (TENORM) (USEPA 2006) and phosphogypsum exceeding 370 Bq kg^{-1} of radioactivity has been banned from all uses by the EPA since 1992. Depending on the quality of the source material (phosphate rock), phosphogypsum can contain as much as 60 times the levels normally found prior to processing. A study was carried out in India to investigate concentration of naturally occurring radionuclides in around a phosphate fertilizer plant (Sahu et al. 2014). ^{238}U, ^{232}Th, ^{226}Ra and ^{40}K activities in soil samples were 21–674 Bq kg^{-1}, 11–44 Bq kg^{-1}, 22–683 Bq kg^{-1} and 51–295 Bq kg^{-1} respectively. The soil from the solid waste disposal area and phosphate rock storage area showed higher levels of natural radionuclides than other locations within the plant premises. The ^{238}U levels in phosphogypsum sample were higher than those compared worldwide. The external gamma radiation had increased level only in these areas as compared to background though such increases were not significantly high. Phosphogypsum also contains fluoride (0.5–1.5%), which may leach down and contaminate groundwater. It contains heavy metals (cadmium, chromium, lead etc.), which if applied in agricultural land, may enter into food chain and damage human and animal health. Extensive reviews (Alcardo and Rechcigl 1993) on the use of phosphogypsum in agriculture, have revealed that the radionuclides, heavy metals, fluorides and other pollutants at concentrations found in phosphogypsum, does not pose any environmental hazards to surface ground water, soils, crops and the ambient atmosphere at the rates normally used in agriculture. However, its unregulated use in agriculture may pose threat to the environment.

9.8 Iron and Steel Industry

India is the world's third-largest producer of crude steel after China and Japan and has been a major contributor to India's manufacturing output. The Government of India is aiming to scale up steel production in the country to 300 Mt. by 2025 from about 90 Mt. in 2015–2016. There are about 11 major integrated steel industries operating in the country in the states of Jharkhand, Odisha, Chattisgarh, West Bengal, Tamil Nadu, Karnataka, Andhra Pradesh and Maharastra. Besides these, several mini steel plants (about 650) having electric arc furnaces are using steel scrap and sponge iron. Most of these industries are located nearby coal mining area.

Slag generated from iron and steel making industries have been found to possess lime supplying potential in acidic soils. Slag materials have been found to contain high amount of Ca (about 30–55% CaO) and Si (12–35% SiO_2) which are essential/ beneficial elements for plants. Acid soils of India are deficient in these two elements, requiring their frequent application in crop production. Considerable researches have also indicated increases in crop yields with the application of slag materials due to increased supply of Ca and Si to plants. India has quite a substantial number of iron and steel industries generating 300 to 500 kg slag per tonne of hot metal produced from blast furnace (BF), electric arc furnace (EAF) or basic oxygen furnace (BOF). Considering projected crude steel production of 140 Mt. (http://steel.gov.in/overview.htm) and average slag generation @ 400 kg t^{-1} of hot metal (http://ibm.nic.in/imyb2010_slagironandsteel.pdf), India has the potential of generating more than 50 Mt. of slag by 2016–2017. Although, a portion of the BF slag find some alternative use (like cement production, road construction etc.), considerable portion of these may be available for enhancing productivity in acid soils. Also, most of the iron & steel industries are located in the acid soil regions, thus minimizing cost of basic slag transportation to agricultural land.

Environmental Threat from Use of Slag The main problem concerning the utilization of steel slags in agriculture consists of the possible leaching of heavy metals. Among heavy metals, steel making slags contain Vanadium (V) and chromium (Cr) in significantly high concentration. Analysis of steel slags from US and Canada indicated that slags generated from BOF and EAF contained very high levels of Cr (23–56 times the avg. background concentration in soil) and V (6–12 times the avg. background concentration in soil) (Proctor et al. 2000). On the basis of TCLP test, however, these materials are not considered hazardous waste and do not likely to contaminated drinking water resources. However, impact of long-term addition of slag materials to agricultural land has not been investigated. In Sweden, death of cattle was reported from vanadium poisoning which occurred due to grazing from slag (containing 3% V) amended land (Frank et al. 1996).

9.9 Aluminium Industry

India is richly endowed with bauxite resources, which is the basic raw material for aluminium production. The total domestic production of aluminium metal during FY 2013-14 was about 1.7 Mt. and in FY14-15, total installed smelting capacity was about 2.9 Mt. The aluminium smelting industries are located in Renukoot (U.P.), Korba (Chhattisgarh), Alupuram (Kerala), Mettur (T.N.), Jharsuguda, Angul, Hirakud (Orissa). Alumina refineries are located in Lanjigarh, Dhamajori (Orissa), Belgaon (Karnataka), Korba (Chhattisgarh), Mettur (T.N.), Renukoot (U.P.), Muri (Jharkhand).

Environmental Threat from Alumina Refineries and Smelters Alumina and aluminium processing creates a range of waste products, the most significant being bauxite residues (also known as red mud). Red mud is a by-product of the Bayer process, the principal means of refining bauxite to produce alumina using NaOH.

Table 9.7 Major constituents of bauxite residues from Indian alumina refineries

Major oxides	Concentration (%)
Al_2O_3	17.5–22
SiO_2	4.8–16
Fe_2O_3	35–54
TiO_2	2.5–19
CaO	0.8–4.5
Na_2O	3.3–6.0

Rai et al. (2012)

Table 9.8 Hazardous elemental concentrations in bauxite residues from alumina refinery

Trace elements	Concentration (mg kg^{-1})
As	267.3
Pb	56.6
Hg	67.3
Cd	27.1
Cr	537.8
Ni	984.9
Ba	212.0
Zn	103.2
Cu	78.2
Mn	187.5

Wang and Liu (2012)

Depending on quality bauxite, approximately 1.5 metric tons of bauxite residues is generated with every metric tonne of alumina produced. Red mud is composed of a mixture of solid and mud of metallic oxides. Iron oxides comprise up to 60% of the mass of the red mud. The mud is highly basic with a pH ranging from 10 to 13. In addition to iron, the other dominant components include silica, residual alumina, and oxides of Ti, Ca and Na (Table 9.7). Besides these, red mud also contains high amount of several hazardous trace elements As, Pb, Hg, Cd, Cr, Ni, and these are present in concentrations several times higher than commonly found in unpolluted soils (Table 9.8). Radioactive elements like ^{226}Ra and ^{232}Th were also found in red mud at significantly higher concentrations. These hazardous elements as well as high pH (10–12.5) are major concern for disposal of red mud wastes.

9.10 Copper Refineries and Smelters

The copper industry in India accounts for 3% of the world copper market at about 500 thousand tonnes by size in terms of annual consumption of refined copper. The country does not have sizable discovered copper reserves and therefore, mined copper production accounts for only 0.2% of the world's global production. Because of the large waste volumes associated with copper production, the processing facilities are usually located near the mines. Copper smelters are located at Maubhandar (Jharkhand) and Khetri (Rajasthan). The smelter at Maubhandar (Ghatshila) receives copper ore from Mosabani, Rakha, Dhobani Rajdah, Tamapahar and Turamdih of Singbhum districts of Jharkhand. Copper smelter at Khetri receives copper ore from Khetri, Kolihan, Chandmari and Dariba areas of Rajasthan as well as from Malanjkhand (Madhya Pradesh). Major fraction of refined copper ore demand for the smelters is met through imports.

Type and volume of wastes as well as their hazardousness depend on quality of ore (chemical composition) and method of processing of ores. Higher grade ores are milled to fine particles and copper minerals are subsequently concentrated by physical separation. The tailings are pumped to the tailings pile and the copper concentrate is transported to a nearby smelter.

Environmental Threat from Copper Production Industries All aspects of copper production—from leaching to milling, smelting, refining and electrowinning—have potential impacts on surface and groundwater quality. Mill tailings from beneficiation process may contain toxic heavy metals (mainly As, Pb, and Cd) and radionuclides due to their natural presence in the ore bodies and thus are a potential source of "Technologically Enhanced Naturally Occurring Radioactive Materials (TENORM)". Mill tailings also contain high amount of pyrite/sulfide minerals which may oxidize to produce sulfuric acid and consequently solubilise heavy metals & radionuclides to contaminate land and water bodies. Leakage from heap leaching with acids may also contaminate environment with heavy metals. Smelting operations in the copper industry emits considerable heavy metals (As, Cd, Cu) in the atmosphere which subsequently precipitates on land and water bodies far away.

9.11 Zinc and Lead Smelting Industry

Zinc and lead smelters India are either located near zinc ore mines (in Rajasthan) or near sea ports (as in Vishakhapatnam and Alwaye, primarily dependent on imported mineral concentrates). Major zinc and lead smelters and refineries in India are located in Rajasthan state (Chanderiya, Debari and Dariba) with total metal production of 880,000 tonnes in year 2014–2015. The ore for these industry is received from mines of Rampura Agucha, Sindesar Khurd, Rajpura Dariba, Kayar and Zawar belonging to the state of Rajasthan. Zinc and lead smelter plant at Vishakhapatnam (A.P.) has an installed capacity of 3000 t Zn, 22,000 t Pb and 75,000 t sulphuric acid and the plant is primarily dependent of imported mineral concentrate. Zinc smelter plant at Alwaye (Kerala) also uses imported zinc concentrates for its operation.

Environmental Threats from Zinc and Lead Smelters The major air pollutants emitted from the smelting processes are particulate matter, and SO_2, along with gaseous forms several metals like As, Hg, F etc. in trace amount. Emitted particulate matters contain oxides of Zn, Pb, As, Hg, Cd, Cu, Sb. All these air pollutants upon settling can contaminate water, soil and food crops. Wastewater effluent from the industry contains dissolved and suspended solids, heavy metals, and oil and grease having adverse effects on land and water bodies. The smelting process typically produces less than 3 tons of solid waste (slag and sludge) per ton of lead/zinc produced. Slag material from the industry may contain 0.5–0.7% lead & zinc along with many other toxic heavy metals, particularly Cd.

9.12 Paper Industry

There are about 759 paper mills in India having a total installed capacity of 12.7 Mt. and was anticipated to produce around 14 Mt. paper, paper board and newsprint per year by the year 2016 according to Govt. of India report. Indian share in total world production is about 2.6%. Among the total paper mills, about 47% is produced through 579 small & medium scale recycle fibre based mills; 31% is produced through 30 wood based large mills; and 22% is produced through 150 medium scale agro-based paper mills. Pulp and paper industry consumes large quantities of fresh water at the rate of 80-150 m^3 per tonne of paper. Generally agro based mills consume much more water than recycle fibre based mills. Paper & pulp industries also use significant amount sodium hydroxide, sodium carbonates, sodium sulfide & bi-sulfites, elemental chlorine & hypochlorite, calcium oxide, hydrochloric acid, etc. during manufacturing of paper and generate larger quantities of effluents containing organic and inorganic salts and toxic pollutants.

Paper mills, and integrated pulp and paper mills also generate significant amount of biodegradable sludge (34 and 105 kg t^{-1} product in large and small paper mills respectively) during the treatment of organic substances coming out of the paper-making and pulp making stages. Sludge is the solid residue formed during waste-water treatment of paper mill effluents. Land filling, land-spreading, composting and energy production are the major disposal/treatment methods for these sludges.

Environmental Threat from Paper Mill Sludge Paper mill effluents contain fugitive fibres, starch, hemi-cellulose and organic acids which cause high COD. High BOD/COD concentration results in depletion of oxygen available to fauna and flora in the downstream of effluent discharge, depleting quality and quantity of aquatic life. A large number of organochlorine compounds, such as, chlorinated derivatives of phenols, acids, dioxins (dibenzo-p-dioxins), furans (Polychlorinated dibenzofurans) 1,3,6,8-TCDD, etc. are generated during paper production, many of which are carcinogens (Thacker et al. 2007). Bleaching process effluents may contain chloroform and carbon tetrachloride, which are also classified as carcinogens.

9.13 Molasses Based Distillery Industry

India is the second largest producer of sugar and fourth largest producer of alcohol in the world. Molasses is used as a raw material in the distillery industries where alcohol is produced by fermentation processes. There are about 356 molasses based distilleries and about 110 grain based distilleries in India with an installed capacity of 4.3 billion liters of alcohol. About 40% of the molasses based distilleries are attached to the sugar industries. About 64% of these industries are located in five states only, viz., Andhra Pradesh, Karnataka, Maharashtra, Tamil Nadu, Uttar Pradesh. Consequently, the wastewater generation from distilleries has gone up to 40 billion liters annually. The effluent generated from distilleries is called raw spent-wash (RSW). Depending upon the technology and quality of molasses, the spent wash generation rate can vary from 7.0 to 15.0 liters per liter of alcohol produced. Spent wash generated in the distilleries is subjected to anaerobic digestion (biomethanation) in many industries for methane recovery and effluent generated after this process is called post-methanation effluent (PME).

The chemical analysis of RSW and PME clearly showed their high manurial potential (Table 9.9). Even application of 1 ha-cm effluent to the soil can meet the N, P, K and S requirements of most of the agricultural crops.

Environmental Threats from Distillery Effluents Distilleries are classified as "Red Category" among the 17 categories of highly polluting industries by CPCB. Disposal of spent wash, a distillery waste, has long been a problem for the industry. Seepage of effluent from storage lagoon pond to contaminate groundwater has been reported at several places which resulted increase in salinity and coloration making

Table 9.9 Chemical composition of untreated spent wash (SW) and treated spent wash (PME)

Parameter	SW	PME
pH	4.2	7.4
EC (dS m^{-1})	25.30	25.06
Total solid (g L^{-1})	98.0	43.0
Organic C (g L^{-1})	37.5	13.5
Total N (g L^{-1})	2.02	1.310
Ammoniacal-N (g L^{-1})	1.254	0.675
Nitrate-N (g L^{-1})	Trace	0.013
Total P (g L^{-1})	0.24	0.132
Organic P (g L^{-1})	0.139	0.068
Total K (g L^{-1})	13.0	11.50
Water soluble K (g L^{-1})	7.9	8.90
Total S (g L^{-1})	2.82	0.852
Sulphate-S (g L^{-1})	1.31	Trace
Chloride (g L^{-1})	4.05	3.49
BOD (mg L^{-1})	36,000	3960
COD (mg L^{-1})	113,240	39,252

Adopted from Biswas et al. (2011)

it unsuitable for drinking purposes. Release of distillery effluent may cause rise in salinity, biological oxygen demand, chemical oxygen demand in surface water bodies deteriorating water quality for aquatic life. Although the effluent may facilitate better crop yields at the beginning due to its richness in nutrients (N, P, K, Ca, S, micronutrients), its continuous or indiscriminate/heavy use may cause accumulation of salts endangering the soil productivity and quality.

9.14 Leather Industry

There are more than 2000 tannery industry existing in the country, of which about 70% are present in Tamil Nadu and West Bengal. Leather industry in India is spread over both organized as well as unorganized sector in India. Unorganized small scale leather industry is dominated by the presence of family units, which accounts for over 90% of the total production. The major production centers for leather and leather products are located in the states of Tamil Nadu, West Bengal and Uttar Pradesh. Tanning is an important leather processing step and tanning industry has grown in clusters of Tamil Nadu, West Bengal, Uttar Pradesh and Punjab. Mineral tanning usually uses chromium in the form of basic chromium sulfate and produces stretchable leather which is excellent for use in handbags and garments. The leather processing requires large quantities of chemicals like sodium chloride, chromium sulphate, calcium and ammonium salts and organic dyes. Wastes and chemicals produced during washing, dehairing and tanning of the leather are released to water system causing pollution concerns for the leather industry.

Environmental Threats from Tannery Industry Effluents The effluents generated from tannery and textile industries are characterized by high value of BOD, COD, Na and dissolved solids (Table 9.10). The tannery effluents are also ranked as the highest polluting waste amongst all industrial waters. The effluent from chrome tannery generally contains much higher suspended solids, dissolved salts, and several heavy metals particularly Cr compared to those generated from vegetable tanning. Use of tannery effluents in crop land increases several plant nutrient contents like N, P and K levels in soil. However it has adverse effect on soil physical properties as well as builds up heavy metals, particularly Cr in soil.

9.15 Textile Industry

Textile sector is one of the oldest industries in India dating back several centuries. This sector is one of the largest contributors (approximately 11% of total) to India's total exports. The organized textile sector has more than 3400 textile mills both in small scale industry (SSI) and the non-SSI sectors, producing about 2500 million m^2 of cloth apart from 2500 million kg of man-made fibres & filament yarns. Most of the medium to large cotton textile industries are located in Maharastra, Gujarat and

Table 9.10 Average total pollution load in tannery industry effluent

Parameter	Average total pollution load	Typical limits for surface water disposal
BOD_5 (mg O_2 L^{-1})	2000	30–40
COD (mg O_2 L^{-1})	4000	125–250
Suspended solids (mg L^{-1})	2000	35–100
Cr^{3+} (mg Cr L^{-1})	150	1.5–2.0
S^{2-} (mg S L^{-1})	160	1.0–2.0
Total N (mg N L^{-1})	160	100
Cl^{-1} (mg Cl L^{-1})	5000	Locally specific
SO_4^{2-} (mg SO_4 L^{-1})	1400	Locally specific
Oil & grease (mg L^{-1})	130	Locally specific
TDS (mg L^{-1})	10,000	Locally specific
pH	6–9	5.5–9.5

UNIDO (2011)

Tamil Nadu; while jute mills and woolen industries are concentrated in West Bengal and Punjab respectively. The textile industry uses high volumes of water throughout its operations, from the washing of fibres to bleaching, dyeing and washing of finished products. On average, approximately 200 L of water are required to produce 1 kg of textiles.

Environmental Threats from Textile Industry Effluents Colour, high contents suspended and dissolved solids, residual chlorine, toxic metals (like Cr, As, Cu) and organic pollutants (mainly dyes) are the major pollutants present in textile mill effluent. All of these have adverse effect on soil physical properties (due to salts), plant growth (due to residual chlorine and Na, sulphate), microbial biodiversity (due to dye compounds and heavy metals) and food quality (due to salts and toxic elements) when used as irrigation. Many of the textile dyestuffs are resistant to aerobic degradation and carcinogenic. Textile mill effluent constituents also have adverse impact on aquatic life when released in surface water bodies.

9.16 Mercury Emission/Discharge in India

Now a days scientific attention on mercury distribution in the environment is being seriously focused; because this element's exposure poses a potential health risks to human and animal. Methyl mercury (CH_3Hg), which is the most toxic organic form of mercury, bioaccumulates in the biota and subsequently gets biomagnified in the aquatic food chain, especially in fish. Chakraborty et al. (2013) quantified the anthropogenic emission and major transfer pathways of mercury to the environment (air, water, soil) in India in the period 2001–2020. Total mercury emissions in India were approximately 415 t in 2001, 310 ts in 2010, and are projected to rise to 540 t in 2020. About 76% of these emissions in 2010 went to the atmosphere. The chlor-alkali industry is the major consumer as well as emitter of mercury and account for 74% of the basic chemicals production in India. The mercury consumption in Indian caustic-chlorine companies is at least 50 times higher than the global best companies. They alone contribute to about 40% of the total mercury pollution in the country. On an average, 47 g mercury is lost in the production of 1 t of caustic soda in the Indian caustic- chlorine industry. This estimate of mercury loss is however, based on the total caustic soda production, irrespective of the production process. The average specific mercury loss from mercury cell plants in India is about 142 g per tonne NaOH produced (http://www.cseindia.org/dte-supplement/who-responsible.htm). Other industries, which release mercury in atmosphere and surface water in India, are coal based industries like thermal power plants, steel industries and cement industries.

Coal, containing about 0.04–0.7 mg kg^{-1} Hg, is consumed as a source of energy in various thermal power plants, steel industries etc. India has consumed about 644 Mt. of coal in 2014 through these industries which might have resulted significant quantity of mercury emission. The total mercury pollution potential from coal in India is estimated to be 77.91 t a^{-1}, considering average concentration of mercury in coal as 0.272 ppm (Rai et al. 2013). About 59.29 t a^{-1} mercury is mobilized from coal-fired thermal power plants alone.

The iron and steel industry was reported to release as high as 56 mg kg^{-1} Hg through dust fall out. The fallout of elemental mercury over the soil-horizon in the vicinity of a steel plant was reported to be in the range of 60.36 to 836.18 g km^{-2} $month^{-1}$ (http://www.cseindia.org/dte-supplement/mercury-menace.htm).

References

Alcaordo LIS, Rechcigl JE (1993) Phosphogypsum in agriculture: a review. Adv Argon 49:55–118

Biswas AK, Ramana S, Singh AB et al (2011) Safe use of distillery effluents in agriculture. IISS Research Bulletin No. 2, Indian Institute of Soil Science (ICAR), Bhopal

Chakraborty LB, Qureshi A, Vadenbo C et al (2013) Anthropogenic mercury flows in India and impacts of emission controls. Environ Sci Technol 47:8105–8113

CPCB (2009) Comprehensive environmental assessment of industrial clusters (Ecological impact assessment Series: EIAS/5/2009-10). Central Pollution Control Board, Delhi. http://cpcb.nic.in/divisionsofheadoffice/ess/NewItem_152_Final-Book_2.pdf. Accessed 3 Oct 2016

Ferraiolo G, Zilli M, Converti A (1990) Fly ash disposal and utilisation. Chem Biotechnol 47:281–305

FICCI (2013) Development of Indian mining industry – the way forward: non-fuel minerals. Mines and Metals Division, Federation of Indian Chamber of Commerce and Industry, New Delhi. http://www.ficci.com/spdocument/20317/Mining-Industry.pdf

Frank A, Madej A, Galgan V et al (1996) Vanadium poisoning of cattle with basic slag. Concentrations in tissues from poisoned animals and from a reference, slaughter-house material. Sci Total Environ 181(1):73–92

Kumar V, Jha GK (2014) Use of fly ash in agriculture: Indian scenario. International Workshop on Agricultural Coal Ash Uses, 27–29 May 2014. Agricultural Center, Bet Dagan, Israel

Maurice J (2001) Tannery pollution threatens health of half-million Bangladesh residents. Bull World Health Organ 79(1), Genebra. DOI: 10.1590/S0042-96862001000100018

MoM (2015) Exploration and mining opportunities in India: an investor guide. Ministry of Mines, Government of India, New Delhi

Mwinyihija M, Strachan NJC, Dawson J et al (2006) An ecotoxicological approach to assessing the impact of tanning industry effluent on river health. Arch Environ Contam Toxicol 50:316–324

Pandey VC, Singh N (2010) Impact of fly ash incorporation in soil systems. Agric Ecosyst Environ 136(1–2):16–27

Pandey VC, Singh JS, Singh RP et al (2011) Arsenic hazards in coal fly ash and its fate in Indian scenario. Resour Conserv Recycl 55(9):819–835

Poole AJ, Allington DJ, Baxter AJ et al (1995) The natural radioactivity of phosphate ore and associated waste products discharged into the eastern Irish Sea from a phosphoric acid production plant. Sci Total Environ 173(174):137–149

Proctor DM, Fehling KA, Shay EC et al (2000) Physical and chemical characteristics of blast furnace, basic oxygen furnace, and electric arc furnace steel industry slags. Environ Sci Technol 34:1576–1582

Rai S, Wasewar KL, Mukhopadhyay J et al (2012) Neutralization and utilization of red mud for its better waste management. Arch Environ Sci 6:13–33

Rai VK, Raman NS, Choudhary SK (2013) Mercury in thermal power plants – a case study. Int J Pure Appl Biosci 1(2):31–37

Ram LC, Masto RE, Srivastava NK et al (2015) Potentially toxic elements in lignite and its combustion residues from a power plant. Environ Monit Assess 187:4148. doi:10.1007/s10661-014-4148-0

Rastogi SK, Pandey A, Tripathi S (2008) Occupational health risks among the workers employed in leather tanneries at Kanpur. Indian J Occup Med 12:132–135

Sahu SK, Ajmal PY, Bhangare RC et al (2014) Natural radioactivity assessment of a phosphate fertilizer plant area. J Radiation Res Appl Sci 7:123–128

Shukla KN, Mishra LC (1986) Effect of fly ash extract on growth and development of corn and soyabean seedling. Water Air Soil Pollut 27:155–167

Singh PK, Tripathi P, Dwivedi S et al (2016) Fly-ash augumented soil enhances heavy metal accumulation and phytotoxicity in rice (*Oryza sativa* L.); A concern for fly-ash amendments in agriculture sector. Plant Growth Regul 78(1):21–30

Song Z, Williams CJ, Edyvean RJ (2000) Sedimentation of tannery wastewater. Water Res 34:2171–2176

Thacker NP, Nitnaware VC, Das SK et al (2007) Dioxin formation in pulp and paper mills of India. Environ Sci Pollut Res Int 14(4):225–226

UNIDO (2011) Introduction to treatment of tannery effluents. United Nations Industrial Development Organization. Austria, Vienna

USEPA (2006) Technical report on technologically enhanced naturally occurring radioactive materials from Uranium mining. Vol 1: Mining and reclamation background. U.S. Environmental Protection Agency, Washington, DC

Wang P, Liu DY (2012) Physical and chemical properties of Sintering red mud and Bayer red mud and the implications for beneficial utilization. Materials 5:1800–1810

World Bank (2013) India- diagnostic assessment of select environmental challenges: an analysis of physical and monetary losses of environmental halth and natural resources (Report No. 70004-IN). Disaster Management and Climate Change Unit, Sustainable Development Department, South Asia Region, Washington, DC

Yao ZT, Xia MS, Sarker PK et al (2014) A review of the alumina recovery from coal fly ash, with a focus in China. Fuel 120:74–85

Chapter 10
Impact of Different Developmental Projects on Soil Fertility

Abstract Soil fertility is highly important for countries where economy is dependent considerably on agriculture and animal husbandry. With the ambition of boosting economic growth many countries like India tap their vast mineral and human resources and expand industrial sector which also have considerable impact on soil fertility and agriculture. In order to have sustainable economy of a country, a harmony between industry and agriculture is highly essential. Though the cost involving protective and remedial measures appears to cut away the profit of industrial sector, these are extremely essential for shielding non-renewable soil resources which anchor the food producing agriculture sector. In order to formulate effective policies on protective and remedial measures, knowledge on the impact of different industries is required. This chapter reviews the research carried out worldwide to study the impact of major industries on soil quality and its consequences on quality of life including human. While some of the industries like thermal power plants, mining and associated ore beneficiation processes have direct impact on soil resources, other industries like textiles, paper & pulp, chemical & petrochemical, distillery etc. affect soil quality via polluted water used for irrigation.

Keywords Soil • Pollution • Mining • Smelting • Industry • Environmental impact • Wastes

10.1 Impact of Mining Activities

Impact of mining projects on environment begins from initiation phase, because it may involve cleaning of wide areas of vegetation to allow entry of machineries for drilling. In the exploratory phase, feasibility of mining project is explored. If feasibility is ascertained, subsequent activities like construction of access roads, cleaning of land, construction of housing facilities for personals and equipment, active mining are undertaken. All of these are associated with huge adverse impact on environment; and the severity of impacts varies depending on whether the mine is operational and abandoned, mining methods used, land topography and the geological conditions (Bell et al. 2001). The Environmental Impact Assessment

J.K. Saha et al., *Soil Pollution - An Emerging Threat to Agriculture*, Environmental Chemistry for a Sustainable World 10, DOI 10.1007/978-981-10-4274-4_10

(EIA) must assess separately, the impacts associated with all these phases of mining operations.

In India, every one million ton of coal extracted by surface mining methods damages a surface area of about 4 ha (Ghosh 1990). Likewise many other mines are excavating other major and minor metallic and non-metallic minerals having severe environmental implications. Many workers around the world have reported various effects of mining which cause massive damage to landscapes and biological communities on the earth (Banerjee et al. 2004; Dutta and Agrawal 2002; Ekka and Behera 2011; Ghose 1996; Singh and Singh 2006). The unscientific mining of minerals poses a serious threat to environment, resulting in the reduction of forest cover, erosion of soil, pollution of air, water and land and reduction in biodiversity. Surface mines produce dust from blasting operations and haul roads; many coal mines produce methane, a greenhouse gas. If mining area is situated near to agricultural land, wind-blown dust generated from different mining activities (blasting, excavation, transportation) is likely to get deposited on the surface of the fertile land (Adewole and Adesina 2011). Such dust particles from excavated mine pits generally contain high amount of heavy metals (Fig. 10.1).

Therefore, high metal contamination in the soils of surrounding agricultural land is frequently observed. Erosion of exposed soils, extracted mineral ores, tailings and fine material in waste rock piles can result in substantial sediment loading to surface waters which may be deposited on the agricultural land of flood plain area. Sediments deposited in layers in flood plains/agricultural lands normally contain sulfide minerals which upon oxidation may reduce pH, mobilize heavy metals, infiltrate subsoil layer and contaminate food chain (Gitt and Dollhopf 1991; Gould et al. 1996). Reduction in soil pH can also alter soil biodiversity and reduce soil fertility. Toxic trace elements, once entered into the soil, remain there for a long time (several thousands of years). Therefore it is not cheap and easy to rehabilitate

Fig. 10.1 Open-cast copper mine at Malanjkhand, Madhya Pradesh emits contaminated dusts during blasting

the contaminated land near mining site. Extent and severity of soil contamination depend on the distance and geographical position of the land with respect to mining area. In the middle and lower reaches of the Mashavera valley in SE Georgia (USA), most of the irrigated soils under different agricultural land use displayed a strong enrichment of heavy metals that could be traced back to irrigation with water polluted by mining wastes (from copper and precious metals mines) over a period of several decades (Henningsen et al. 2010). The concentrations of total amounts of Cu, Zn and Cd increased with intensity of land use and amount of irrigation. There are four major types of mining impacts on water and soil quality.

10.1.1 Acid Mine Drainage (AMD) and Its Impact on Soil

During the mining process, if rocks containing large quantities sulfide minerals are excavated and exposed to open air, it reacts with water and oxygen to produce sulphuric acid. This reaction is mediated and further aggravated by a naturally occurring bacterium called *Thiobacillus ferroxidans* which survives under acidic condition. When water is acidified, these organisms get involved and accelerate the oxidation and acidification processes that result in leaching of even more trace metals from the mine wastes. Such leaching process continues as long as sulfides in the rock are exposed to air and water and/or until all sulfides are leached out. This may last hundreds, even thousands of years according to type of source rock and local conditions. Acid is carried off the mine site by rainwater or surface drainage and deposited into nearby streams, rivers, lakes and groundwater. Acid mine drainage severely degrades water quality, and can kill aquatic life and can make water virtually unusable.

In coal mining districts, AMD is generally caused by the oxidation and hydrolysis of pyrites (FeS_2), an accessory mineral occurring in coal and associated strata (Gray 1996). Coal mining exposes pyrites to atmosphere which subsequently enhances dissolution of pyrites. This appears to be the main mechanism for the occurrence of AMD exacerbating the problem of acid rock drainage (Blahwar et al. 2012). According to an estimate, about 19,300 km of streams and rivers, and about 720 km^2 of lakes and reservoirs worldwide are adversely affected by mine effluents (Johnson and Hallberg 2005). The problem of AMD is reported in India as well, especially in the coalfields of India; however, quantitative estimate about the extent of impact of AMD on water bodies is not available.

10.1.2 Heavy Metal Contamination and Leaching

Metals in the underground bed rocks are exposed to air and water by mining activities. When these rocks and minerals are exposed in the aerobic environment, associated heavy metals are released upon weathering, which contaminate the surrounding ecosystem. The major heavy metals and metalloid introduced through mining activities are As, Co, Cu, Cd, Pb, Cr, Hg and Zn. After mineralization, these are leached out and are carried downstream by water. Although these metals are immobile under neutral pH conditions, low pH of the water due to acid mine drainage, accelerates the leaching of these metals. In BacKan province of North Vietnam, mining activities has been found to contaminate soil, water bodies and plants with heavy metals (Bui et al. 2016). Soils from vegetable fields in the mining areas were contaminated with Cd, Pb and As; while irrigation water was contaminated with Pb. About 71% and 44% of the vegetable samples grown in the area contained Pb and As respectively more than the safe limits.

10.1.3 Pollution due to Processing Chemicals

The mining companies use large quantity of chemical agents such as cyanide or sulphuric acid to remove the waste and separate the target mineral from the ore during beneficiation/ enrichment process. During this process, these chemicals spill, leak, or leach from the mine site and carry metals to contaminate the nearby water bodies causing pollution. Most of these chemicals/metals are highly toxic to humans and wildlife.

10.1.4 Erosion and Sedimentation

Construction and maintenance of roads, open pits, and waste impoundments during mine development disturbs soil and rock. Due to poor management, prevention and control strategies, the incidence of soil and sediment erosion is a major problem in the mine area. The erosion of the exposed earth may carry substantial amounts of sediment into streams, rivers and lakes. These sediments can clog riverbeds and dams, smother watershed vegetation, wildlife habitat and aquatic organisms. Pandey et al. (2014) studied the impact of dumping of mine overburden near the banks of perennial Damodar River. They claimed that it caused positive topographic changes (up to 20 m) indicating adverse effects on its hydrological regime. Due to the extraction of coal, the negative relief change in the form of deep depression of up to 66 m was created within coal mines. These depression areas within the abandon mines became the water accumulation zones and caused pollution of surface and ground water resources.

10.2 Impacts due to Smelting Industry

Smelting is the process of separating a metal from the other impurities generally by heating processed ore concentrates in the presence of added reducing chemical compounds. The smelting operation produces metal and generates slag, a solid waste product. Air pollution is the major impact of smelting industry as it emits sulphur dioxide (SO_2) and toxic trace elements into the atmosphere. Emitted SO_2 into atmosphere acidifies soil through acid rain. Acidification of soil near smelting industries is commonly observed and this increases availability of heavy metals to contaminate food chain (Gunn 1996; Dudka and Adriano 1997; Hamilton et al. 2016). Easily vaporized trace elements (As, Cd, Hg) are also released from the smelters and get deposited in terrestrial ecosystem to contaminate agricultural land. Smelters of different metals emit about 77%, 76%, 74%, 71% of total atmospheric emission of Zn, As, Cd, Cu respectively through anthropogenic activities (Pacyna and Pacyna 2001) (Table 10.1). Copper smelting results maximum As and Cu emission into the atmosphere. After being emitted to the atmosphere, trace metals are subject to transport within air masses and a large part of this gets deposited on the land mass not only around the source but also in area far away, even crossing boundaries of the country/ continent (Pacyna and Pacyna 2001). The elemental contamination and acidification lower soil fertility and reduce soil biodiversity (Maxwell 1991). Contribution of copper smelting on SO_2 emission and consequently to acid rain is higher as compared to other metal smelting due to having higher sulfide: metal ratio in the ore. Although smelting industries also produce solid and liquid wastes, their extent of impact on land is smaller as compared to air emission. Installation of electrostatic precipitator technology in modern smelters has reduced gaseous emission of metals through capturing of dust particles.

10.3 Impacts due to Textile Industries

The textile industries are classified on the basis of the types of textile fiber they use, viz., cellulose fibres, protein fibres and synthetic fibres. The type of dyes and chemicals used to dye different types of fabrics are as below (Ghaly et al. 2014):

- *For dyeing cellulose fibres*: Reactive dyes (remazol, procion MX and cibacron F), direct dyes (congo red, direct yellow 50 and direct brown 116), naphthol dyes (fast yellow GC, fast scarlet R and fast blue B) and indigo dyes (indigo white, tyrian purple and indigo carmine).
- *For dyeing Protein fibres*: Acid dyes (azo dyes, triarylmethane dyes and anthraquinone dyes) and lanaset dyes (Blue 5G and Bordeaux B).
- *For dyeing synthetic fibres*: Dispersed dyes (Disperse yellow 218 and disperse navy 35), basic dyes (basic orange 37 and basic red 1) and direct dyes.

Table 10.1 An assessment of global emissions of trace metals (% of total) to the atmosphere from different anthropogenic activities worldwide

Metal	Combustion of fuels in stationer sources	Vehicular traffic	Copper production	Lead production	Zinc production	Pig iron & steel production	Cement production	Munic waste incineration	Sewage sludge incineration	Gold production
As	16	0	64	0	5	7	5	2	1	0
Cd	23	0	44	1	27	2	1	1	0	0
Cr	69	0	0	0	0	19	9	0	3	0
Cu	27	0	69	0	1	1	0	2	0	0
Hg	66	0	3	0	4	1	6	5	0	15
Pb	10	75	5	5	2	2	0	0	0	0
Ni	90	0	9	0	0	0	0	0	0	0
Zn	17	0	7	0	66	4	5	1	0	0
V	100	0	0	0	0	0	0	0	0	0
Se	89	0	7	1	2	0	0	0	0	0

Pacyna and Pacyna (2001)

Table 10.2 Characteristics of typical untreated textile wastewater

Parameter	Range
pH	6–10
EC (mS cm^{-1})	4.43–8.71
BOD (mg L^{-1})	80–6000
COD (mg L^{-1})	150–12,000
Total suspended solids (mg L^{-1})	15–8000
Total dissolved solids (mg L^{-1})	2900–3100
Chloride (mg L^{-1})	1000–6000
Free Chlorine (mg L^{-1})	<10
Sodium (mg L^{-1})	975–2185
Potassium (mg L^{-1})	11–19
Fe (mg L^{-1})	0.017–0.163
Pb (mg L^{-1})	0.011–0.061
Cu (mg L^{-1})	0.006–0.311
Cr (mg L^{-1})	0.015–7.854
F (mg L^{-1})	0.7–2.2
Ca (mg L^{-1})	12–28
Mg (mg L^{-1})	13–29
Mn (mg L^{-1})	0.001–0.022
Oil & grease (mg L^{-1})	10–30
NO_3-N (mg L^{-1})	<5
SO_4 (mg L^{-1})	600–1000
Total Nitrogen (mg L^{-1})	70–80
Color (Pt-Co)	50–2500

Hussain et al. (2004) and Eswaramoorthi et al. (2008)

Textile industry requires large volume water as well as uses a wide variety of chemicals during various processes. On an average, 200 L of water are required to produce 1 kg of textiles. Quite a large fraction of these chemicals and water are discharged as wastewater causing severe damage to the environment. It is estimated that about 10% of the dye compounds used in the dying process don't bind the fibres/fabrics and are released into the environment through waste water. General characteristics of untreated wastewater generated from textile industries are presented in Table 10.2.

Hazardness of the textile industry waste water arises due to presence of significantly higher concentration of salt, surfactants, ionic metals and their complexes, toxic organic chemicals and dye chemicals. Many of dyes released and their breakdown products are toxic, carcinogenic or mutagenic to life forms mainly because of carcinogens, such as benzidine, naphthalene and other aromatic compounds (Suteu et al. 2009; Zaharia et al. 2009). In absence of adequate treatment, these dyes can remain in the environment for a long period of time. For instance, the half-life of hydrolysed Reactive Blue 19 is about 46 years at pH 7 and 25 °C (Hao et al. 2000). Several heavy metals are constituents of certain group of azo-dyes used for coloring which are often highly toxic and carcinogenic. Copper, Cr, Ni and Pb compounds are common inorganic pigments used.

Appropriate effluent treatment plants effectively remove considerable amount of organic compounds and most of the heavy metals. However, salts of alkali metals as well as certain types of azo-dyes (recalcitrant to microbial breakdown) are released into environment even after wastewater treatment. As a result of irrigation with effluent from dye and textile industries, considerable accumulation of total organic dyes in cultivated soil had been observed and such accumulation has also been transported in plant tissue (Zhou 2001).

Most of the effluents find its way into surface water bodies (rivers, lakes etc.) which are used in many cases, for irrigation by farmers across the downward streamline affecting soil quality. Increase in salinity, sodicity, heavy metal contents are common direct impact on soil fertility. Azo-dyes have been found to affect adversely the activities of beneficial microorganisms involved in nutrient transformation, mineralization and cycling processes related to soil fertility. Certain azo-dye had been found to affect adversely the growth of atmospheric N fixing cyanobacterium *Anabaena* sp. (Hu and Wu 2001). Sulfonated azo dye has significant adverse effect on urease activity, ammonification and nitrification rates in soil and therefore reduces nitrogen use efficiency in crop production (Khalid et al. 2009). Phytotoxicty of some of the dyes has also been reported by many researchers (Kalyani et al. 2008; Ayed et al. 2011). Development of soil salinity results in high demand of irrigation water, poor crop growth (Saha 2005) and crop cover which indirectly enhances soil erosion during high rainfall events. Increase in sodicity (due to high Na) results in degradation of soil physical conditions like compaction, decrease in infiltration rate (causing more soil erosion) besides Na injury to crops. Such impacts are more severe in areas receiving less rainfall and in soils with high clay content (i.e. heavy texture).

10.4 Impacts of Paper and Pulp Industry on Soil Quality

Paper and pulp industry is another category of highly polluting industry due to discharge of high volumes of densely colored toxic waste affluent. Worldwide paper is mostly produced from cellulose fibres. Less than two third comes from wood, one third comes from recycled paper and 5% comes from non-wood sources. Principal industrial processes in the paper production are pulping, bleaching and paper making.

Some components of the additives used in the paper making process (i.e. acid or alkali, chlorine, hypochlorite), in maintaining equipment (i.e. PCB in electrical equipment) or of the raw materials used in the industry (i.e. fuel oil) can pollute the soil at the plant site and/or at the dump sites used for disposing solid wastes. Waste water from the industry is rich in dissolved solids such as chloride and sulfate of

Table 10.3 Physico chemical parameters of paper industry effluent

Parameters	Mean value+SD
Colour	Light white
Odor	Un pleasant
PH	8.56 + 0.33
Temp°C	30.33 + 1.52
Ec(μs)	1300 + 26.0
TDS(mg/l)	2950 + 38.0
BOD(mg/l)	380 + 36.55
COD(mg/l)	1830 + 81.85
Na(mg/l)	260.99 + 7.06
Ca(mg/l)	299.66 + 19.09
Mg(mg/l)	184.2 + 12.48

Kuzhali et al. (2012)

Na, K, Ca and varying amount of suspended organic materials, including significant amount of organo-chlorine compounds (Table 10.3). The BOD/COD ratio in paper industry effluent is normally 1:4 or less, which indicates low biodegradability of the organic substances. Depending on the type of chemicals used for bleaching, the waste stream from the bleaching process may contain chlorine compounds and organics. The mixture of chemicals may result in the formation of a number of toxic chemicals (such as dioxins, furans, and chlorinated organics). Although the effluent is generally subjected to a waste water treatment plant, several chemicals simply "pass through" (i.e. the treatment plant does not significantly reduce the concentrations of these pollutants) and accumulate in the rivers and lakes which receive the treatment plant discharges.

In most of short term research studies, diluted paper mill effluent has been found beneficial to many crops, as it supplies significant amount of plant nutrients (Chhonkar et al. 2000). However, it has been found to affect soil health through development of salinity, alkalinity and degradation of soil physical characteristics like water holding capacity and bulk density (Giri et al. 2014; Sharma et al. 2014). Germination of seed and seedling growth have been found adversely affected due to irrigation with such effluent water (Chhonkar et al. 2000). Magnitude of adverse impact of paper mill industry effluent on soil depends on climate, soil type as well as dilution of effluent water in surface water bodies before irrigation. While dry climatic region promotes accumulation of salts, high rainfall in humid region facilitates washing of the salts away from the surface layer and improves the soil health. Similarly, sandy and sandy loam soils with low CEC and high porosity can be reclaimed easily by soil washing with rain water or good quality irrigation as compared to heavy textured black soils with less macro porosity.

10.5 Impact of Pesticide Industry on Agroecosystem

Globally, India is the fourth largest producer of agrochemicals, after United States, Japan and China. More than 60 technical grade pesticides are being manufactured by more than 125 industries in India. The principal manufacturing steps are (a) preparation of process intermediates; (b) introduction of functional groups; (c) coupling and esterification; (d) separation processes, such as washing and stripping; and (e) purification of the final product. Each of these steps may generate air emissions, liquid effluents, and solid wastes. The principal air pollutants are volatile organic compounds (VOCs) and particulate matter (PM). Wastewater from pesticide manufacturing industries originate from cleaning activities after batch operation during the synthesis processes. Composition/properties of wastewater from pesticide industries vary widely: chemical oxygen demand (COD) in the range of 0.4–73, 000 mg L^{-1}; oil and grease content in the range of 1–13,000 mg L^{-1}; total suspended solids in the range of 4–43,000 mg L^{-1}. They may also contain toxic organics and pesticide residues which pose threats to the quality of surface and groundwater. Wastewater from agrochemical/pesticide manufacturing industries was reported to have pesticide contamination levels as high as 500 mg L^{-1} (Chiron et al. 2000). Major solid wastes of concern include process and effluent treatment sludge, spent catalysts, and container residues. Approximately 200 kg of waste is generated per metric ton of active ingredient manufactured.

Pesticides entering through wastewater are harmful to many organisms including microorganisms and consequently, impact on environment in many ways.

- Loss in biodiversity: Continuous pesticides entry into the soil system causes suppression of sensitive microbes and emergence/promotion of growth of resistant groups. It is known that a range of specific soil microorganisms are playing an important role in the suppression of soil-borne plant diseases as well as in plant growth promotion (Kennedy and Smith 1995). Systems with reduced microbial diversity generally suffer from impaired functioning in respect of the sustainability of microbial nutrient transformation and plant disease suppression (Read 1998).
- Impaired nutrient cycling processes: Urea hydrolysis, nitrification, N_2 fixation, organic matter (and associated N, P, S, micronutrients) mineralization processes in soil are inhibited (Shaw and Robinson 1960; Potera 2007; Lo 2010). As a result, nutrient use efficiency from manures and fertilizers decreases necessitating additional application of fertilizer nutrients to maintain higher yields.
- Development of resistance among targeted organisms towards pesticides: Due to this, many groups of pesticides lose efficacy in controlling diseases/pests, requiring either higher doses of application and even newer formulation.

10.6 Impact of Thermal Power Plants

India's power requirement is largely met through thermal power production. India has quite a significant number (>85) of thermal power plants operating all over the country. Coal based power plants significantly impact the local environment. This sector uses huge amount of coal as fire. However major problem associated with Indian coal is its high ash content. Direct impacts resulting from construction and ongoing operations of thermal power plants include:

- Air Pollution due to thermal power plants occurs due to atmospheric release of particulates, ozone, sulphur oxides, nitrous oxides, volatile organic compounds (VOC), toxic metals and other hazardous chemicals. Atmospheric fall-out of heavy metals originated from these plants into the agricultural land can be sometimes quite significant. Considering average emission factors proposed by Pacyna and Pacyna (2001), thermal power plants in India also emit annually 67.8 t of As, 33.9 t of Cd, 576.3 t of Cr, 474.6 t of Cu, 678 t of Ni, 339 t of Pb, 339 t of V and 271.2 t of Se. Ozone, SO_2 and NO_2 are also known phytotoxicants. Also, such air pollution, if occurs with severe intensity, may result in acid rain causing damage to plants and soil.
- Water Pollution occurs due to discharge of effluents in local water streams, rivers and groundwater and percolation of hazardous materials from the stored fly ash.
- Land Degradation occurs due to alterations of land used for storing fly ash. During the power generation huge quantity of ash is disposed and dumped on the valuable land area. Approximately an acre of land is required per MW of installed thermal capacity for ash disposal. Many of India's thermal power plants are located on non-forest area, and therefore, resettlement and rehabilitation problems have not occurred much. Even then the adverse impacts of emission and ash dumping on flora and fauna, wild life and human life have been found quite significant and sometimes alarming. The most serious land related problem of thermal power plants is its requirement for ash disposal along with percolation of hazardous elements to ground water through disposal in ash ponds.
- Soil contamination with heavy metals due to activities of thermal power plant running on coal has been reported. Özkul (2016) reported varying degree of contamination of surface soils with As, Hg, Ni, and Cr around thermal power plants as evidenced by high geoaccumulation index and enrichment factor.
- Aerial Deposition of emitted trace and toxic metals on agricultural land has been widely reported causing adverse effect on soil ecology, nature and extent of which depends on the distance from power plants, duration of deposition and their concentrations in coal (Pacyna and Pacyna 2001; Rind et al. 2013). Fly-ash generated from thermal power plants disturbs microbial activities in the soil by decreasing their population and also by decreasing the enzymatic activities like phosphatase, sulphatase, dehydrogenase and invertase responsible for several nutrient transformation processes (Pitchel and Hayes 1990; Wong and Wong 1986). Addition of fly ash particularly to sandy soil greatly inhibited the

microbial respiration, enzymatic activity and soil N cycling processes like nitrification and N mineralization (Garau et al. 1991). Depending on the coal type, fly ash may contain significant variable concentrations of essential plant nutrient elements as well as toxic heavy metals (Table 10.4). Fly ash has been found to contaminate soils with heavy metals (Swamy et al. 2010); however, their uptake by plants has rarely been observed. In Shaoguan City of Guangdong Province (China), substantial build-up of As, Cd, Cu, Pb, and Zn in soil had been recorded in the predominant wind direction of the power plant (Wang et al. 2016). Average Cd (0.69 mg kg^{-1}) and Pb (0.39 mg kg^{-1}) contents in rice grain grown in the area had exceeded their maximum permissible limits (0.2 mg kg^{-1} for both the element) in foods of China indicating potential health risk to the local residents.

- Coal contains a little amount of radioactive elements (on an average about 1 ppm Uranium and 2 ppm thorium) which gets concentrated in ash and also emitted in the atmosphere. It has been reported that threat due to radioactivity nearby coal based thermal power plants is higher than that near pressurized-water or boiling-water reactors of nuclear power plants (McBride et al. 1978). Levels of radioactivity (due to naturally occurring radioactive metals ^{232}Th, ^{228}Ra, ^{238}U, ^{226}Ra, and ^{210}Pb) in the ash were up to five times higher than in normal soil, and up to 10 times higher than in the parent coal itself as combustion concentrates radioactivity (Lauer et al. 2015). Therefore, concern on build up of naturally occurring radioactive materials in soil has often been raised with repeated application of fly ash as soil amendment during crop husbandry. Large number of demonstration experiments conducted under Fly Ash Mission of Govt. of India revealed that radioactivity levels in the soils were within the safe limits and the grains, straws and vegetables produced from fly ash amended soils are safe for human and animal consumption (Table 10.5). However, information on impact of repetitive fly ash application on radionuclides accumulation in soils is practically absent. Such information is required before recommending this waste as soil amendment in crop production.

10.7 Impact of Nuclear Power Plants

A nuclear power plant starts disturbing the environment during its construction. New roads, increasing traffic flow in the existing roads, excavations, cutting trees and other plants, are some of the adverse impacts on land to be expected during the construction of a power plant. Site selection for nuclear power plants should be carefully made to avoid, or minimize to the extent possible, most of those impacts. The environmental impact of nuclear power plants results from the nuclear fuel cycle, operation, and the effects from nuclear accidents.

Table 10.4 Physical and chemical properties of fly ash

Properties	Contents[a] (global review)	Contents in fly ash generated in India[b]
Physical properties		
Bulk density (g cm^{-3})	<1.0	0.85–1.2
Water-holding capacity (%)	35–40	45–60
Porosity (%)	50–60	45–55
Plant nutrients contents (%)		
N	–	–
P	0.004–0.8	–
K	0.15–3.5	0.03–0.7
Ca	0.11–22.2	0.14–5.7
Mg	0.04–7.6	0.006–0.3
S	0.1–1.5	
Al	0.1–17.3	14.2–23.3
Na	0.01–2.03	0.05–0.31
Fe	36–1333	3.3–6.4
Si		17.8–29.4
Trace element contents (mg kg^{-1})		
Mn	58–3000	500–750
Zn	10–3500	50–150
Cu	14–2800	40–80
B	10–618	17–38
As	2.3–6300	1.0–4.0
Cd	0.7–130	5–10
Co	7–520	10–50
Cr	10–1000	50–150
Hg	0.02–1.0	BDL
Mo	7–160	2.2–6.7
Ni	6.3–4300	50–150
Pb	3.1–5000	10–70
Se	0.2–134	0.6–2.6
Radioactivity levels in fly ash (Bq kg^{-1})		
^{226}Ra	–	30–110
^{228}Ac	–	30–110
^{40}K	–	180–500

Source: [a] Basu et al. (2009); [b] NTPC (2015)

Radioactive effluents released from nuclear power plants are normally monitored and controlled to minimize exposure of the public to ionizing radiation. The total amount of radioactivity released through gaseous and liquid radiological effluents depends on the technologies used during installation of power plant, the regulatory requirements, and the plant's performance. As far as contamination through liquid release is concerned, tritium (^{3}H) is by far the most dominant

Table 10.5 Daily intake ratio of radio nuclides through eatables grown with fly ash

Food items	Maximum radionuclide level (Bq kg^{-1})		
	^{40}K	^{226}Ra	^{228}Ac
Grain	95	0.75	1.0
Vegetable	85	0.50	0.8
Daily intake of radionuclide (Bq kg^{-1}) based on average dietary intake			
Grain	38	0.30	0.4
Vegetable	17	0.10	0.16
Total	55	0.40	0.56
Safe limit (Bq kg^{-1})	173.34	1.00	3.67

Kumar and Jha (2014)

radionuclide released routinely. In terms of radioactivity, liquid releases of all other radionuclides combined correspond only about 1% of that released by tritium discharges. The movement of radionuclides in the environment depends on several factors, like point of release, rate of release, chemical form, natural constituents of the surrounding environment, among others. Tritium (half-life of 12.3 years) emits radioactive beta particles. Once tritium is inhaled or swallowed, its beta particles can bombard cells. If a particle zaps a DNA molecule in a cell, it can cause genetic mutation. The primary concern for tritium release is its presence in drinking water, irrigation water and other aquatic environment. Biological magnification of tritium in crops and animals consumed for food can occur when such contaminated water is used for irrigation and drinking.

A nuclear power plant should store gas and liquid radioactive materials generated during normal operation in tanks for a period of time to allow for the radioactivity level to decrease before being treated and/or released in a planned, monitored way. This is required to keep the amount of radioactive material in releases low and within regulatory limits. However, adverse impact of nuclear power plants becomes significant under emergency situation where accidental release of large amount of harmful radionuclides like iodine (^{131}I), and caesium (^{134}Cs, ^{137}Cs) occur. Radioactive ^{131}I has a half-life of 8 days and decays naturally within weeks. It may only be present in the environment for several weeks after the release has occurred. Radioactive caesium (^{134}Cs has a half-life of approximately 2 years and ^{137}Cs has a half-life of approximately 30 years) can persist in the environment for longer time. For example traces of caesium-137 (^{137}Cs) have been found in the environment for several years after the accident at the power station at Chernobyl and Fukushima Daiichi Power Plant.

The radioactive material in the air can settle or deposit on soil and local vegetation. Cow milk can also get contaminated when contaminated pasture grasses are fed to cattle population. Therefore, samples of air, soil, food products, milk etc. from nearby areas of the power plants are required to be analyzed periodically. All water bodies like lakes, ponds, and streams nearby nuclear power plants are required to be inspected periodically for the presence of radioactive material; and more so is the water samples from the liquid waste stream released from the plant.

10.8 Impact of Hydropower Industry

Hydroelectric power plants include both massive hydroelectric dams and small run-of-the-river hydroelectric plants. The size of the reservoir created by a hydroelectric project can vary widely, depending largely on the capacity of the hydroelectric generators and the topography of the land. Hydroelectric plants in flat areas require much more land than those in hilly areas or canyons where deeper reservoirs can hold more volume of water in a smaller space. Flooding land for a hydroelectric reservoir has an extreme environmental impact as it destroys forest, wildlife and human habitat, agricultural and other land.

Estimates for life-cycle GHGs emissions from hydroelectric plants built in tropical areas or temperate peat lands are quite high. After the reservoir is flooded, the vegetation and soil organic matter in the area decompose and release both carbon dioxide and methane. The exact amount of emissions depends greatly on site-specific characteristics. Current estimates suggest that life-cycle emissions can be over 225 gram of carbon dioxide equivalent per kilowatt-hour (IPCC 2011). Thus, estimates of life-cycle global warming emissions for natural gas generated electricity are between 272 and 900 gram of CO_2 equivalent per kilowatt-hour and estimates for coal-generated electricity are between 635 and 1633 gram of CO_2 equivalent per kilowatt-hour (IPCC 2011).

10.9 Impact of Distillery Industry

Distillery is the agro-based industry and uses mainly molasses (with few using grain or fruits) to produce alcohol. However, this industry produces considerable amount of liquid effluent (called spent wash) which is estimated at the rate 12–15 times of volume of alcohol produced. It is one of the most complex, troublesome and strongest organic & acidic (pH 4.3–5.3) industrial effluents, having extremely high COD, BOD values, total dissolved and suspended solids, high contents of K, Cl^-, and SO_4^{-2}. Therefore, such spent wash has significant environmental impact, particularly on land and water resources. Treatment of spent wash is generally accomplished by two different routes viz.; (a) concentration in solar ponds followed by incineration, and (b) anaerobic digestion with biogas recovery (biomethanation) followed by concentration in solar pond. Even after biomethanation, organic matter and mineral content in the effluent remains very high and at more than the permissible limits. Evaporation pond with quite a large area is required for producing sludge, thereby destroying fertile land. Moreover, in several cases, effluents percolate to contaminate groundwater with salts and sometimes with organics, extent of which depends upon nature of geological formation. Light texture of soil lining the pond and fractured underground rocks facilitate groundwater contamination with distillery effluent from lagoon pond. During rainy season, high rainfall events can wash out distillery effluents and sludge to contaminate surface water bodies (thus impacting aquatic life) and land masses. Use of contaminated water (both surface and groundwater) for irrigation to agricultural land may have adverse impact on soil quality by increasing salinity and sodicity.

10.10 Impact of Infrastructure Projects

Airports, road, bridges, transportation, communication, sewage treatment and disposal, drinking water and electricity distribution systems are all examples of infrastructure. These infrastructural projects normally require high-cost investments; and of course, are vital to a country's economic development and prosperity. However, infrastructure projects also have considerable impact on soil resources, the nature and extent of which depend on the type of projects and associated activities.

Soil degradation is one of the most significant impacts of highway construction, which can alter the soil developmental level and soil structure, damage soil texture, lead to loss of the organic matter and inorganic nutrients in soil, reduce soil fertility and water holding capacity, deteriorate soil quality, cause soil erosion, and limit soil productivity (Zhao 2002). Soil is disturbed heavily by highway construction activity, such as slope fill and slope cut, invasion of a large number of artificial soils, and destruction of original soil and humus layer (Xu et al. 2006). Studies have shown that soil pH and bulk density are higher, soil moisture and nutrient levels are lower near road, and soil layer thickness is altered, which is related to destruction of soil layer from road construction, dust origin from traffic and destruction of vegetation (Auerbach et al. 1997). Road construction also causes sheet, rill and gully erosions, which are mainly related to soil type, slope length and change in the direction of water flow due to construction. Therefore, soil quality changes significantly near highway: the contents of clay particles, soil organic matter and nutrient content declines while pH increases with different degrees and ranges (Zhao et al. 2007). Parking lots treated with coal tar based seal coat are a major source of cancer causing contaminants that can pollute air, soil, water and wildlife, posing a significant health risk to humans who may breathe, drink or consume them through fish and other foods. Coal-tar based sealants contain high levels of polycyclic aromatic hydrocarbons (PAHs). The persistent cancer causing PAHs widely pollute the environment, wildlife and people.

Construction of buildings using bricks also has high impact on precious soil resources. In India, bricks are prepared from top fertile soils and thereby destroying soil productivity which has been built-up through millions of years of soil forming processes. As top soil layers are most fertile, impact of brick-making on loss of soil productivity is huge. Construction of road, buildings, industries make a concrete sealing of land surface. As land surface facilitates percolation of rainwater through profile to recharge groundwater and reduces runoff, it's sealing due to infrastructure and housing projects leads to impeded underground recharge and enhanced partitioning of rainwater towards surface flow causing floods. Destruction of vegetation to facilitate different infrastructure projects leads to clearing of land and facilitation of soil loss through runoff water during monsoon season.

References

Adewole MB, Adesina MA (2011) Impact of marble mining on soil properties in a part of Guinea savanna zone of southwestern Nigeria. Ethiopian J Environ Stud Manag 4(2):1–8

Auerbach NA, Walker MD, Walker DA (1997) Effects of roadside disturbance on substrate and vegetation properties in Arctic Tundra. Ecol Appl 7:218–235

Ayed L, Mahdhi A et al (2011) Decolorization and degradation of azo dye Methyl Red by an isolated *Sphingomonas paucimobilis*: biotoxicity and metabolites characterization. Desalination 274:272–277

Banerjee SK, Mishra TK, Singh AK et al (2004) Impact of plantation on ecosystem development in disturbed coal mine overburden spoils. J Trop For Sci 16:294–307

Basu M, Pande M, Bhadoria PBS, Mahapatra SC (2009) Potential fly-ash utilization in agriculture: a global review. Prog Nat Sci 19:1173–1186

Bell FJ, Bullock SET, Hälbich TFJ et al (2001) Environmental impacts associated with an abandoned mine in the Witbank Coalfield, South Africa. Int J Coal Geol 45:195–216

Blahwar B, Srivastav SK, Smeth JB (2012) Use of high resolution satellite imagery for investigating acid mine drainage from artisanal coal mining in North-Eastern India. Geochem Int 27:231–247

Bui ATK, Nguyen HTH, Nguyen MN et al (2016) Accumulation and potential health risks of cadmium, lead and arsenic in vegetables grown near mining sites in Northern Vietnam. Environ Monit Assess 188(9):525. doi:10.1007/s10661-016-5535-5

Chhonkar PK, Datta SP, Joshi HC, Pathak H (2000) Impact of industrial effluents on soil health and agriculture – Indian experience: Part 1 – Distillery and paper mill effluents. J Sci Ind Res 59:350–361

Chiron S, Fernandez-Alba A, Rodriguez A, Garcia-Calv E (2000) Pesticide chemical oxidation: state-of-the-art. Water Res 34:366–377

Dudka S, Adriano DC (1997) Environmental impacts of metal ore mining and processing: a review. J Environ Qual 26:590–602

Dutta RK, Agrawal M (2002) Effect of tree plantations on the soil characteristics and microbial activity of coal mine spoil land. Trop Ecol 43:315–324

Ekka JN, Behera N (2011) Species composition and diversity of vegetation developing on an age series of coal mine spoil in an open cast coal field in Orissa, India. Trop Ecol 52:337–343

Eswaramoorthi S, Dhanapal K, Chauhan D (2008) Advances in textile waste water treatment: the case for UV-ozonation and membrane bioreactor for common effluent treatment plants in Tirupur, Tamil Nadu, India. Environmental technology awareness series 1–17

Garau MA, Dalmau JL, Felipo MT (1991) Nitrogen mineralization in soil amended with sewage sludge and flyash. Biol Fertil Soils 12:199–201

Ghaly AE, Ananthashankar R, Alhattab et al (2014) Production, characterization and treatment of textile effluents: a critical review. J Chem Eng Process Technol 5:182. doi:10.4172/2157-7048.1000182

Ghose MK (1996) Damage of land due to coal mining and conservation of topsoil for land reclamation. Environ Ecol 14:466–468
Ghosh AK (1990) Mining in 2000 AD – challenges for India. J Inst Eng (India) 39:1–11
Giri J, Srivastava A, Pachauri SP, Srivastava PC (2014) Effluents from paper and pulp industries and their impact on soil properties and chemical composition of plants in Uttarakhand, India. J Environ Waste Manag 1: 26–30
Gitt MJ, Dollhopf DJ (1991) Coal waste reclamation using automated weathering to predict lime requirement. J Environ Qual 20:285–288
Gould AB, Hendrix JW, Ferriss RS (1996) Relationship of mycorrhizal activity to time following reclamation of surface mine land in western Kentucky. I Propagule and spore population densities. Can J Bot 74:247–261
Gray NF (1996) Field assessment of acid mine drainage contamination in surface and ground water. Environ Geol 27:358–361
Gunn JM (1996) Restoring the smelter-damaged landscape near Sudbury, Canada. Restor Manag Notes 14:129–136
Hamilton JG, Farrell RE, Chen N et al (2016) Characterizing Zinc speciation in soils from a smelter-affected boreal forest ecosystem. J Environ Qual. doi:10.2134/jeq2015.03.0145
Hao OJ, Kim H, Chang PC (2000) Decolorization of wastewater. Crit Rev Environ Sci Technol 30:449–505
Henningsen FP, Urushadze TF, Steffens D, Kalandadze BB, Narimanidze EI (2010) Uptake of heavy metals by food crops from highly-polluted Chernozem-like soils in an irrigation district south of Tbilisi, eastern Georgia. Agron Res 8:781–795
Hu T, Wu SC (2001) Assessment of the effect of azo dye RP2B on the growth of a nitrogen fixing cyanobacterium – Anabaena sp. Bioresour Technol 77:93–95
Hussain J, Hussain I, Arif M (2004) Characterization of textile wastewater. J Ind Pollut Control 20:137–144
IPCC (2011) IPCC special report on renewable energy sources and climate change mitigation. In: Edenhofer O, Pichs-Madruga R et al (eds) Working Group III of the Intergovernmental Panel on Climate Change. Cambridge University Press, Cambridge/New York, p 1075
Johnson DB, Hallberg KB (2005) Acid mine drainage remediation options: a review. Sci Total Environ 338:3–14
Kalyani DC, Patil PS et al (2008) Biodegradation of reactive textile dye Red BLI by an isolated bacterium Pseudomonas sp. SUK1. Bioresour Technol 99:4635–4641
Kennedy AC, Smith KL (1995) Soil microbial diversity and the sustainability of agricultural soils. Plant Soil 170:75–86
Khalid A, Arshad M, Crowley DE (2009) Biodegradation potential of pure and mixed bacterial cultures for removal of 4-nitroaniline from textile dye wastewater. Water Res 43:1110–1116
Kumar V, Jha GK (2014) Use of fly ash in agriculture: Indian scenario. International workshop on agricultural coal ash uses, 27–29 May 2014. Agricultural Center, Bet Dagan, Israel
Kuzhali SS, Manikandan N, Kumuthakalavalli R (2012) Physico chemical and biological parameters of paper industry effluent. J Nat Prod Plant Resour 2:445–448
Lauer NE, Hower JC, Hsu-Kim H, Taggart RK, Vengosh A (2015) Naturally occurring radioactive materials in coals and coal combustion residuals in the United States. Environ Sci Technol 49:11227–11233
Lo CC (2010) Effect of pesticides on soil microbial community. J Environ Sci Health 45:348–359
Maxwell CD (1991) Floristic changes in soil algae and cyanobacteria in reclaimed metal-contaminated land at Sudbury, Canada. Water Air Soil Pollut 60:381–393
McBride JP, Moore RE, Witherspoon JP, Blanco RE (1978) Radiological impact of airborne effluents of coal and nuclear plants. Scientia 202:1045–1050
NTPC (2015) Use of ash in agriculture – grey to green revolution, Fly Ash Utilization Programme, TIFAC, DST. Published by NTPC Limited, New Delhi
Özkul C (2016) Heavy metal contamination in soils around the Tunçbilek Thermal Power Plant (Kütahya, Turkey). Environ Monit Assess 188:284

Pacyna JM, Pacyna EG (2001) An assessment of global and regional emissions of trace metals to the atmosphere from anthropogenic sources worldwide. Environ Rev 9:269–298

Pandey B, Agrawal M, Singh S (2014) Coal mining activities change plant community structure due to air pollution and soil degradation. Ecotoxicology 23:1474–1483

Pitchel JR, Hayes JM (1990) Influence of fly ash on soil microbial activity and population. J Environ Qual 19:593–597

Potera C (2007) Agriculture: pesticides disrupt nitrogen fixation. Environ Health Perspect 115: A579

Read DJ (1998) Plants on the web. Nature 396:22–23

Rind AM, Mastoi GM, Hullio AA (2013) Impacts of Jamshoro thermal power station on soil of the surrounding area. Indian J Sci Res Technol 1:65–71

Saha JK (2005) Changes in salinity and sodicity of soils with continuous application of contaminated water near industrial area. J Indian Soc Soil Sci 53:612–617

Sharma V, Garg UK, Arora D (2014) Impact of pulp and paper mill effluent on physico-chemical properties of soil. Arch Appl Sci Res 6:12–17

Shaw WM, Robinson B (1960) Pesticide effects in soils on nitrification and plant growth. Soil Sci 90:320–323

Singh AN, Singh JS (2006) Experiments on ecological restoration of coal mine spoil using native trees in a dry tropical environment, India: a synthesis. New For 31:25–39

Suteu D, Zaharia C, Bilba D, Muresan A, Muresan R, Popescu A (2009) Decolorization wastewaters from the textile industry – physical methods, chemical methods. Ind Text 60:254–263

Swamy N, Dash N, Nahak G et al (2010) Effect of coal fly ash on growth, biochemistry, cytology and heavy metal content of *Allium cepa*. NY Sci J 3:10–16

Wang X, Zeng X, Chuanping L et al (2016) Heavy metal contaminations in soil-rice system: source identification in relation to a sulfur-rich coal burning power plant in Northern Guangdong Province, China. Environ Monit Assess 188:460–471

Wong MH, Wong JWC (1986) Effects of flyash on soil microbial activity. Environ Pollut A 40:127–244

Xu X, Zhang K, Kong Y et al (2006) Effectiveness of erosion control measures along the Qinghai–Tibet highway, Tibetan plateau, China. Transportation Res Part D 11:302–309

Zaharia C, Suteu D, Muresan A, Muresan R, Popescu A (2009) Textile wastewater treatment by homogenous oxidation with hydrogen peroxide. Environ Eng Manag J 8:1359–1369

Zhao JQ (2002) Highway traffic and environmental protection (in Chinese). China Communications Press, Beijing, pp 34–39

Zhao SQ, Cui BS, Gao LN, Liu J (2007) Effects of highway construction on soil quality in the Longitudinal Range-Gorge Region in Yunnan Province. Chin Sci Bull 52:192–202

Zhou Q (2001) Chemical pollution and transport of organic dyes in water soil crop system of the Chinese Coasts. Bull Environ Contam Toxicol 66:784–793

Chapter 11
Status of Soil Pollution in India

Abstract Industrial sector in India is witnessing rapid growth since the last decade of twentieth century with reforms in economic laws and with establishment of special economic zones (SEZ). Such rapid industrial growth has also increased threat to the environment. In spite of great difficulty in its remediation in comparison with polluted air and water, soil pollution as a threat to human life is by and large ignored at national level in India due to lack of comprehensive information on the subject. Though coordinated effort on assessment of soil pollution is absent at national level, sporadic information has been generated by several researchers on various aspects of pollution affecting soil quality. This chapter analyses these information and attempts to assess the quantum of threat being faced by agroecosystem in the country. It indicates that soil resources are facing threats from deliberate use of contaminated organics, amendment materials and irrigation water or from atmospheric depositions, spillage of effluents etc. Nature pollutants varies from salts, toxic metals, metalloids, persistent organics with varying degree of toxicity and may be of both industrial and geogenic origins.

Keywords Soil • Pollution • India • Heavy metals • Groundwater • Organic pollutants • Salinity

Over the years agriculture has been the major source of livelihood of the Indian population. During the period from 1940–1970, there was restrictive growth of private sector and Gross Domestic Product (GDP) grew at a rate of 1.4% per annum. In 1994–1995 the industrial sector registered remarkable 8.4% growth and its contribution in GDP increased thereafter. However, rapid industrial growth was also associated with surge in release of toxic effluents in the environment, including land and water bodies. Entry of pollutants directly (release of effluents on land) or indirectly (use of polluted water as irrigation to crops) has been reported to contaminate vast area of soil resources and groundwater bodies, affecting crop production as well as human and animal health through food contamination. As per the latest available estimate, about 33,900 million litres per day (MLD) urban waste water and 23,500 MLD industrial waste water was generated in our country during 2009, polluting water and soil resources of India. Small scale industries with less or no effluent treatment plants are considered bigger polluter than big industries. In another category of soil pollution, pollutant chemicals enter into the soil body

J.K. Saha et al., *Soil Pollution - An Emerging Threat to Agriculture*, Environmental Chemistry for a Sustainable World 10, DOI 10.1007/978-981-10-4274-4_11

through processes uncontrollable by land owners. For example, air borne pollutants enter the soil body from the emissions from several industries, power plants, vehicles, radioactive and toxic chemical fallouts during disasters. Gas-dust releases into the atmosphere under high temperature technological processes (e.g. power plants, metal smelting, the burning of raw materials for cement, etc.), waste incineration, vehicular activities and fuel combustion; and get deposited on land surface far away from source of generation. All these pollutant activities call for a need to develop a database on pollutant element contamination in soil and water so that remedial measures can be undertaken to protect our soil resources.

11.1 Land Use Pattern in India

India has a total land area of approximately 328 million hectares. Land utilization statistics are available for almost 93% of the entire area that is around 306 million hectares. The land use pattern of India is as below:

Agricultural land	: 43.6%
Permanent pastures and meadows	: 4.6%
Cultural wastelands	: 12.2%
Forests	: 10.7%
Barren and uncultivable land	: 8.4%
Urban land	: 5.3%
No information available	: 5.2%

The mounting population and advanced standards of living have resulted in an ever increasing demand for residential land, both in villages and towns. Land is also needed to develop industry, commerce, transport and recreational facilities. Area under non-agricultural use includes all lands occupied by buildings, roads, railways, industrial establishments or that under water, e.g. rivers and canals etc. The increasing area under non-agriculture use driven by population growth is of major concern. Land under non-agricultural uses increased by 11.73 million hectares during 1950–2000. The increasing population and change in life style would further aggravate the pressure on agricultural land in the future.

11.2 Soil Degradation in India

Soil degradation is the decline in quantity and quality of soil. It includes: (a) Erosion by wind and water; (b) Biological degradation (the loss of humus and plant/animal life); (c) Physical degradation (loss of structure, changes in permeability); and (d) Chemical degradation (acidification, declining fertility, changes in pH, salinisation and chemical toxicity). Soil degradation encompasses several issues at various spatial time scale. The major types and underlying causes of soil degradation is presented Table 1.2 in Chap. 1. This indicates that soil degradation estimates in India is almost entirely focused on loss of soil and its productivity due to either natural processes or accelerated natural processes of erosion and ionic movements through faulty soil and water management. There is no denying fact that arresting soil loss and its productivity through above degrading forces are extremely important for ensuring food security in the country. Quality of human health is often linked with soil quality (Parr et al. 2009). Therefore, soils around cities and industrial areas having high population density play an important role on human health, as considerable fraction of the food requirement are met from the agricultural activities surrounding the area. An estimate of area under soil degradation (erosion, inundation, salinity build up, acidity) has been made mostly using satellite images and remote sensing data as well as through extensive network of organizations & Institutes like Soil and Land Use Survey of India and ICAR-National Bureau of Soil Survey and Land Use Planning. However such quantitative assessment of area under pollution by industrial, urban and mining activities is only possible through information on geographical locations of such activities as well as through direct estimation of pollutants accumulation in soil and their transfer to food and organisms around the intense activity zones. Information generated on the pollutants accumulation in environmental samples related to agriculture and foods are described in the following sections.

11.3 Soil Pollution Due to Anthropogenic Activities

Although surge in India's economic growth aided by higher levels of industrialization has remained a subject of pride, there is also a huge concern for the environmental degradation that slowly but loudly being voiced out. CPCB identified critically polluted industrial areas and clusters or potential impact zone based on its Comprehensive Environmental Pollution Index (CEPI) rating. Forty three critically polluted zones were reported in the 16 states which have CEPI rating more than 70 (Table 11.1). Among the 43 sites, 21 sites exist in only four states namely Gujarat, Uttar Pradesh, Maharashtra and Tamil Nadu.

Information on soil pollution has been generated by research organizations in several of these critically polluted areas and such information is unevenly distributed. In some areas having very high CEPI rating like Haldia, Bhiwadi, Chandrapur, Singrauli, Bhiwadi, published information on the soil pollution in nearby agricultural areas is practically absent (Table 11.1). Information on type of pollutants and location of sites covered were highly skewed towards nearness (to the polluted area), facilities available and mandate of the organizations responsible for the study.

Table 11.1 Critically polluted industrial areas in India

State	Critically polluted industrial area	CEPI rating	Information available on soil pollution[a]
Andhra Pradesh	Vishakhapatnam	70.82	*
	Patancheru – Bollaram	70.07	***
Chhattisgarh	Kobra	83.00	***
Delhi	Nazafgarh drain basin	79.54	**
Gujarat	Ankleshwar	88.50	*
	Vapi	88.09	*
	Ahmedabad	75.28	–
	Vatva	74.77	*
	Bhavnagar	70.99	–
	Junagarh	70.82	–
Haryana	Faridabad	77.07	–
	Panipat	71.91	–
Jharkhand	Dhanbad	78.63	**
Karnataka	Mangalore	73.68	–
	Bhadravati	72.33	–
Kerala	Greater Cochin	75.08	–
Madhya Pradesh	Indore	71.26	–
Maharashtra	Chandrapur	83.38	–
	Dombivalli	78.41	–
	Aurangabad	77.44	–
	Navi Mumbai	73.77	–
	Tarapur	72.01	–
Orissa	Angul Talcer	82.09	***
	Ib Valley	74.00	–
	Jharsuguda	73.34	*
Punjab	Ludhiana	81.66	**
	Mandi Gobind Garh	75.08	–
Rajasthan	Bhiwadi	82.91	–
	Jodhpur	75.19	–
	Pali	73.73	**
Tamil Nadu	Vellore (North Arcot)	81.79	***
	Cuddalore	77.45	*
	Manali	76.32	*
	Coimbatore	72.38	***
Uttar Pradesh	Ghaziabad	87.37	**
	Singrauli	81.73	–
	Noida	78.90	–
	Kanpur	78.09	***
	Agra	76.48	*
	Varansi-Mirzapur	73.79	***

(continued)

Table 11.1 (continued)

State	Critically polluted industrial area	CEPI rating	Information available on soil pollution[a]
West Bengal	Haldia	75.43	–
	Howrah	74.84	*
	Asansol	70.20	–

CPCB (2009)
[a]'–': no information; '*': very few information; '**': few information; '***': moderate information

11.3.1 Entry of Sodium into Ecosystem and Increase in Soil Salinity and Sodicity

Industries, particularly those associated with chlor-alkali, textiles, glass, rubber production, animal hide processing and leather tanning, metal processing, pharmaceuticals, oil and gas drilling, pigment manufacture, ceramic manufacture, soap & detergent production are the major consumers of salts (mainly NaCl) produced in the world today. When released into the environment, salt ions present in the industrial effluents percolate through the soil profile and contaminate the groundwater due to their high mobility in the matrix. Most of the effluent treatment plants don't remove salts from the effluent water. As a result of this, salinity of groundwater has been found elevated in and around many industrial clusters of India; deteriorating drinking and irrigation water quality (Table 11.2). As crop production in most of the countries rely considerably on groundwater and surface water, salinity build-up in soil is inevitable around areas of high industrial activity through 'industry → effluent → soil →groundwater → soil' route (Fig. 11.1).

Soils of agricultural land surrounding industrial areas of several cities recorded high electrical conductivity (EC) and exchangeable Na indicating considerable accumulation of salts due to irrigation with contaminated surface and groundwater (Saha 2005, Panwar et al. 2010).

11.3.2 Entry of Heavy Metals in Soil

Soil health assessment surveys in India have been carried out discretely by several researchers. These indicate clearly that contamination of soils with heavy metal in the impact zone is quite prevalent nearby industrial areas. However, such surveys are highly inadequate keeping in view of the extent of polluting activities going on in the country and many of the studies had in the past been carried out using less sensitive instrumental techniques. Inadequate numbers of competent laboratories with limited organizational strengths have generated some indicative information only on extent of soil contamination with heavy metals. A study conducted by ICAR-Indian Institute of Soil Science, Bhopal has indicated built-up of heavy

Table 11.2 Impact of industries on electrical conductivity (dS m^{-1}) of groundwater

Location	Nature of industries	Polluted area			Unpolluted area		
		Range	Mean	Reference	Range	Mean	Reference
Gajraula (district: Jyotiba Phule Nagar, UP)	Distillery	0.71–0.88	0.79	Jain et al. (2005)	0.46–0.48	0.47	Jain et al. (2005)
Kancheepuram (district: Kancheepuram, TN)	Textile dyeing	1.76–4.21	3.34	Balakrishnan et al. (2008)	0.23–3.42	1.28	CGWB[a]
Mettupalayam (district: Coimbatore, TN)	Textiles, paper and pulp	0.14–10.38	3.91	Mukherjee and Nelliyat (2007)	0.76–4.15	1.87	Mukherjee and Nelliyat (2007)
Dindigul (district: Dindigul, TN)	Tannery industry	0.52–21.2	3.72	Mondal et al. (2005)	0.125–6.52	1.9	CGWB[a]
Chennai (district: Chennai, TN)	Tannery, textile	0.48–3.76	2.02	Somasundaram et al. (1993)	1.17–1.93	1.75	CGWB[a]
Tiruppur (district: Tirupur, TN)	Textile, dye	0.41–15.95	4.02	Sellamuthu et al. (2011)	0.66–4.08	1.93	CGWB[a]
		5.9–8.4	6.78	Panwar et al. (2010)	2.4–2.9	2.62	Panwar et al. (2010)
Patancheru (district: Medak, AP)	Miscellaneous	0.70–10.2	2.6	Panwar et al. (2010)	0.18–4.85	1.78	CGWB[a]
Ratlam (district: Ratlam, MP)	Chemicals, dye, pharmaceuticals, distillery	1.49–4.50	2.84	Saha and Sharma (2006)	0.28–3.08	1.03	CGWB[a]
Udaipur (district: Udaipur, Rajasthan)	Zn smelter	3.14–6.79	5.43	Panwar et al. (2010)	0.36–6.44	1.33	CGWB[a]
Pali (district: Pali, Rajasthan)	Textile	5.91–9.54	7.36	Panwar et al. (2010)	0.35–7.95	2.73	CGWB[a]

[a]Range and mean values have been compiled for the district in which the study areas belong (Source: CWGB 2013), assuming most of the information belonged to unpolluted area.

Adopted from Saha et al. (2013b)

Fig. 11.1 Salinity build-up due to irrigation with polluted river water at Nagda

metals in soils due to industrial activities and nature of contamination varied with type of industries operating (Table 11.3). The study had also indicated contamination of groundwater with heavy metals at different locations of India due to industrial activities (Table 11.4).

11.4 Instances of Pollution from Industrial Effluents

Significant part of the pollutant loaded effluents, generated particularly from small scale industries are released untreated into land and water bodies. In most of cases, metals are present in dilute and small quantities in polluted water bodies and may not cause any harm to plant growth immediately when used for irrigation. However, their immobility and consequent persistence imply that concentrations may become elevated in the long run to such an extent that they begin exhibiting toxic effect on plant, soil microorganisms and food chain. Long-term exposure to heavy metals has been reported to affect human and animal health adversely (ATSDR 2005). Among the heavy metals, Ni, Co, Cr and Cu are relatively more toxic to plants and As, Cd, Pb and Hg are relatively more toxic to higher animals (McBride 1994). Build up of different pollutants in Indian soils and their impact on soil quality, agricultural productivity and food quality as well as impact on organisms were investigated by researchers which are described below.

Table 11.3 Heavy metals accumulated in soils around different industrial areas

Location	Nature of industries	Heavy metals accumulated
Pithampur (Dhar), Madhya Pradesh	Automobile manufacturing, food processing, chemical processing, distilleries, textile industries and other manufacturing industries	Cr, Zn, Co
Debari (Udaipur), Rajasthan	Zinc smelter	Zn, Cd, Pb
Korba, Chhattisgarh	Thermal power plant, Metallurgical (Al), Textiles, Engineering workshops, Tyre retreading, and others	Cd, Cr
Coimbatore, Tamil Nadu	Electroplating, Textile, Dye	Ni, Pb, Cd, Cr
Kanpur-Unnao (UP)	Textile, leather tanning, fertilizer, miscellaneous small scale chemical factories	Ni, Zn, Cr, Sn

Panwar et al. (2010)

Table 11.4 Impact of industrial activities on soil and groundwater bodies

Industrial area	Soil quality parameters affected	Surface water quality parameters affected	Groundwater quality parameters affected
Ratlam industrial Area, Madhya Pradesh	EC, ESP	–	TDS, SAR, Coloration, Pb, Cd
Nagda industrial area, Madhya Pradesh	EC, ESP	TDS, high Na, Cl, SO_4^{-2}	
Pithampur (Dhar) industrial area, Madhya Pradesh	EC, SAR, Co, Cr		EC, SAR, Cr
Patancheru industrial area, Medak district, Andhra Pradesh	–	EC, As	High pH, EC, Ni, As
Zinc smelting area in Udaipur, Rajasthan	Zn, Cd	Zn, Cd, F	Zn, Cd
Textile industries in Pali, Rajasthan	EC		EC, Na, Cu, Pb, Cr, As
Korba industrial area (Chhattisgarh)	Acidic pH, EC, Cr, Cu		Cd, Co, Cr, Ni, Zn
Tiruppur Industrial Area	EC, ESP	EC	Pb, Cr
Coimbatore industrial area, Coimbatore	Ni, Pb, Cd, Cr		EC, Na, Cl, SO_4^{-2}

Adopted from Panwar et al. (2010) and Saha and Sharma (2006)

11.4.1 Ratlam Industrial Area, Madhya Pradesh

In Ratlam, about 2000 m^3 effluent per day generated from industrial area was used to pour through Dosinala which runs through the city southwards, finally draining into river Mahi. Majority of the industries were shut down by 1996 due to Supreme Court order on the Public Interest Litigation (PIL) filed on pollution related issue.

Polluted groundwater at Ratlam | Polluted river water at Nagda

Fig. 11.2 Pollution of groundwater at Ratlam and river water at Nagda due to industrial activity

As a result of industrial activities, groundwater of nearby villages has turned red and the impact is still visible even after 20 years (Fig. 11.2).

The groundwater at about 60–80 m depth in several villages had been polluted with salts due to contamination with percolating industrial effluent and was being used for irrigation to winter crops (Saha 2005). While mean EC values of ground water of unaffected villages were in the range of 0.85–0.92 dS m^{-1}, the same in affected villages ranged from 1.49 to 4.50 dS m^{-1} with an overall mean of 2.84 dS m^{-1}. Contents of sodium, sulphate and chloride in groundwater of affected villages were, respectively, in the range of 14.5–30.9 m*M* (mean 22.61 m*M*), 0.92–3.55 m*M* (mean 2.02 m*M*) and 6.16–35.88 m*M* (mean 2.83 m*M*). These values were, respectively, 348%, 288% and 364% more than the similar values obtained in groundwater samples of surrounding unaffected villages. The values of sodium absorption ratio (SAR), a measure of Na hazard, were considerably higher (range 3.36–12.29; mean 8.52) in ground water of affected villages as compared to the values (range 2.57–6.19; mean 3.88) obtained for unaffected villages. About 40% of the water samples in the polluted area were categorized as having very high salinity (>2.25 dS m^{-1}) and sodium hazard (SAR > 9) and about 71% of the samples had potential for severe Cl^- hazard (>10 meq Cl^- L^{-1}) permitting their use as irrigation only to tolerant crops. Use of such bad quality irrigation water caused considerable decline in area under vegetable cultivation in the groundwater pollution area. Groundwater samples of polluted area contained, on an average, 9.1 μg Pb L^{-1}, 4.1 μg Cd L^{-1} and 18.5 μg Cu L^{-1}; which were more by about 162, 26 and 83%, respectively over those in groundwater samples of unpolluted area. Considering World Health Organization (WHO) limits for groundwater, samples from Bhajankheda, Jadwasa khurd and Dosigaon villages of polluted area contained unsafe levels of Pb and Cd.

There were significant increases in salinity as well as exchangeable sodium percentage (ESP) levels in soils due to irrigation with polluted water in the affected villages. Soil samples collected from groundwater polluted villages in the month of

Table 11.5 Mean chemical properties of soils of polluted and unpolluted area in Ratlam

Parameter	Polluted area			Unpolluted area		
	Mean	Range	Standard deviation	Mean	Range	Standard deviation
pH (1:2 in water)	7.70	7.11–8.27	0.28	7.83	7.28–8.01	0.14
EC (dS m^{-1})	1.83	0.49–5.01	1.00	0.41	0.25–0.76	0.21
SAR	3.98	0.69–27.12	4.72	0.86	0.69–2.65	0.86
OC (%)	0.70	0.44 0–0.98	0.15	0.45	0.15–0.89	0.19
Available P (mg kg^{-1})	18.70	16.0–22.4	1.65	26.64	11.5–43.7	7.95
Available K (mg kg^{-1})	230.38	171.5–332.5	49.30	188.58	128.0–293.5	43.76

Saha (2005)

February had, on an average, higher EC (4.5 times), SAR (4.6 times), organic C (1.5 times) and available K (22% more) as compared to the soils from unpolluted area (Table 11.5). Significant increases in concentrations of Na^+, Cl^-, SO_4^{-2} and HCO_3^- and decreases in the concentrations of K^+ and NO_3^- in the soil solution were observed in the polluted soils of Ratlam, which indicates that salt loaded polluted groundwater/river water had direct impact on the crop yields as well as soil parameters in the polluted area. Monsoon rainfall decreased EC as well as concentrations of Na^+, Cl^- and SO_4^{-2} indicating reduction in soil salinity. However, a significant increase in the soil pH was observed after rainy season, due to washing out of Cl^- and SO_4^{-2} from surface soil layer and their replacement with CO_3^{-2} and HCO_3^- in the soil matrix.

Groundwater pollution had adversely affected the economic condition through changes in cropping pattern, reduction in crop yield, as well as reduced longevity of irrigation infrastructure (Saha and Sharma 2006). Area under vegetable and pulse cultivation was severely reduced and the same under fallow had increased during rabi season as many farmers preferred to keep their land fallow instead of using polluted groundwater as irrigation. The yields of the traditional crops such as soybean, gram wheat, methi and garlic were less on the fields irrigated with polluted ground water (Table 11.6). Onion grown using polluted groundwater had a short keeping quality and started rotting within 10–15 days after harvest; thus inflicting heavy economic losses to the farmers. Poppy (*Papaver somniferum*) cultivation, previously grown widely in the area, was abandoned because of drastic reduction in yield and consequently the government denotified the area for its cultivation. The iron pipes of the tubewell got rusted within 5 years; and therefore, the farmers incurred losses on account of repair and maintenance of their irrigation infrastructure. The drinking water source of the eight villages became unfit for human use. People were either forced to consume polluted drinking water or to spend much effort and time in fetching potable water from a distance up to 3–4 kms.

Table 11.6 Estimation of losses due to reduction of crops yields

Crop	Average yield (q ha^{-1}) Normal conditions	Polluted conditions	Yield reduction (q ha^{-1})	Price (Rs q $^{-1}$)	Loss (ha $^{-1}$)
Wheat	36	29	7	650	4550
Gram	12.5	9[a]	3.5	1400	4900
Soybean	15	11.5	3.5	1200	4200
Methi	18	12	6	1500	9000
Garlic	60	45	15	2000	30,000

Saha and Sharma (2006)
[a]When irrigation is given at later stages

Approximately 300 man-days were used each day for bringing potable water from distant sources.

11.4.2 Nagda Industrial Area, Madhya Pradesh

In Nagda, a textile industry complex located on the North-West of town manufactures a variety of products including viscous rayon, caustic soda, liquid chlorine, carbon disulfide, etc. A few other ancillary industrial units have also been established to make various chlorinated products utilizing the chlorine produced by the textile industry complex. All these industrial activities result in a high volume of wastewater generation, which is carried away by a natural surface drain, leading to river Chambal.

Water of Chambal river became severely polluted with effluents from textile industry containing salts namely, Na^+, Cl^- and SO_4^{-2} and was being used for irrigation to winter crops in nearby areas of several villages (Saha 2005). Irrigation water (from Chambal river) near affected villages had EC ranging from 2.38 to 4.11 (mean 3.56 dS m^{-1}) and contained 31.6–57.5 m*M* Na, 0.44 to 1.66 m*M* K, 3.54–4.32 m*M* Ca, 1.07–2.72 m*M* Mg, 2.93–6.76 m*M* SO_4^{-2} and 25.6–25.8 m*M* Cl^-, which were on an average, 4.7, 9.9, 7.7, 5.8, 1.8, 9.0, and 6.9 times more than the corresponding mean values obtained in irrigation water (groundwater) in unaffected villages. The mean SAR value in Chambal river water was 5.1 times higher than the similar value obtained in groundwater of unaffected villages. Long-term application of these polluted water to soil resulted significant accumulation of salts in the root zone layer (Table 11.7). There were significant increases in salinity as well as ESP levels in soils due to irrigation with polluted water. Magnitude of salinity development in soil was much more in the soils irrigated with polluted river water at Nagda as compared to polluted groundwater irrigated lands in Ratlam. Polluted soils of Nagda recorded much higher concentrations of major cations and anions (except NO_3^-) in soil solution. Monsoon rainfall decreased EC, SAR, ESP as well

Table 11.7 Mean chemical properties of soils of polluted and unpolluted area in Nagda

	Polluted area			Unpolluted area		
Parameters	Mean	Maximum	Standard deviation	Mean	Maximum	Standard deviation
pH (1:2 in water)	7.69	7.10–8.50	0.30	7.96	7.83–8.32	0.20
EC (dS m^{-1})	4.03	0.35–12.90	3.10	0.44	0.25–1.32	0.52
SAR	19.31	0.45–75.97	20.70	1.57	0.53–2.43	0.69
OC (%)	0.67	0.29–0.99	0.20	0.53	0.33–0.75	0.11
Available P (mg kg^{-1})	18.48	14.9–23.0	2.16	11.35	5.9–22.3	4.24
Available K (mg kg^{-1})	270.00	145.5–454.0	88.03	177.10	110.0–300.5	45.95

Saha (2005)

as concentrations of Na^+, Cl^- and SO_4^{-2} in solution of soils irrigated with polluted water. However, the impact on the soil properties lessened considerably during rainy season, probably due to the presence of appreciable amount of divalent cations, Ca^{+2} and Mg^{+2} in polluted irrigation water. Available Cu contents in soils of polluted area were higher as compared to the soils of unpolluted area. Concentrations of Zn and Cu were also considerably more in the in the wheat plant tissue of polluted area as compared to those of unpolluted area.

11.4.3 Pithampur (Dhar) Industrial Area, Madhya Pradesh

Pithampur is the second largest industrial area in Asia having both large and small scale industries. Majority of the automobile companies of India have their factories in Pithampur. Also, this area is housing food processing, chemical processing, distilleries, manufacturing, and textile industries.

Water of open wells and tube wells in Cheerkhani and Silotia villages near the industrial area tested high salinity (EC 1.91–4.07 dS m^{-1}) and sodium hazard (SAR > 10) and about 82% of the samples had potential for severe Cl^- (>10 me L^{-1}) hazard permitting their use as irrigation only in tolerant crops. The EC of some of the tube well water samples from polluted villages had gone up by more than 2.5 mS cm^{-1} indicating that effluents contaminated the ground water. The ground water samples of polluted area contained, on an average, 84.2 μg Cr L^{-1}, 3.7 μg Pb L^{-1} and 1.2 μg Cd L^{-1}. Several groundwater samples of polluted area had Cr concentrations more than the WHO permitted level for drinking water. Surface soil samples of Cheerkhani and Silotia villages had, on an average, higher EC (3.4 times) and SAR (3.1 times) due to considerable accumulation of Na^+ and Cl^-. The soils receiving polluted ground water was higher in Co (7.5 times) and Cr (1.5 times) contents as compared to soils of unpolluted area (Table 11.8).

Table 11.8 Impact of industrial activity in Pithampur on total heavy metal contents in soil

	Unpolluted area			Polluted area		
Heavy metal (mg kg^{-1})	Range	Mean	Median	Range	Mean	Median
Cd	0.05–0.1	0.1	0.1	0.05–0.1	0.1	0.1
Co	5.2–6.7	5.9	5.9	26.8–63.9	49.9	51.5
Cr	18.4–47.1	34.6	35.5	68.1–252.0	87.3	96.9
Cu	382.4–445.3	417.5	424.9	82.4–252.4	179.4	170.3
Ni	32.1–38.1	35.3	34.5	21.4–39.2	37.8	37.1
Pb	5.2–6.1	5.7	5.7	1.1–21.0	7.6	6.2
Zn	27.2–29.4	28.5	28.9	76.6–763.0	145.2	96.3

Panwar et al. (2010)

11.4.4 Patancheru Industrial Area, Medak District, Andhra Pradesh

Pattancheru-Bollaram cluster is an agglomeration of different industrial areas. Production in these areas is dominated by bulk drug manufacturing. It is located at the north-western outskirts of Hyderabad. Starting with the first pharmaceutical production facilities in the mid 1970s, it is now one of the biggest pharmaceutical industrial areas of India, with more than 90 manufacturers (Beijer et al. 2013). Since 1989, wastewater generated from these facilities is being treated in a common effluent treatment plant (CETP). Until 2009, most of the CETP pretreated effluents were being drained into rivers and lakes in the Patancheru area (Larsson 2007). Patancheru-Bollaram cluster was banned for further projects in 2013 by Ministry of Forests and Environment (MoEF) due to investigations conducted by the Central Pollution Control Board (CPCB).

Arsenic levels were found to be high in effluent water from the industrial area. In effluent water, Ni concentration varied from 4.7 to 57.4 μg L^{-1} (average of 23.4 μg L^{-1}), Pb varied from 0.3 to 14.2 μg L^{-1} (average of 2.0 μg L^{-1}) and Zn varied from 32.9 to 293.9 μg L^{-1} (average of 81 μg L^{-1}) (Panwar et al. 2010). Some samples showed high values of Fe, Ni, Pb and Zn, which are near the vicinity of industrial areas. The groundwater in some places near the study area has also been found contaminated with salts (high pH and EC) and some metals like As, Ni, Cr and Zn (Table 11.9).

11.4.5 Zinc Smelting Area in Udaipur, Rajasthan

The zinc smelter plant near Udaipur has smelting capacity of about 49,000 tonne per annum (TPA). With the expansion of the smelter plant, a number of other production units was commissioned which includes production units for sulphuric acid (87,000 TPA), cadmium metal (190 TPA), phosphoric acid (26,000 TPA), single superphosphate (72,000 TPA) and zinc dust (36,000 TPA). Since its

Table 11.9 Descriptive data on groundwater analysis (μg L^{-1}) from Pattancheru industrial area, Medak district

Parameter	Mean	Median	Minimum	Maximum
pH	8.1	8.0	6.9	9.4
EC (dS m^{-1})	2.6	1.7	0.7	10.2
Cr (μg L^{-1})	13.3	6.9	2.0	69.7
Mn (μg L^{-1})	75.8	75.6	2.4	11,384
Fe (μg L^{-1})	106.7	87.3	34.6	497.2
Ni (μg L^{-1})	69.0	43.0	3.9	264.8
Zn (μg L^{-1})	89.4	57.8	21.4	310.8
As (μg L^{-1})	129.4	119.3	4.2	1139.0
Pb (μg L^{-1})	2.10	0.95	0.30	7.20
Sr (μg L^{-1})	1079	875	134	2681

Panwar et al. (2010)

Fig. 11.3 Accumulation of pollutants on soil surface receiving Zn-smelter industrial effluent from Udaipur industrial area

inception the effluent from the plant was being discharged into a stream which flowed through 3 kms to the east and merged into Berach river. The effluent of zinc smelter was being discharged in a stream, employed for irrigating the crops in the vicinity of the smelter plant (Fig. 11.3).

The concentrations of zinc and fluoride in the groundwater were higher than permissible limit of 5 and 2 mg L^{-1}, respectively (BIS 2012). Well water samples had high concentrations of Zn (2.2–9.7 mg L^{-1}) and Cd (0.004–0.081 mg L^{-1}) which indicates contamination of groundwater due to industrial activity. The concentrations of these (Zn and Cd) heavy metals were high in effluent irrigated soils nearer to the discharge point (Gorla and Bichhari village) and decreased with the distance from the effluent discharge point. A large variation in the content of total zinc (65–1860 mg kg^{-1}), total cadmium (0.07–10.4 mg kg^{-1}) and total lead

Table 11.10 Chemical parameters and total heavy metals content (mg kg^{-1}) in soil of agricultural land irrigated with effluent water from smelter industries in Udaipur

Parameter	Range	Mean	Median
pH	7.2–10.4	8.2	8.3
EC (dS m^{-1})	0.2–6.5	1.0	0.6
Zn	65–1860	619	590
Pb	27.5–180.0	51.4	38.0
Cu	21.2–70.0	36.0	34.9
Cd	0.07–10.37	1.56	0.86
Co	7.6–18.5	11.9	11.8
Ni	19.8–50.2	34.9	34.3

Panwar et al. (2010)

(27.5–180 mg kg^{-1}) was recorded in the soils of the area (Table 11.10). While comparing with the safe concentration limits determined for soils (Saha et al. 2010b; Saha et al. 2013a), most of the soils nearby Zn smelting area accumulated toxic levels of Zn and Cd. With few exceptions, total Zn, Cd and available Zn content of the soil decreases with an increase in the lateral distance from the stream and river.

11.4.6 Soil and Water Pollution by Textile Industries in Pali, Rajasthan

Industrial area at Pali is having more than 800 textile units and is indicated as one of the critically polluted area by Central Pollution Control Board (CPCB). The textile printing and dyeing industries were discharging industrial effluents into the river Bandi, a non-perennial river with no flow in the lean season, thus severely contaminating both the river as well as the groundwater. The industries discharged a variety of chemicals, dyes, acids and alkalis besides heavy metals and other toxic compounds. The effluents were multi-colored and highly acidic and/or alkaline. Groundwater from downstream villages was highly saline with salts of Na as compared to upstream villages. Copper concentration was more than drinking water standards in all the wells in downstream villages; while Pb was high in Kerla, Sukarlai and Nehada; Cr level is high in Kerla, Sukarlai, Gadhwara and Phikaria; As is high in Jewadiya, Kerla and Phikaria (Table 11.11).

The well water was not suitable for irrigation due to high salinity (>4 dS m^{-1}). The Nahada dam built for storing water, has become a industrial storage tank and thus led to groundwater contamination. The soils under cultivation using contaminated well waters also showed high salinity due to high salt content of irrigation water (Fig. 11.4).

Table 11.11 Average heavy metal (μg mL^{-1}) contents in groundwater in selected villages towards upstream and downstream side of effluent discharge point in Bandi river of Pali

	Village	Cu	Zn	Pb	Ni	Cr	As
Downstream	Jewadiya	0.17	0.07	0.03	0.25	0.04	0.28
	Kerla	0.52	0.12	0.56	1.42	0.11	0.49
	Sukarlai	0.05	12.16	0.08	0.78	0.27	0.03
	Nehada	0.07	0.06	0.08	0.16	0.05	0.02
	Phikaria	0.13	0.03	0.02	0.22	0.05	0.13
	Gadhwara	0.16	0.09	0.04	0.26	0.06	0.07
Upstream	Iycea	0.03	0.04	0.03	0.08	0.03	0.01
	Hemavas	0.02	0.03	0.01	0.11	0.01	ND[a]

[a]*ND* not detected
Panwar et al. (2010)

Land area of unpolluted area

Land area of polluted area

Fig. 11.4 Soils of land receiving unpolluted groundwater and polluted industrial effluent

11.4.7 Korba (Chhattisgarh) Industrial Area

Korba city is the power capital of central India with the National Thermal Power Corporation Limited's (NTPC) super thermal power plant working at 90% plant load factor. Korba is also having aluminium industry Bharat Aluminum Company Limited (BALCO), textiles, engineering workshops, hardware (aluminum & iron), detergents, plastic toys, PVC cable pipes, cement products, electricity transformer, bakelite, distemper, clay insulator manufacturing units and other small industries, generating large quantity of acidic effluents which contaminated surrounding land areas (Fig. 11.5).

Groundwater samples collected from villages nearby industrial area contained high levels of heavy metals Cd, Co, Cr, Ni, and Zn; mean values of which were considerably higher (Cd: 19 times, Co: 67 times, Cr: 6 times, Ni: 5 times, and Zn: 10 times) as compared to those collected from far away villages. Majority of the groundwater samples from polluted area had heavy metals more than the levels

Fig. 11.5 Same agricultural land near Korba industrial area before (year 2008) and after (year 2010) contamination with industrial effluent

permitted for drinking purpose. Soil irrigated with industrial effluent turned highly acidic (Table 11.12). Contents of soluble salts, soil organic carbon, DTPA extractable heavy metals and total heavy metals contents increased due to use of industrial waste/effluent or contaminated water as compared to the non-polluted soils. The total as well as DTPA extractable heavy metals particularly, Cr and Cd were in toxic range in most of the polluted soils (Table 11.12).

11.4.8 Tiruppur Industrial Area, Tamil Nadu

Tiruppur has been identified as one of the critically polluted area by CPCB. Industrial area discharges more than 90 MLD into Noyyal river (tributary of Cauvery river). Industrial effluent passes through Tiruppur and is stored up in the Orathapalayam Dam to be used in agriculture and drinking purposes for the downstream villages in the Tiruppur and Erode district. The Industrial area is having 729 bleaching and dyeing units. Due to pollution, drinking water, fisheries and the agriculture in Tiruppur area and downstream villages of Noyyal river has been affected.

The river water was injurious (EC >3 dS m^{-1}) to agriculture in an area of about 146.3 km^{-2} and critical (EC 1.1 to 3 dS m^{-1}) in about 218.3 km^{-2}. The groundwater in some villages was having high values of Pb and Cr which may be attributed to the industrial activities (Panwar et al. 2010). Majority of the samples were not suitable for domestic purposes and were far from drinking water standards. Irrigation of cropping land with polluted water transformed the productive soils into saline soil (>4 dS m^{-1}); the dominant cations and anions being Na^+ and Cl^- and $SO_4{}^{-2}$, respectively. Irrigation with polluted Noyyal river water resulted build-up of salinity (EC > 4 dS m^{-1}) in soils of agricultural land (Table 11.13).

Table 11.12 Soil physico-chemical properties and heavy metal status of soils irrigated with industrial effluent in Korba

Parameter	Mean	STDEV
pH	4.61	±0.093
EC (dS m^{-1})	2.21	±0.196
OC (%)	0.46	±0.062
DTPA extractable heavy metals (mg kg^{-1})		
Cd	0.056	±0.009
Co	1.726	±0.029
Cr	0.096	±0.006
Cu	2.356	±0.038
Ni	1.828	±0.028
Pb	5.944	±0.042
Zn	1.350	±0.062
Total heavy metals (mg kg^{-1})		
Cd	8.10	±0.031
Co	25.1	±0.42
Cr	172.3	±6.41
Cu	40.5	±3.86
Ni	14.1	±0.86
Pb	15.3	±2.16
Zn	29.4	±2.31

Panwar et al. (2010)

Table 11.13 Soil properties in selected villages around Noyyal river in Tiruppur district

	Downstream villages			Upstream villages		
Parameters	pH	EC (dS m^{-1})	OC (%)	pH	EC (dS m^{-1})	OC (%)
Minimum	7.24	4.29	0.19	7.27	1.86	0.21
Maximum	8.37	8.31	0.57	8.34	3.62	0.69
Mean	8.07	6.59	0.39	7.52	2.39	0.34
Median	7.92	6.47	0.36	7.63	2.53	0.47

Panwar et al. (2010)

11.4.9 Coimbatore Industrial Area, Tamil Nadu

Coimbatore industrial area is the 2nd largest industrial area in Tamil Nadu. The industrial area is having about 500 textile industry, 200 electroplating industry, 100 foundries and 300 dyeing industries. All the industrial effluent/sewage finds its way to Ukkadam river, which is source of irrigation in nearby area. Heavy metal contents in the city sewage water were quite high and varied widely with season. ICAR-Indian Institute of Soil Science, Bhopal investigated changes in soil properties in agricultural land nearby different industrial clusters; namely electroplating industry, textile industry, dye industry and city sewage irrigated areas (Panwar et al. 2010). Groundwater near industrial area has developed salinity due to contamination mainly with salts of Na^+ and Cl^-; magnitude of contamination was more near textile and dye industries. Sulphate contamination was the highest in the groundwater near electroplating industries.

Soils of agricultural land near textile and dye industries have developed severe salinity (EC > 6 dS m^{-1}) and slight alkalinity (pH >8.0). Soils of agricultural land near industrial areas contained 47–178 mg Ni kg^{-1}, 47–214 mg Pb kg^{-1}, 0.5–4.2 mg Cd kg^{-1} and 43–241 mg Cr kg^{-1} (Table 11.14).

Most of the soils had all the heavy metals more than the safe concentration limits determined by Saha et al. (2013a), which indicate that these soils may pose threat to the environment. Nickel and Pb concentrations were high in soils near the electroplating and sewage from industrial area; Cd concentration was higher in soils irrigated with mixed effluents, and sewage; Cr concentration was higher in soils irrigated with textile, dye and sewage effluent. DTPA extractable heavy metal contents were also very high as compared to those normally observed in unpolluted soils, which indicates soils of agricultural land near industrial area of Coimbatore are likely to impart considerable threat to living organisms (Table 11.15).

11.4.10 Katedan Industrial Development Area in South of Hyderabad, Andhra Pradesh

Katedan industrial area of Hyderabad is hosting more than 300 industries which are involved in dyeing, edible oil production, battery manufacturing, metal plating, chemicals production etc. Dumping of the waste materials around this area is commonly observed and is a major cause of soil pollution which occurs due to spreading hazardous material through rainwater and wind. Govil et al. (2008) reported the prevalence of very high concentrations of Pb, Cr, Ni, Zn, As and Cd throughout the industrial area. Hazardous metals like As and Cr, Cu, Pb and Zn had also been observed in some of the residential area nearby industrial complex.

The wastes and effluents generated from different industrial activities in and around this area are discharged into the ponds without adequate treatment. It resulted in contamination of groundwater in and around Katedan area with salts and several heavy metals (Cd, Cr, Ni and Cu). All these heavy metals exceeded the permissible limits in the groundwater, except Pb. Continuous use of the contaminated groundwater for agricultural production increased the contents of Pb, Ni and Cr in the soil (Bhupal Raj et al. 2009).

11.4.11 Industrial Area at Thane Region of Maharashtra

There are about 5449 industries in Thane region, which includes textile industries, dye manufacturing industries, match box factories, canning factories of various food stuff, pharmaceutical and chemical industries, paper mill, paint industry,

Table 11.14 Total heavy metal contents (mg kg^{-1}) in soils of agricultural land nearby different industrial clusters at Coimbatore

	Ni		Pb		Cd		Cr	
Source of contamination	Range	Median	Range	Median	Range	Median	Range	Median
Sewage	78–140	107	116–214	158	1.2–4.2	2.17	75–176	124
Electroplating	84–178	122	74–167	136	1.1–2.7	1.63	48–152	109
Textile	47–126	72	52–159	114	0.6–3.1	1.28	86–241	168
Dye	62–111	89	58–147	82	0.5–2.4	1.34	79–187	127
Mixed effluents	58–132	96	47–171	92	1.3–3.8	2.39	43–158	96

Table 11.15 DTPA-extractable heavy metal contents (mg kg^{-1}) in soils of agricultural land nearby different industrial clusters at Coimbatore

	Ni		Pb		Cd		Cr	
Source of contamination	Range	Median	Range	Median	Range	Median	Range	Median
Sewage	0.9–9.7	5.34	1.6–9.6	5.26	0.5–1.7	0.79	0.2–4.6	2.69
Electroplating	1.0–8.8	4.71	0.8–3.4	2.81	0.1–0.8	0.34	1.2–4.7	3.92
Textile	0.3–1.2	0.79	1.3–2.6	2.31	0.1–0.4	0.18	2.4–6.8	4.61
Dyeing	0.5–3.6	2.26	1.4–2.3	1.79	0.05–0.3	0.09	1.9–4.2	2.87
Mixed effluents	1.5–4.9	2.14	1.8–4.3	2.74	0.1–0.5	0.19	0.6–4.3	1.81

insecticide industries, etc. Waste effluents from the industrial area contaminated surface water and groundwater bodies. The random dumping of hazardous waste was the cause of contamination (Bhagure and Mirgane 2011). Ground water in this region contained very high concentration of total dissolved solids, total hardness, total alkalinity, chemical oxygen demand, chloride, as well as heavy metals like As (12–500 μg L^{-1}), Cd (4–21 μg L^{-1}), Hg (1–12 μg L^{-1}), and Ni (5–38 μg L^{-1}), most of which were more than WHO limits for drinking water (10, 3, 1 and 20 μg L^{-1} respectively). Similarly, the soils samples collected from residential, commercial and industrial areas were heavily contaminated by As, Cd, Hg, and Ni (as per the Swedish soil guideline values for polluted soil), mainly because of local dumping of hazardous wastes.

11.4.12 Manali Industrial Area in Chennai, Tamil Nadu

The Central Pollution Control Board (CPCB) has identified Manali industrial area as one of the critically polluted areas in the country. The industrial town is situated to the north of Chennai near the Buckingham canal. It encompasses all types of processing industries, including chemical, plastic, petrochemicals, refineries, and fertilizers industries. The industrial area houses about 300 industries generating hazardous wastes. Soils in the industrial area of Manali had very high concentrations of Cr (149.8–418.0 mg kg^{-1}), Cu (22.4–372.0 mg kg^{-1}), Ni (11.8–78.8 mg kg^{-1}), Zn (63.5–213.6 mg kg^{-1}) and Mo (2.3–15.3 mg kg^{-1}) (Krishna and Govil (2008). The enrichment factors for Cr in soils ranged between 5.88 and 51.85, which categorized these under the class of extremely high enrichment. The source of Cr appears to be anthropogenic from some industries producing steel, textiles in the area. The assessment of the overall contamination of soil was based on the degree of contamination (C_{deg}). On the basis of the contamination factor (C_f^i), the soils were classified as slightly contaminated with As and Ba, moderately contaminated with Co, V and Zn, considerably contaminated with Ni, Mo and Pb and highly contaminated with Cr and Cu. Chromium contributed most (22.38%) to the degree of contamination index (C_{deg}). Copper accounted for 21.6%, Ni-8.57%, Mo and Pb- 7.31%, Zn- 6.33% and Co- 5.9% (Krishna and Govil 2008).

11.4.13 Kanpur-Unnao Industrial Area of Ganga Plain, Uttar Pradesh

Kanpur-Unnao region is an industrial hub for leather processing and manufacturing of leather goods and includes several multinational leather industries. The industrial hub is situated on the bank of Ganges and is considered to be the hot spot region of pollution in the Gangetic Plain. The outskirts land of both the cities is mainly utilized for agricultural purposes. These land area are flooded with wastewater either by over flooding of city drains or by sheet flow during heavy monsoon rainfall. Additionally, farmers utilize the wastewater of the city drains for irrigation, as they do not have any other option. The studies conducted by Ansari et al. (1999) registered the elevated contents of Cd, Co, Cr, Cu, Hg, Ni, Pb, Sn, Zn and organic carbon in sediments and soils of this region. Very high contents of OC (upto 5.9%), Cr (upto 2.16%), Sn (upto 1.21%), Zn (upto 975 mg kg^{-1}) and Ni (upto 482 mg kg^{-1}) were found in top 20 cm soils during the pre-monsoon period in 1994. In relation to the natural background values, the contribution of anthropogenic inputs of the toxic metals in soils were about 90% of Cr and Sn; about 75% of Cd; and 25% of OC, Cu, Ni and Zn. The Enrichment Factors were 10.7 for Cr, 9.0 for Sn, 3.6 for Cd, 1.8 for Ni and 1.5 for Cu and Zn in soils, respectively. The considerable Cr accumulation in soils and other environmental samples were also reported by other workers from this area. Concentration of this metal in soils was found quite high (1323 mg kg^{-1}) in the area having large number of tanneries (Rawat et al. 2009). Analysis of surface soil samples from Jajmau and Unnao industrial areas using X-ray fluorescence spectrometer indicated that these were significantly contaminated with heavy metals such as Cr (161.8–6227.8, average 2652.3 mg/kg) (Gowd et al. 2010). The plant available Cr in soil (extracted by 0.01 M $CaCl_2$) was not detected at control (unpolluted) site while its level at wastewater irrigated soils was quite high (33.26 to114.26 μg g^{-1} dw) (Sinha et al. 2006).

11.4.14 Chromium Pollution in Soils Around Vellore Tannery Industries, Tamil Nadu

Leather industry is among the major sources of pollution in the state of Tamil Nadu. It has been estimated that more than 50,000 ha of productive agricultural lands in Vellore district have been contaminated with Cr alone due to the disposal of tannery wastes, where more than 60% of Indian tanneries are located (Rangasamy et al. 2015). Effluent generated from the tanneries had highly variable characteristics and had pH 6.17–8.17 and contained very high soluble salts (EC 10.4–23.0 dS m^{-1}; sodium 2.04–9.0 g L^{-1}) and Cr (0.62–26.2 mg L^{-1}) (Mahimairajah et al. 2000). Soils surrounding tannery industries were severely contaminated with Cr (16731–79,865 mg kg^{-1}). More than 90% of the soil samples from agricultural land in 65 locations of *six Taluks* (Walajapet, Arcot, Vellore, Thirupattur, Vaniyambadi, and Gudiyatham) had high concentration of Cr (>200 mg kg^{-1} and upto 1646 mg kg^{-1}) due to use of tannery wastes (Rangasamy et al. 2015). Since $Cr_2(SO_4)_3$ is predominantly used in tanning process, the tannery effluent and sludge are rich in Cr III. As a result, contaminated soil contained mostly trivalent Cr (Mahimairajah et al. 2000). Chromium and salts had also leached through soil profile and contaminated ground water. The Cr concentration in groundwater ranged from trace to 36.7 mg L^{-1}. About 28% of the samples had relatively higher concentration of Cr, exceeding the safer limit of drinking water (0.05 mg L^{-1}) and irrigation water (2 mg L^{-1}), prescribed by the WHO and FAO, respectively.

11.4.15 Mercury Pollution in the Vicinity of Chlor-Alkali Plant at Ganjam, Orissa

Chlor-alkali plants are major consumer of mercury for its operation; and considerable concentration of this extremely toxic metal is found in liquid effluent and in solid wastes generated from there. Sediments of effluent carrying channel and low lying area and solid waste deposits had been found to contain high concentration of Hg (41–2550 mg kg^{-1}) (Lenka et al. 1992). Aquatic plants growing in effluent carrying channel and low lying area as well as vegetables grown in soils nearby solid waste dumping site accumulated high levels of Hg. However such impact (Hg contamination) due to chlor-alkali plant was highly localized as rice crops in the surrounding agricultural land did not show any Hg accumulation.

11.4.16 Fluoride Contamination in Soil and Plant in the Vicinity of Aluminium Smelter Plant at Angul, Orissa

The Angul Talcher area in Angul district of Orissa has been declared by CPCB as one of the hot spot of pollution in Orissa with CEPI rating index of 82.09. The industrial area houses about 184 industries which includes thermal power plant, aluminium smelters, steel plants as well as coal mines. The activity of aluminum industry is one of the major environmental concerns in this area. The most highlighted pollutant from the smelter plant is fluoride. During smelter process, F is volatilized from molten cryolite at 1000 °C as gaseous fluoride such as HF, SiF_4 and flurosilicic acid (H_2SiF_6). Tiny particles in the form of different compounds such as various aluminum fluoride, apetite, CaF_2 and NaF are mechanically blown out through the stacks. Such particles ultimately settle down on the natural vegetation and water bodies. In the effluent treatment plant, most of the fluoride is caught as sodium fluoro-silicate, cryolite, AIF_3 etc., which are again recycled. But due to poor effluent treatment facilities and lack of efficient techniques to catch the whole fluoride, it escapes into the environment and create nuisance. Although fluoride is beneficial for dental health in low dosage, its chronic exposure in large amounts causes gastro-intestinal problems and interferes with bone formation. A study was conducted around smelter plant to determine the fluoride content of water, soil and plant samples (Jena et al. 2003). The fluoride content of soil, water and leaf samples varied from 0.52 to 5.52 (water soluble fluoride), 0.2 to 3.24 and 25 to 390 mg kg^{-1}, respectively. Similar samples collected from places around

Bhubaneshwar (180 km away from Angul) showed fluoride content in the range of 0.36–0.44 mg L^{-1} (water soluble fluoride), 0.10–0.19 mg L^{-1} and 10–30 mg L^{-1}, respectively. The fluoride content of soil, water and soil samples decreased with increasing distance except in Santiri and Purukia village, which might have contaminated due to the inflow of effluent water from aluminum industry.

11.4.17 Lead Pollution in Some Industrial Cities of Chhattisgarh

Lead pollution has become global health issue due to its toxicity to human and widespread use and leakage into the environment. In industrial areas, Pb enters the environment through particles generated by coal burning in power plants and roasting of minerals in smelters. The elevated levels of Pb in blood of children and dogs of Indian mega cities have been reported (Kaul et al. 2002; Balagangatharathilagar et al. 2006). Lead levels in various environmental compartments (air, rain water, runoff water, surface soil, sludge and plant) of different industrial cities of Chhatisgarh states (viz., Raipur, Bhilai and Korba) was investigated (Patel et al. 2010a). Different medium and large industries are located in Raipur (cement, steel and ferro-alloy), Bhilai (steel and others) and Korba (thermal power plants and others). These industrial areas recorded considerably higher levels of Pb in air as compared to far away small residential cities. Soils of these cities, particularly from coal burning area of Korba city contained very high amount of this metal.

11.4.18 Heavy Metal Contamination in Agricultural Soils and Plants in Peri Urban Areas of Some Cities in Gujarat

Ankleshwar, Vatva, Nandesari in Gujarat have high industrial activity and is considered to have significant impact on environment. Patel et al. (2010b) conducted study to assess impact of the industrial activities on environment. Soil and sewage effluent samples from peri-urban areas of Ankleshwar (Bharuch), Vatva (Ahmedabad) and Nandesari (Vadodara) cities were analyzed for different quality parameters. The effluents were generally alkaline and contained high salts with EC ranging from 1.90 dS m^{-1} in Koyali (Vadodara) to 12.0 dS m^{-1} in Kasbativad (Ankleshwar). The concentrations of different micronutrients in the effluents samples from major industrial areas of Gujarat were generally high while that of pollutant elements were quite low, except for Cd and Co. Among the pollutant elements, only Co and Cr were above the threshold level in several areas and contaminations of other heavy metals viz. Cd, Ni and Pb was low in soil samples from these areas.

11.4.19 Impact of Industrial Effluent Form Visakhapatnam City on Soil and Plants

Industrial effluent from Visakhapatnam industrial area (having chemical, petrochemical, metallurgical industries) contained high soluble salts and heavy metals (Cu, Pb, Mi, Cr, Zn) (Bhupal Raj et al. 2009). Soils of agricultural land receiving effluents from different industries developed salinity. All the soil samples collected from surrounding areas contained high levels of Pb, Zn, Cu, Cd, Ni and Co. About 100%, 96%, 91%, 70%, 65% and 44% of the plant samples from the polluted area contained high levels of Cr, Zn, Pb, Ni, Co and Cu, respectively.

11.4.20 Effect of Cement Kiln Dust Pollution of Heavy Metal Accumulation in Soils

Cement industry is one of the 17 most polluting industries listed by the CPCB. Cement dust contains heavy metals like chromium, nickel, cobalt, lead and mercury, which are having impact on vegetation, human health, animal health and ecosystem. Effect of dust pollution from cement plant in Dindigul district (Tamil Nadu) on soil with reference to EC, pH, total Pb, Ni and available Zn, Cu, Fe and Mn content was seen up to 1 km distance and the effect was more pronounced in soil samples collected at 0.5 km distance. The soil reaction tended towards alkalinity while no effect was seen on salt concentration. An increase in total Pb and Ni content was also seen in the samples collected from 0.5 km (Stalin et al. 2010).

11.5 Soil Pollution in Mining Areas

Most of the country's mining activities (about 92%) are concentrated in the states of Gujarat, Andhra Pradesh, Jharkhand, Madhya Pradesh, Rajasthan, Karnataka, Odisha, Tamil Nadu, Maharashtra, Chhattisgarh and West Bengal. Geological Survey of India (2007–2008) estimated an affected area of 1394 km^2 through large-scale mapping (Ministry of Mines 2008). A number of studies have been carried out around mining areas in India to evaluate the extent of soil pollution (Goswami et al. 2008, 2010a, b; Swain et al. 2011). The changes in soil quality were found to be drastic and continuously deteriorating in and around mining areas.

11.5.1 Coal Mines Impact in Eastern India

India is rich in coal mines and excavation processes have devastating impact on terrestrial ecosystem including nearby agricultural land area. Coal fires are common in coal mine areas which may start by natural cause like forest fire or by human activity. In Jharia (Jharkhand), coal in the mines is burning for more than 100 years. Hazardous effects from coal fires include the emission of noxious gases and particulate matter into the atmosphere, and their condensation on the land and water surfaces leading to water and soil pollution (Stracher and Taylor 2004). Soil samples analyzed from an opencast coal mine (OCM) and a coal fire affected area (CFA) in Jharia coalfield revealed that Cr and Ni contents were elevated in soils nearby both CFA and OCM; V and Zn were enriched in soils nearby CFA. However, the levels of Cr, Ni, and Zn in these soils are below the USEPA soil screening limits (Masto et al. 2011). Using statistical tools (principal component analysis combined with multiple linear regression analysis), Pandey et al. (2016) identified coal mining activities (including mine fires) as major factor for build-up of Ni, Cu and Cr in soils of the area; while wind-blown dust was the major contributor Pb and Cd. Chemicals released from the coal mines; overburden and tailings contained high concentration of metals such as Cu, Cd, Fe, Hg and Zn; which also affected the organisms adversely.

A core committee (along with its sub-committees) was constituted by National Green Tribunal (NGT) to quantify industrial pollution and impact assessment of water, air, soil and health in and around Singrauli. The committee observed that groundwater in the villages was contaminated with high fluoride and mercury. The mercury concentration in groundwater was found exceeding the limit of 0.001 mg L^{-1} in the samples collected from Kirwani, Parasi, Harrahwa, Naktu, Sirsoti, Chilkadand, Parsavar-raja, Govindpur, Kusmaha, Khairahi, Jayant Colony, Jaitpur, MPCC colony, and Dibulganj villages around thermal power plants and mines in Singrauli and Sonebhadra area (Business Standard 2015).

A study was conducted to investigate changes on soil fertility near open cast coal mining area Godda district of Jharkhand (Ghose 2004). Soils around the area had lower soil fertility (in terms of available major plant nutrients) and microbial population as compared to unmined soils. Similarly other workers also reported that organic matter content and available nutrients like N, P, K in soils were much lower while heavy metals content were higher in mining areas as compared to normal soils (Maharia et al. 2010; Yellishetty et al. 2009; Juwarkar et al. 2003).

11.5.2 Copper Mines Impact

Soil samples collected from Khetri copper mine area were found to contain abnormally high Cu concentration (763 mg kg^{-1}), which was 30-folds higher (phytotoxic level) than that of uncontaminated soil (26.4 mg kg^{-1}). Also concentrations of Cr, Fe, Zn and Pb in soil were elevated as compared to unpolluted soils (Maharia et al. 2010). The abandoned copper mines in Mosaboni (Jharkhand, India) left huge amount of untreated tailings containing high concentration of toxic, environmentally available (equal to total metal except silicate matrix bound metal) Cu (154 mg kg^{-1}), Ni (136 mg kg^{-1}) and Pb (9.9 mg kg^{-1}) which became a source of metal pollutants. About 12.5%, 0.8% and 8% of environmentally available fractions of Cu, Ni and Pb were in bio-available forms (DTPA extractable), respectively (Shyamsundar et al. 2014). In all the samples, concentration of total Cu and Ni were found exceeding the toxicity threshold limit as indicated by Kabata-Pendias and Pendias (1984).

11.5.3 Chromite Mining Impact

About 95% of India's chromite minerals are deposited in the state of Orissa with approximately 183 million tons of deposits located in the region's Sukinda and Baula-Nuasahi mining belts (Ministry of Mines 2010). Sukinda chromite valley in the district of Jajpur, Orissa has one of the largest chromite deposits of the country and produces 8% of total chromite, mainly through opencast mining method. High Cr(VI) concentration in ground and surface water, mine effluents and seepage water samples in the area was reported (Tiwary et al. 2005). Chromium (VI) concentration was found to be varying between 0.02 and 0.12 mg L^{-1} in mine effluents and 0.03–0.8 mg L^{-1} in shallow hand pump waters and 0.05–1.22 mg L^{-1} in quarry seepage. The concentration of Cr(VI) in the surface water source (Damsal creek) was in the range of 0.03–0.14 mg L^{-1}, which increased in the downstream due to mining activities.

Dhal et al. (2010) also assessed the environmental impact of chromite mining belts in Baula-Nuasahi area and they reported hostile conditions for organisms in the surrounding environment. This study also revealed that most of the water quality parameters exceeded national/international permissible standards. The soils in and around the overburden region had low nutrient (N, P and K) and the microbial population. Also hazardous metals were found to be leached and accumulated in nearby agriculture lands and caused them less fertile for crop production. The main source of Cr pollution in this region was found to be overburden dumps and seepage water.

A case study at South Kaliapani, Chromite Mine Area, Orissa on mine waste water irrigated rice grown soil indicated that Cr(VI) concentration (0.65 mg L^{-1}) in the mine wastewater used for irrigation was beyond the toxic limit i.e., > 0.008 mg L^{-1} and total Cr content in soil irrigated with mine waste water was very high (11,170 mg kg^{-1}) compared to normal soil (Mohanty et al. 2011). Soils of agricultural land near abandoned chromite-asbestos mine area of Chaibasa (Jharkhand state) had accumulated high Cr and Ni as indicated by values of contamination factor and geoaccumulation index (Kumar and Maiti 2015). Concentrations of Zn, Mn, Co, Cu, Pb, and Cd were found low and within toxicity limit. Metal grouping and site grouping cluster analysis also revealed that Cr and Ni were closely linked with each other and chromite-asbestos mine waste was the major source of contamination.

11.5.4 Arsenic Toxicity Near Gold Mining Area of Karnataka

Prevalence of arsenicosis and As related cancers among human population had been reported from several villages of Raichur, Yadgir and Gulbarga districts in north-eastern Karnataka and a study conducted jointly by Govt. of Karnataka and UNICEF indicated a unsafe levels ($>10\ \mu g\ L^{-1}$ As) of As in drinking water samples (groundwater) of 69 villages in these districts. A comprehensive study to investigate the cause of As like symptoms among villagers of the area was conducted in Kiradalli Tanda village of Yadgir district (Chakraborti et al. 2013). The village is only 4 km away from a gold mine which had been closed for mining operations since 1994. Arsenical skin lesions (as confirmed through histopathological analysis) were observed among 58.6% of a total 181 screened individuals. Analysis of hair and nail samples from all of these As affected individuals had elevated As contents. About 79% of the tube-well water samples had As above 10 $\mu g\ L^{-1}$. Top soil samples from the residential area contained As in the range 99–9136 mg kg^{-1}, which were very high considering the its commonly reported range of 2.2–25 mg kg^{-1} for unpolluted soils (McBride 1994). This indicate that inhalation of soil dust might be another route of As entry into human being. Arsenic concentrations in the food grains were however, found considerably low in the area.

11.6 Aerial Deposition of Heavy Metals on Land

The rapid industrialization and urbanization have resulted atmospheric deposition of heavy metals. Several case studies have indicated that industrial, mining and urban activities generate considerable dust in the atmosphere and these dust particles are normally enriched with heavy metals (Patel et al. 2010a; Mishra et al. 2013; Pal et al. 2014). A study conducted by CPCB (2011) indicated that suspended particulate matter (SPM) was more in air near industrial areas as compared to that in city residential areas of both Raipur and Raigarh in Chhatisgarh. SPM of these areas are loaded with heavy metals (0.43–0.89% in Raigarh and 1.17–1.87% in Raipur). Moradabad in UP is known as brass city due to large number of brassware industries. Pal et al. (2014) found that SPM (PM_{10}) was highest in industrial area followed by commercial area and least in residential area. Metal (Cd, Cr, Cu, Fe, Ni, Pb and Zn) concentrations in the air due to suspended particles were also considerably higher near industrial area as compared to commercial and residential areas. Coal mining areas in Jharia also contained high SPM (PM_{10}) in air (20.8 μg m^{-3}) (Mishra et al. 2013). Mean concentrations of all SPM were around 2 times higher than that of non-mining area. Level of these pollutants for coal mining areas was found higher than that of most of the cities nearby.

Tiwari et al. (2008) investigated atmospheric deposition of heavy metals in urban and sub-urban area of Varanasi city. Atmospheric deposition was maximum for Mn (387.3 g ha^{-1} $year^{-1}$) followed by Zn (336.7 g ha^{-1} $year^{-1}$), Cr (124.4 g ha^{-1} $year^{-1}$), Pb (71.0 g ha^{-1} $year^{-1}$), Ni (51.2 g ha^{-1} $year^{-1}$), Cu (39.8 g ha^{-1} $year^{-1}$) and Cd (6.9 g ha^{-1} $year^{-1}$). Their deposition was the maximum in heavy traffic zone followed by commercial, residential and sub-urban areas in the decreasing order. Another study (Sharma et al. 2008) in Varanasi indicated deposition rate of Cd as 13.8 g ha^{-1} $year^{-1}$, Zn as 525 g ha^{-1} $year^{-1}$, Cu as 66.8 g ha^{-1} $year^{-1}$ and Pb as 9.8 g ha^{-1} $year^{-1}$. An earlier study (Tripathi et al. 1993) in Mumbai showed deposition rate for Cd, Zn, Cu, and Pb as 0.4, 371.6, 94.3 and 32.4 g ha^{-1} $year^{-1}$, respectively.

11.6.1 Risk to the Peri-Urban Agriculture with Atmospheric Deposition

Most the atmospheric emissions of heavy metals from the centers of anthropogenic activities are likely to end up as their depositions over nearby agricultural land directly or through washing-off of the vegetations. If the mean metal concentration in biomass of rice and wheat grown in uncontaminated soils (Kabata-Pendias and Pendias 1992) are considered, annual uptake of heavy metals by rice-wheat cropping system should not be more than 1.3 g Cd ha^{-1}, 2 g Cr ha^{-1}, 65 g Cu ha^{-1}, 6 g Ni ha^{-1}, 5 g Pb ha^{-1} and 440 g Zn ha^{-1} under Indian condition (assuming wheat and rice grain yield of 5 and 4 t ha^{-1}). On the other hand, if means of the annual heavy metal deposition rates, as found out by different workers above are assumed (7 g Cd ha^{-1}, 124 g Cr ha^{-1}, 67 g Cu ha^{-1}, 51 g Ni ha^{-1}, 38 g Pb ha^{-1} and 410 g Zn ha^{-1}), the annual deposition of Cd, Cr, Pb and Ni can be several times higher than their uptake by above ground biomass (Table 11.16). However, even at this rate of soil contamination due to atmospheric deposition, heavy metal build-up in the soil would be very slow. For example, it will take 200 years for Cd, 460 years for Cr, 2900 years for Ni and 1050 years for Pb to get their total contents become double than the present level in plough layer at Bhopal city as an example.

However, Sharma et al. (2008) have reported that atmospheric deposition had substantially contributed towards the heavy metals accumulation in vegetables and consumption of which lead to potential health risks to the consumers. This study revealed that both Cu and Cd posed health risk to human via all the tested vegetables consumption, whereas Pb only through cauliflower. Another study in China revealed that Pb and Cd in the grain grown nearby highway were due to the foliar uptake from atmosphere which, accounted for about 46% of Pb and 41% of

Table 11.16 Estimation of impact of atmospheric deposition on heavy metal contents in soil

Heavy metal	Annual mean atmospheric deposition rate (g ha^{-1})	Annual mean removal by rice-wheat system (g ha^{-1})	Annual increase in total content in soil (mg kg^{-1})
Cd	7	1.3	0.003
Cr	124	2	0.061
Cu	67	65	Nil
Ni	51	6	0.022
Pb	38	5	0.016
Zn	410	440	Nil

Cd of the total uptake and there was no significant contribution of atmosphere to Cr, Zn and Cu in grain. The study concluded that atmospheric Pb and Cd around highway can directly contaminate food (Feng et al. 2011). Smolders (2001) concluded from his research work that in rural areas with low atmospheric Cd deposition, of less than 2 g Cd ha^{-1} y^{-1}, airborne Cd in the plants has only a marginal influence on the crop Cd concentrations. On the other hand, when the atmospheric Cd deposition is well above 10 g Cd ha^{-1} y^{-1} can be a major source of crop Cd and dietary Cd. This kind of conditions may occur in and around the pyrometallurgic smelters with high Cd emissions.

11.7 Pollution Around Municipal and Hazardous Waste Dumpsites

Municipal solid wastes (MSW) are considered to be loaded with hazardous material and have severe environmental consequences if it not properly treated. Unorganized, indiscriminate and unscientific dumping of municipal wastes is very common disposal method in many Indian cities which causes adverse impacts to the environment. Environmental impact of unscientific land-filling/dumping of MSW usually results from the run-off of the toxic compounds into nearby land area, surface water and groundwater which eventually lead to water pollution as a result of percolation of leachate. For example, examination of soils in three municipal waste dumpsites of Allahabad, Uttar Pradesh, showed elevated total metal concentrations of Cr, Cu, Fe, Ni, Pb and Zn (32.46–108.85 mg kg^{-1}) (Tripathi and Misra 2012). Among the sites investigated, Daraganj dumpsite was highly contaminated while Phaphamau dumpsite was least contaminated. The order of metal contamination in dumpsites was Pb > Zn > Fe > Ni > Cu > Cr > Cd. It indicates that the heavy metal contamination at unscientific dumpsites is higher and of great concern for their surrounding environment and organisms. The Thane-Belapur industrial area, in Maharashtra produces 5 tonnes of waste every day, which is co-disposed with municipal waste in municipal waste dumpsites. The water bodies in the vicinity of this dumpsite area are polluted and sediment in the Ulhas river has registered high levels of mercury and arsenic (Vision 2025 of Planning

Commission, Government of India). Considerable leachate migration from municipal dumpsite in Ariyamangalam of Tiruchirappalli district (Tamil Nadu) resulted heavy metals contamination in soils of nearby area (Kanmani and Gandhimathi 2013).

Hazardous waste disposal sites are one of the major sources of elevated levels of metals in the soil environment around industrial area. Parth et al. (2011) studied and reported the degree of contamination of soil in respect of heavy metals accumulation in and around hazardous waste disposal sites located in the north-western part of Hyderabad, India. It was estimated that annual hazardous/industrial waste of approximately 50,000 tonnes was abandoned as landfill over 200 acres of area in the city outskirts. These hazardous wastes were contaminating soil resource. Soils in the vicinity of dumpsite and the downstream were considerably contaminated with metals. The heavy metals such as As, Cr, Pb in the soils was found to exceed the threshold and natural background values. The highest concentrations of Cu, Ni and Zn exceeded the prescribed threshold limit. The soil pH was acidic to alkaline and was one of the major factors affecting mobility/solubility of metals in soil environment. In Kolkata, Adhikari et al. (1993) found significant accumulation of Cr, Pb and Cd in upland surface soils of agricultural land due to long-term application of sewage sludge.

11.8 Soil Contamination Due to Agricultural Activities

Although contribution of agriculture to GDP of India has fallen over time, absolute production has increased almost continuously over the years, mainly due to technological innovations, and intensive use of agricultural inputs. Many a times, safety of environment has been compromised with quality of these inputs due to pressure of enhancing productivity of land. Fertilizers have been considered as an essential input to agriculture as these play an important role in achieving the total food grain production worldwide to feed the ever increasing population and to meet their daily needs of food, fuel and fiber. Consumption of chemical fertilizers and organic manure bear a direct relationship with food grain production. The total nutrient consumption ($N + P_2O_5 + K_2O$) was about 24 million tonnes during 2013–2014, with about 141 kg fertilizer per ha (Annual Report 2014-‘15, Department of Fertilizers, Govt. of India). Excessive use of fertilizers and pesticides, antibiotics and hormones in livestock and irrigating farms with contaminated wastewater are agricultural factors affecting soil pollution. Many of these agricultural inputs have been reported to be contaminated with heavy metals (Table 11.17).

Table 11.17 Agricultural sources of trace element contamination in soils (mg kg^{-1})

Element	Sewage sludge	Phosphate fertilizers	Limestone	Nitrogenous fertilizers	Municipal solid waste compost	Manures	Pesticides
As	2–26	2–1200	0.1–24	2–120	–	3–150	22–60
Cd	2–1500	0.1–170	0.04–0.1	0.05–8.5	Tr.–8.4	0.3–0.8	–
Cr	20–40,600	66–245	10–15	3–19	14–401	5.2–55	–
Cu	50–3300	1–300	2–125	1–15	25–865	2–60	12–50
Hg	0.1–55	0.01–1.2	0.05	0.3–3	–	0.09–26	0.8–42
Mn	60–3900	40–2000	40–1200	–	–	30–550	–
Ni	16–5300	7–38	10–20	7–38	8.6–190	7.8–30	–
Pb	50–3000	7–225	20–1250	2–1450	11–647	6.6–15	60
Zn	700–49,000	50–1450	10–450	1–42	82–946	15–250	1.3–25

Adopted from Kabata-Pendias (2000) and Saha et al. (2010a)

11.8.1 Impact of Fertilizer Use on Heavy Metal Build-Up in Agricultural Land

Generally fertilizers are manufactured from the raw materials that are collected from underground through mining like rock phosphate, sulphates etc. Therefore fertilizers contain highly variable amount of heavy metals as impurities (Table 11.17). Among the fertilizers, use of rock phosphates and phosphatic fertilizers in agriculture are considered as environmental concern due to their potential for enhancing heavy metal levels in soil and contaminating food crops (Mortvedt 1996; Gupta et al. 2014). For instance, concentrations heavy metals in phosphate fertilizers were reported as 0.5–20 ppm (mean 11.3) As, 0.1–250 ppm (mean 65 ppm) Cd, 63–896 ppm (mean 173.2 ppm) Cr, 0.2–1170 ppm (56.6 ppm) Cu and 0.5–151 ppm (27.5 ppm) Ni (USEPA 1999). Concerns are raised on accumulation of heavy metals and their contamination to food crops due to continuous application of phosphatic fertilizers in soil. However, long-term experiments in different countries indicated no significant contamination of food due to continuous P fertilizer application (Smilde and van luit 1983; Rothbaum et al. 1986; Mortvedt 1987). On the contrary, uptake of Cd by herbage was found higher where P fertilizer was being applied continuously over several decades under a long-term experiment (Jones and Jonston 1989).

In India, analysis of soils and plant samples from long term (>39 years) fertilizers experiments at Barrackpore, Jabalpur, Bangalore, Ranchi, Palampur, Pantnagar and New Delhi indicated that application of 150% and/or 100% recommended doses of NPK for this period resulted build-up of the heavy metals like Cd, Pb, Co, Cr and Ni. However, contamination level didn't reach the unsafe levels. Also, heavy metals in edible plant parts showed that risk to human health is very little (Adhikari et al. 2012).

11.9 Soil Pollution Through Use of Geogenically Contaminated Groundwater

11.9.1 Use of Arsenic (As) Contaminated Groundwater for Irrigation

The prevalence and consistent increase in evidences of As contamination of groundwater has been reported from various countries. In India also, its contamination in water and soil have been recognized a serious clinical problem in several states. In West Bengal, extensive As contamination in groundwater and soil has been reported from nine districts, particularly in intensive cropping areas within the upper delta plain along the Bhagirathi and other rivers (CGWB 1999). About 10% of total human population in the state was exposed to the risk of As toxicity by consuming As contaminated ground water for drinking purpose (Elangovan and Chalakh 2006). Analysis of 140150 water samples from tube wells in all 19 districts of West Bengal for arsenic indicated that 48.1% samples had As >10 μg L^{-1} (WHO guideline value), 23.8% had >50 μg L^{-1} (Bureau of Indian Standard) and 3.3% had >300 μg L^{-1} (concentration predicting overt arsenical skin lesions) (Chakraborti et al. 2009). Arsenic contamination of groundwater, soil and food had also been reported from other regions of India such as Uttar Pradesh, Jharkhand, Bihar, Madhya Pradesh, Chhattisgarh, Assam, Manipur, Tripura, Arunachal Pradesh, Punjab and Andhra Pradesh (Chakraborti et al. 1999, 2003; Chetia et al. 2011; Govil et al. 2001; Mukherjee et al. 2006; Rao et al. 2001; Singh et al. 2011; Sharma et al. 2016). In a study in the state of Punjab, the concentration of As in 200 groundwater and surface soil samples were analyzed (Singh et al. 2011). In southern states of Punjab, As in groundwater varied from 5.33 to 17.27 μg As L^{-1}, and in soil it varied from 1.09 to 2.48 mg As kg^{-1}. About 40% of ground water samples exceed the permissible limit (10 μg As L^{-1}).

Reductive dissolution of Fe-hydroxides (FeOOH) stimulated by microbial activity and organic materials is regarded as the most important mechanism of releasing As into the aquifer (Mukherjee and Bhattacharya 2001; Ravenscroft et al. 2001; Smedley et al. 2003; McArthur et al. 2004). Continuous use of As-contaminated groundwater may elevate the soil arsenic level, thereby increasing the possibility of arsenic entering into the food chain. Arsenic contamination or build-up in fertile alluvial soils of Malda, Dinajpur (North and South), Murshidabad, Nadia, Burdwan, 24 Parganas (North and part of South), Hoogly districts has arisen from use of arsenic loaded ground water as a source of irrigation. Beside underground source of drinking water, As can also enter into human body through consumption of food that is grown using contaminated groundwater for irrigation.

Impact of As contaminated groundwater irrigation in different vegetables and crops (Table 11.18), and their dietary intake was studied (Santra et al. 2013). The results revealed that tubers accumulated higher amount of arsenic than leafy vegetables followed by fruit vegetables. The As accumulation was high in potato, brinjal, arum, amaranthus, radish, lady's finger and cauliflower; and was at moderate level in beans, green chilli, tomato, bitter guard, lemon and turmeric. Its accumulation in mustard was in the range 0.339–0.373 mg kg^{-1}. Among the pulses, pea showed the highest As content (1.30 mg kg^{-1}) and the lowest As concentration was found in moong bean (0.314 mg kg^{-1}).

In Gangetic West Bengal, huge amount of the groundwater is used as irrigation for production of winter (*boro*) and summer (*aush*) rice during winter and summer (November to May). In the affected districts, use of As-contaminated groundwater as irrigation in paddy fields caused accumulation of As in rice irrespective of its varieties (Halder et al. 2014). As a result, As accumulation was found more in *boro* rice than that grown in rainy-season (*aman*, grown predominantly with rain water). Further, high yielding rice varieties accumulated more arsenic than local varieties. Garari et al. (2000) predicted that the left over roots after harvest of crops have contributed substantially to the As accumulation in soils. However, the toxicity due to As to human and crops depends on forms or species of As rather than the total As

Table 11.18 Arsenic levels in soil, crops and vegetables grown in West Bengal, (India) and adjoining countries

Country	Arsenic in soil (mg kg^{-1})	Arsenic in crops and vegetable (mg kg^{-1})		References
		Rice	Vegetables	
Bangladesh	NA	0.358	0.034	Chowdhury et al. (2001)
West Bengal, India	11.35	0.245	<0.0004–0693	Roychowdhury et al. (2002)
Bangladesh	NA	NA	0.306–0.489	Alam et al. (2003)
Bangladesh	NA	NA	0.011–0.103	Farid et al. (2003)
Bangladesh	7.31–27.28	0.04–0.27	0.2–3.99	Das et al. (2004)
West Bengal, India	7.0–38.0	0.30	NA	Norra et al. (2005)
China	6.04	0.117	0.003–0.116	Huang et al. (2006)
Bangladesh	14.5	0.5–0.8	NA	Rahman et al. (2007)
Nepal	6.1–16.7	0.180	<0.010–0.550	Dahal et al. (2008)
West Bengal, India	1.34–14.09	0.16–0.58	NA	Bhattacharya et al. (2009)
West Bengal, India	5.70–9.71	0.334–0.451	0.030–0.654	Bhattacharya et al. (2010)
Bihar, India	0.027	0.019	0.011–0.015	Singh et al. (2011)
West Bengal, India	NA	0.156–0.194	0.069–0.78	Samal et al. (2011)
West Bengal, India	NA	0.01–0.64	0.03–0.35	Halder et al. (2013)

Santra et al. (2013)
NA not available

content. Generally As exists in AsO_4^{3-} and AsO_3^{3-} forms in soil and the later is considered being more toxic to animal and human. Soil properties like texture, mineralogy, redox potential (Eh) and pH control the speciation and mobility of As in soil.

Average daily intakes of As by adult and children through their diet were computed as 560 μg and 393 μg, respectively, on the basis of average dietary habit and concentrations of the toxicant in common food items. Further the people having poor nutrition were found to be more vulnerable to As toxicity than the people having adequate nutrition (Santra et al. 2013). Contamination of food chain and daily intake of As by human through food and drinking water was estimated in Nadia district where unsafe levels of the element in groundwater in widely prevalent (Samal et al. 2011). Average concentrations of As in drinking water and commonly grown food in the area were 16 μg L^{-1} in drinking water, 156–194 μg kg^{-1} in rice, 69–780 μg kg^{-1} in vegetables and 24.7 μg kg^{-1} in pulse (lentil). Total intake of As through foodstuffs was computed as 560 μg day^{-1} by adults and 393 μg day^{-1} by children in the area which were quite alarming. After adjusting the excretion through urine, investigators indicated considerable potential risk of As exposure to local inhabitants through continuous consumption of As-contaminated foodstuffs and drinking water. Such intake rates of As are alarmingly higher keeping in view of maximum tolerable daily intake 2.1 μg kg^{-1} body weight by humans from all sources (WHO 1988). In Ropar district of Punjab, consumption of As contaminated wheat grains was found to pose higher risk of cancer and non-cancer health disorders as compared to intake of As contaminated groundwater by both adults and children (Sharma et al. 2016).

11.9.2 Use of Selenium Contaminated Ground Water for Irrigation

Selenium is an essential element in human and animals; and however, it shows toxicity symptoms when taken in larger amount. Long-term ingestion of excess Se may result in chronic disease, called 'selenosis' with symptoms of nausea, diarrhea, joint pain, loss of nail and teeth and skin rashes appearing according to severity of the disease. Upper intake level of Se is 400 μg d^{-1} for adult and is 45–280 μg d^{-1} (varies according to age) for children (ATSDR 2001). Selenium toxicity is also observed in animals with symptoms of vision loss, random walking, loss of hair, deformed and sloughing hooves, joint erosion, paralysis etc.

Geology of a region affects or influences the distribution of Se in soils. Soils containing >0.5 mg Se kg^{-1} are considered as seleniferous as the forages produced on such soils absorb Se more than the maximum permissible level for animal consumption. Pockets of seleniferous soils have been identified in north-eastern parts (mainly in Hoshiarpur and Nawanshahar districts) of Punjab, India. Dhillon and Dhillon (2003) examined the Se content of soils, irrigation water, plants and animal tissues in the region. These seleniferous soils occupied more than 1000 ha, but toxic sites were reported only in 4–16 ha that were distributed sporadically in the study area. The Se content in surface (2.12 ± 1.13 mg kg^{-1}) and sub surface (1.16 ± 0.51 mg kg^{-1}) soils in the toxic sites was 4–5 times higher than that of non-seleniferous areas. The development of seleniferous pockets was mainly because of the deposition of seleniferous materials transported by seasonal rivulets from higher reaches of the Siwalik hills and use of groundwater for frequently irrigating crops like lowland rice. Some parts of Rajasthan and Southern parts of the Haryana also had soils with selenium levels above normal soil. Selenium in contaminated soil and water exists mainly as highly mobile toxic inorganic species such as selenate (SeO_4^{2-}, Se^{6+}) and selenite (SeO_3^{2-}, Se^{4+}). As a result, these transfer efficiently through the soil-plant-animal- human system. Rice crop grown on a seleniferous soil (2.85 mg kg^{-1} Se) from Nawanshahar, Punjab recorded reduction in growth and delayed flowering. Selenium accumulation increased by about 3 to 20-folds in leaves and grains of rice grown on seleniferous soil as compared to normal soil containing 0.135 mg Se kg^{-1} (Sharma et al. 2014) (Fig. 11.6).

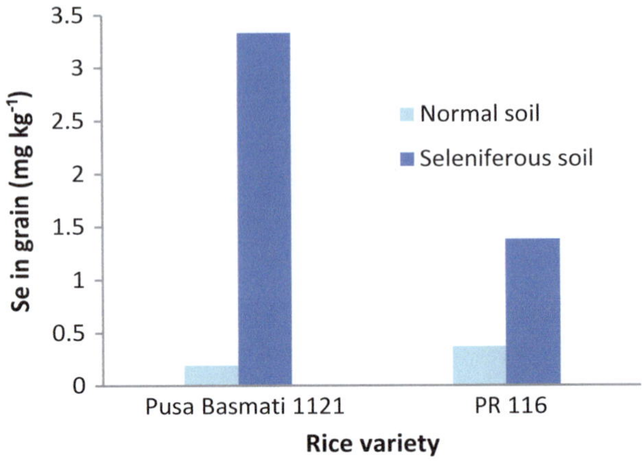

Fig. 11.6 Selenium accumulation in rice grain grown on seleniferous soil (Adopted from Sharma et al. 2014)

11.10 Environmental Risks of Organic Pollutants in Environmental Samples

11.10.1 Indiscriminate Use of Pesticides and Insecticides

In addition to fertilizers, a large amount of pesticides are used in agriculture to ensure a good yield of crops. Most part of the applied pesticide, irrespective of crops, ultimately finds its way into soil. Though a large part of these are degraded by soil microorganisms or inactivated by soil matrix through absorption, these affect adversely the functioning of non-target microbes and other soil organisms before inactivation. Prakash et al. (2004) studied the presence of HCH isomers residues in 45 surface (0–15 cm) and subsurface (15–30 cm) soils samples from agricultural sites of Delhi, Haryana, and Uttar Pradesh and around the HCH manufacturing plant of Indian Pesticide Limited. Thirty nine soil samples contained residues of *b*-HCH (2.5–463 mg kg^{-1}) and the remaining six samples showed the presence of *g*-HCH (0.08–43.00 mg kg^{-1}). And residues of *a*-HCH (0.04–98.00 mg kg^{-1}) and *d*-HCH (0.07–458.00 mg kg^{-1}) were detected less frequently. Random monitoring of pesticides in water had also detected residues of persistent organo-chlorines in many rivers like Ganga, Yamuna, Cooum, Ulsoor, Mandori, Hoogly inflicting damage to aquatic life and health of fish consuming human population (UNEP 2002).

In India, several cases of residues like parathion, endosulfan, DDT etc. were reported in food samples. The presence of pesticide residues in samples of fruits, vegetables, cereals, pulses, grains, wheat flour, oils, eggs, meat, fish, poultry, bovine milk, butter and cheese in India was reported by several investigators. Analysis of 16,948 samples of vegetable, fruits, spices, cereals, pulses, milk, animal feed, fish/crustacean, tea, honey, meat, egg, soil and ground water by 21 participating laboratories across the country during the period of April 2011 to March 2012, for the possible residues of agrochemicals like organo-chlorine, organophosphorous, synthetic pyrethroids, carbamates, herbicides etc. revealed that about 290 (1.7%) samples were found to contain these chemicals above maximum residue limit (MRL) as prescribed under Prevention of Food Adulteration Act (PFA)/Food Safety Standard Authority of India (FSSAI) (Kulshrestha 2013).

11.10.2 Environmental Risks of Other Organic Pollutants in Environmental Samples

Though persistent organic pollutants (other than pesticides) are less investigated in agricultural soils, concerns are expressed on their entry through several inputs. Devanathan (2012) have studied the contamination status of organo-halogen compounds (OHCs), including polychlorinated biphenyls (PCBs), polybrominated diphenyl ethers (PBDEs) and hexabromocyclododecanes (HBCDs) in human milk, fish and dust samples collected from different locations in India. Higher levels of OHCs were found in the dust samples near e-waste recycling areas and improper e-waste recycling and dismantling are considered to be the major sources of these contaminants. Farm-raised fishes contain relatively high levels of PCBs and PBDEs than wild fishes. However, the concentration of these contaminants and dietary intake from fish was much lower than the guideline values which indicate less risk. Municipal dumpsites in India have been found to contain dioxins and related compounds like PCDD/DFs high amount (Subramanian and Tanabe 2007, Subramanian et al. 2015). Toxic persistent organic pollutants (PCDD/DFs, PCBs, and PAH) have also been found in significant amount in composts prepared from municipal solid wastes which can be a potential medium for entry of these pollutants in agricultural land (Grossi et al. 1998). Fish and human milk samples particularly from surrounding areas of the municipal waste dumping sites of Kolkata and Chennai in India contain significantly higher levels of PCBs which suggest that there is a greater risk for infants living near these sites (Someya et al. 2010). The hazard quotients (HQs) values were above one for PCBs in the infants and toddlers living near the municipal dumping and e-waste recycling areas and it indicated high risk with toxic organic pollutants among human population living in the city and industrial area. Hence regular monitoring is necessary for having a more real assessment on status of these toxic organic pollutants and taking appropriate measures for reducing the pollutants level in the environment.

References

Adhikari S, Gupta SK, Banerjee SK (1993) Heavy metals content of city sewage and sludge. J Indian Soc Soil Sci 41:170–172

Adhikari T, Wanjari RH, Biswas AK et al (2012) Final Report of the project entitled "Impact assessment of continuous fertilization on heavy metals and microbial diversity in soils under

long-term fertilizer experiment" (Submitted to Ministry of Forest and Environment, New Delhi), p 175
Alam MG, Snow ET, Tanaka A (2003) Arsenic and heavy metal contamination of vegetables grown in Samta village, Bangladesh. Sci Total Environ 308:83–96
Ansari AA, Singh IB, Tobschall HJ (1999) Status of anthropogenically induced metal pollution in the Kanpur-Unnao industrial region of the ganga plain, India. Environ Geol 38:25–33
ATSDR (2001) Toxicological profile for selenium. Agency for Toxic Substances and Disease Registry, U.S. Department of Health And Human Services, Public Health Service, Division of Toxicology/Toxicology Information Branch, Atlanta, Georgia
ATSDR (2005) Toxicological profile for nickel. Agency for Toxic Substances and Disease Registry, U.S. Department of Health And Human Services, Public Health Service, Division of Toxicology/Toxicology Information Branch, Atlanta, Georgia
Balagangatharathilagar M, Swarup D, Patra RC, Dwivedi SK (2006) Blood lead level in dogs from urban and rural areas of India and its relation to animal and environmental variables. Sci Total Environ 359:130–134
Balakrishnan M, Antony SA et al (2008) Impact of dyeing industrial effluents on the groundwater quality in Kancheepuram (India). Indian J Sci Technol 1:1–8
Beijer K, Gao K, Jonsson ME et al (2013) Effluent from drug manufacturing affects cytochrome P4501 regulation and function in fish. Chemosphere 90:1149–1157
Bhagure GR, Mirgane SR (2011) Heavy metal concentrations in groundwaters and soils of thane region of Maharashtra, India. Environ Monit Assess 173:643–652
Bhattacharya P, Samal AC, Majumdar J, Santra SC (2009) Transfer of arsenic from groundwater and Paddy soil to Rice plant (*Oryza sativa* L.): a micro level study in West Bengal, India. World J Agric Sci 5:425–431
Bhattacharya P, Samal AC, Majumdar J, Santra SC (2010) Arsenic contamination in rice, wheat, pulses, and vegetables: a study inan arsenic affected area of West Bengal, India. Water Air Soil Pollut 213:3–13
Bhupal Raj G, Singh MV, Patnaik MC, Khadke KM (2009) Four decades of research on micro- and secondary- nutrients and pollutant elements in Andhra Pradesh. Research Bulletin. AICRP Micro- and Secondary-Nutrients and Pollutant Elements in Soils and Plants, IISS, Bhopal, pp 1–132
BIS (2012) Indian standard: drinking water-specification, 2nd revision, ICS 13.060.20. Bureau of Indian Standards, New Delhi
Business Standard (2015) NGT flays UP, MP government for pollution in Singrauli. http://www.business-standard.com/article/pti-stories/ngt-flays-up-mp-government-for-pollution-in-singrauli-115100600689_1.html#. Accessed 6 Oct 2015
CGWB (1999) High incidence of arsenic in groundwater in West Bengal. Central Ground Water Board, India, Ministry of Water Resources, Government of India
Chakraborti D, Biswas BK, Chowdhury TR et al (1999) Arsenic groundwater contamination and sufferings of people in Rajnandgaon, Madhya Pradesh, India. Curr Sci India 77:502–504
Chakraborti D, Mukherjee SC, Pati S et al (2003) Arsenic groundwater contamination in middle ganga plain, Bihar, India: a future danger? Environ Health Perspect 111:1194–1201
Chakraborti D, Das B, Rahman MM et al (2009) Status of groundwater arsenic contamination in the state of West Bengal, India: a 20-year study report. Mol Nutr Food Res 53:542–551. doi:10.1002/mnfr.200700517
Chakraborti D, Rahman MM, Murrill M et al (2013) Environmental arsenic contamination and its health effects in a historic gold mining area of the Mangalur greenstone belt of northeastern Karnataka, India. J Hazard Mater 262:1048–1055
Chetia M, Chatterjee S, Banerjee S et al (2011) Groundwater arsenic contamination in Brahamputra river basin: a water quality assessment in Golaghat (Assam), India. Environ Monit Assess 173:371–385
Choudhury UK, Rahaman MM, Mondal BKGK et al (2001) Groundwater arsenic contamination and sufferings of people in West Bengal, India and Bangladesh. Environ Sci 8:393–415

CPCB (2009) Comprehensive Environmental Assessment of Industrial Clusters. Ecological Impact Assessment Series: EIAS/5/2009–2010. Central Pollution Control Board, Ministry of Environment and Forest, Government of India

CPCB (2011) Report on SPM characterization for heavy metals concentration: Study areas-Raipur & Raigarh in Chhatisgarh state 2010–2011. Central Pollution Control Board, Bhopal. http://cpcb.nic.in/SPMCharacterization.pdf Accessed 19 Aug 2016

CWGB (2013) Central Ground Water Board, Govt. of India. http://gis2.nic.in/cgwb/Gemsdata.aspx. Accessed 16 Jan 2013

Dahal BM, Fuerhacker M, Mentler A (2008) Arsenic contamination of soils and agricultural plants through irrigation water in Nepal. Environ Pollut 155:157–163

Das HK, Mitra AK, Sengupta PK et al (2004) Arsenic concentrations in rice, vegetables, and fish in Bangladesh: a preliminary study. Environ Int 30:383–387

Devanathan G, Subramanian A, Sudaryanto A et al (2012) Brominated flame retardants and polychlorinated biphenyls in human breast milk from several locations in India: potential contaminant sources in a municipal dumping site. Environ Int 39:87–95

Dhal B, Das NN, Pandey BD, Thatoi HN (2010) Environmental quality of the boula-nuasahi chromite mine area in India. Mine Water Environ 30:191–196

Dhillon KS, Dhillon SK (2003) Distribution and management of seleniferous soils. Adv Agron 79:120–184

Elangovan D, Chalakh ML (2006) Arsenic Pollution in West Bengal. Technical Digest, National Bank for Agriculture and Rural Development, Issue 9, pp 31–35. https://www.nabard.org/pdf/issue9td-8.pdf Accessed 12 Aug 2016

Farid ATM, Roy KC, Hossain KM, Sen R (2003) A study of arsenic contaminated irrigation water and it's carried over effect on vegetable. Fate of arsenic in the environment. Bangladesh University of Engineering and Technology, Dhaka, pp 113–121

Feng J, Wang Y, Zhao J et al (2011) Source attributions of heavy metals in rice plant along highway in eastern China. J Environ Sci 23:1158–1164

Garari TK, Das DK, Sarkar S (2000) Effect of iron and zinc application on the availability of native and applied arsenic simulating low land rice condition. Paper presented at the International Conference on managing natural resources for sustainable agricultural production in the 21st Century, held at the New Delhi 14–18 February 2000

Ghose MK (2004) Effect of opencast mining on soil fertility. J Environ Indus Res 63:1006–1009

Goswami S, Das M, Guru BC (2008) Environmental impact of Siljora opencast manganese mining, Keonjhar: an overview. Vistas Geol Res 7:121–131

Goswami S, Mishra JS, Das M (2010a) Environmental impact of manganese mining: a case study of Dubna opencast mine, Keonjhar district, Orissa, India. J Ecophysiol Occup Health 9:189–197

Goswami S, Das M, Guru BC (2010b) Environmental degradation due to exploitation of mineral resources: a scenario in Orissa. The Bioscan 2:295–304

Govil PK, Reddy GL, Krishna AK (2001) Contamination of soil due to heavy metals in the Patancheru industrial development area, Andhra Pradesh, India. Environ Geol 41:461–469

Govil PK, Sorlie JE, Murthy NN et al (2008) Soil contamination of heavy metals in the Katedan industrial development area, Hyderabad, India. Environ Monit Assess 140:313–323

Gowd SS, Reddy MR, Govil PK (2010) Assessment of heavy metal contamination in soils at Jajmau (Kanpur) and Unnao industrial areas of the ganga plain, Uttar Pradesh, India. J Hazard Mater 174:113–121

Grossi G, Lichtig J, Krauβ P (1998) PCDD/F, PCB and PAH content of Brazilian compost. Chemosphere 37:2153–2160

Gupta DK, Chatterjee S, Datta S et al (2014) Role of phosphate fertilizers in heavy metal uptake and detoxification of toxic metals. Chemosphere 108:134–144

Halder D, Bhowmick S, Biswas A et al (2013) Risk of arsenic exposure from drinking water and dietary components: implications for risk Management in Rural Bengal. Environ Sci Technol 47:1120–1127

Halder D, Biswas A, Šlejkovec Z, Chatterjee D et al (2014) Arsenic species in raw and cooked rice: implications for human health in rural Bengal. Sci Total Environ 497-498:200–208
Huang RQ, Gao SF, Wang WL (2006) Soil arsenic availability and the transfer of soil arsenic to crops in suburban areas in Fujian Province, Southeast China. Sci Total Environ 368:531–541
Jain N, Bhatia A et al (2005) Impact of post-Methanation distillery effluent irrigation on groundwater quality. Environ Monit Assess 110:243–255
Jena D, Nayak MK, Acharya N, Singh MV (2003). Fluoride Distribution in Soil, Water and Plant in the Vicinity of NALCO Smelter Plnat at Angul in Orissa. In: Environmental Pollution-Proceedings of the International Conference on Water and Environment (WE-2003) 15–18 December 2003, Bhopal, India, pp 188–194
Jones KC, Johnston AE (1989) Cadmium in cereal grain and herbage from long-term experimental plots at Rothamsted, UK. Environ Pollut 57:199–216
Juwarkar A, Singh SK, Dubay K, Nimje M (2003) Reclamation of Iron Mine Spoil Waste Dumps Using Integrated Biotechnological Approach. In: Proceedings of national seminar on status of environmental management in mining industry, Varanasi, January 17–18, pp 197–212
Kabata-Pendias A (2000) Trace element in soils and plants, Third edn. CRC Press, Baton Raton, p 432
Kabata-Pendias A, Pendias K (1984) Trace elements in soils and plants. CRC press, Boca Raton, p 315
Kabata-Pendias A, Pendias H (1992) Trace elements in soils and plants. CRC Press, Baton Raton, p 365
Kanmani S, Gandhimathi R (2013) Assessment of heavy metal contamination in soil due to leachate migration from an open dumping site. Appl Water Sci 3:193–205
Kaul PP, Srivastava R, Srivastava SP et al (2002) Relationships of maternal blood lead and disorders of pregnancy to neonatal birth weight. Vet Human toxicol J 44:321–323
Krishna AK, Govil PK (2008) Assessment of heavy metal contamination in soils around Manali industrial area, Chennai, southern India. Environ Geol 54:1465–1472
Kulshrestha S (2013) Report of the expert committee to frame a policy for monitoring of pesticide residues in Fruits & Vegetables. Ministry of Health & family Welfare, Nirman Bhawan
Kumar A, Maiti SK (2015) Assessment of potentially toxic heavy metal contamination in agricultural fields, sediment, and water from an abandoned chromite-asbestos mine waste of Roro hill, Chaibasa. India Environ Earth Sci. doi:10.1007/s12665-015-4282-1
Larsson DJG, de-Pedro C, Paxeus N (2007) Effluent from drug manufactures contains extremely high levels of pharmaceuticals. J Hazard Mater 148:751–755
Lenka M, Panda KK, Panda BB (1992) Monitoring and assessment of mercury pollution in the vicinity of a chloralkali plant. IV. Bioconcentration of mercury in in-situ aquatic and terrestrial plants at Ganjam, India. Arch Environ Contam Toxicol 22:195–202
Maharia RS, Dutta RK, Acharya R, Reddy AVR (2010) Heavy metal bioaccumulation in selected medicinal plants collected from Khetri copper mines and comparison with those collected from fertile soil in Haridwar, India. J Environ Sci Health, Part B: Pesticides Food Contam Agr Wastes 45:174–181
Mahimairaja S, Sakthivel S, Divakaran J et al (2000) Extent and severity of contamination around tanning industries in Vellore district. In: Naidu R, Willett IR, Mahimairajah S, Kookana R, Ramasamy K (eds) Towards better management of soils contaminated with tannery wastes, ACIAR publication no 88. Australian Centre for International Agricultural Research, Canberra, pp 75–82
Masto RE, Lal CR, Joshy G, Vetrivel AS et al (2011) Impacts of opencast coal mine and mine fire on the trace elements' content of the surrounding soil Vis-à-Vis human health risk. Toxicol Environ Chem 93:223–237
McArthur JM, Banjeree DM, Hudson-Edwards KA et al (2004) Natural organic matter in sedimentary basins and its relation to arsenic in anoxic ground water: the example of West Bengal and its worldwide implications. Appl Geochem 19:1255–1293
McBride MB (1994) Environmental chemistry of soils. Oxford University Press Inc, New York

Ministry of Mines (2008) National mineral policy of India. Government of India
Ministry of Mines (2010) National mineral policy of India. Government of India
Mishra AK, Maiti SK, Pal AK (2013) Status of PM_{10} in bound heavy metals in ambient air in certain parts of Jharia coal field, Jharkhand, India. Int J Environ Sci 4:141–150
Mohanty M, Pattnaik MM, Mishra AK, Patra HK (2011) Chromium bioaccumulation in rice grown in contaminated soil and irrigated mine wastewater-a case study at south Kaliapani chromite mine area, Orissa, India. Int J Phytoremed 13:397–409
Mondal NC, Saxena VK et al (2005) Assessment of groundwater pollution due to tannery industries in and around Dindigul, Tamilnadu, India. Environ Geol 48:149–157
Mortvedt JJ (1987) Cadmium levels in soils and plants from some long-term soil fertility experiments in the United States of America. J Environ Qual 16:137–142
Mortvedt JJ (1996) Heavy metal contaminants in inorganic and organic fertilizers. Fert Res 43:55–61
Mukherjee A, Bhattacharya P (2001) Arsenic in groundwater in the Bengal Delta plain: slow poisoning in Bangladesh. Environ Rev 9:189–220
Mukherjee S, Nelliyat P (2007) Groundwater pollution and emerging environmental challenges of industrial effluent irrigation in Mettupalayam Taluk, Tamil Nadu. In: Comprehensive Assessment of Water Management in Agriculture Discussion Paper 4, International Water Management Institute, Colombo, Sri Lanka, p 51
Mukherjee A, Sengupta MK, Hossain MA et al (2006) Arsenic contamination in groundwater: a global perspective with emphasis on the Asian scenario. J Health Popul Nutr 24:142–163
Norra S, Berner ZA, Agarwala P et al (2005) Impact of irrigation with arsenic rich groundwater on soil and crops: a geochemical case study in West Bengal delta plain, India. Appl Geochem 20:1890–1906
Pal R, Mahima A, Tripathi A (2014) Assessment of heavy metals in suspended particulate matter in Moradabad, India. J Environ Biol 35:357–361
Pandey B, Agrawal M, Singh S (2016) Ecological risk assessment of soil contamination by trace elements around coal mining area. J Soils Sediments 16:159–168
Panwar NR, Saha JK, Adhikari T et al (2010) Soil and water pollution in India: some case studies, IISS Technical Bulletin. Indian Institute of Soil Science, Bhopal
Parr JF, Papendick RI, Hornick SB, Meyer RE (2009) Soil quality: attributes and relationship to alternative and sustainable agriculture. Am J Altern Agric 7:5–11. doi:10.1017/S0889189300004367
Parth V, Murthy NN, Saxena PR (2011) Assessment of heavy metal contamination in soil around hazardous waste disposal sites in Hyderabad city (India): natural and anthropogenic implications. J Environ Res Manage 2:27–34
Patel KS, Ambade B, Sharma S et al (2010a) Lead Environmental Pollution in Central India. In: Ramov B (ed) New Trends in Technologies, InTech, Available from: http://www.intechopen.com/books/new-trends-in-technologies/lead-environmental-pollution-in-central-India
Patel KP, Singh MV, George V, Ramani VP (2010b) Four decades of research on management of micro- and secondary- nutrients and pollutant elements in crops and soils of Gujarat. Indian Institute of Soil Science, Bhopal
Prakash O, Suar M, Raina V et al (2004) Residues of hexachlorocyclohexane isomers in soil and water samples from Delhi and adjoining areas. Curr Sci India 87:73–77
Rahman MA, Hasegawa H, Rahman MM (2007) Accumulation of arsenic in tissues of rice plant (*Oryza sativa* L.) and its distribution in fractions of rice grain. Chemosphere 69:942–948
Rangasamy S, Purushothaman G, Alagirisamy B, Mahimairaja S (2015) Chromium contamination in soil and groundwater due to tannery wastes disposals at Vellore district of Tamil Nadu. Int J Environ Sci 6:114–124
Rao VVSG, Dhar RL, Subrahmanyam K (2001) Assessment of contaminant migration in groundwater from an industrial development area, Medak district, Andhra Pradesh, India. Water Air Soil Pollut 128:369–389

Ravenscroft P, McArthur JM, Hoque BA (2001) Geochemical and palaeohydrological controls on pollution of groundwater by arsenic. In: Chappel WR, Abernathy CO, Calderon R (eds) Arsenic exposure and health effects IV. Elsevier Science Ltd, Oxford, pp 53–78

Rawat M, Ramanathan AL, Subramanian V (2009) Quantification and distribution of heavy metals from small scale industrial areas of Kanpur city, India. J Hazard Mater 172:1145–1149

Rothbaum HP, Goguel RL, Johnston AE, Mattingly GEG (1986) Cadmium accumulation in soils from long-continued applications of superphosphate. J Soil Sci 37:99–107

Roychowdhury T, Uchino T, Tokunaga H, Ando M (2002) Survey of arsenic in food composites from an arsenic-affected area of West Bengal, India. Food Chem Toxicol 40:1611–1621

Saha JK (2005) Changes in salinity and sodicity of soils with continuous application of contaminated water near industrial area. J Indian Soc Soil Sci 53:612–617

Saha JK, Sharma AK (2006) Impact of the use of polluted irrigation water on soil quality and crop productivity near Ratlam and Nagda industrial area. Agricultural Bulletin IISS-1. Indian Institute of Soil Science, Bhopal, India

Saha JK, Panwar N et al (2010a) An assessment of municipal solid waste compost quality produced in different cities of India in the perspective of developing quality control indices. Waste Manag 30:192–201

Saha JK, Panwar N, Singh MV (2010b) Determination of lead and cadmium concentration limits in agricultural soil and municipal solid waste compost through an approach of zero tolerance to food contamination. Environ Monit Assess 168:397–406

Saha JK, Panwar N et al (2013a) Risk assessment of heavy metals in soil of a susceptible agro-ecological system amended with municipal solid waste compost. J Indian Soc Soil Sci 61:15–22

Saha JK, Rao AS, Mandal B (2013b) Integrated management of polluted soils for enhancing productivity and quality of crops. In: Gaur RK, Sharma P (eds) Approaches to plant stress and their management. Springer, New Delhi, pp 1–21

Samal AC, Kar S, Bhattacharya P, Santra SC (2011) Human exposure to arsenic through foodstuffs cultivated using arsenic contaminated groundwater in areas of West Bengal, India. J Environ Sci Health A Tox Hazard Subst Environ Eng 46(11):1259–1265

Santra SC, Samal AC, Bhattacharya P et al (2013) Arsenic in food chain and community health risk: a study in Gangetic West Bengal. Proc Environ Sci 18:2–13

Sellamuthu KM, Mayilswami C et al (2011) Effect of textile and dye industrial pollution on irrigation water quality of Noyyal River basin of Tamil Nadu. Madras Agric J 98:129–135

Sharma RK, Agrawal M, Marshall FM (2008) Atmospheric deposition of heavy metal (copper, zinc, cadmium and lead) in Varanasi city, India. Environ Monit Assess 142:269–278

Sharma S, Goyal R, Sadana US (2014) Selenium accumulation and antioxidant status of rice plants grown on seleniferous soil from northwestern India. Rice Sci 21:327–334

Sharma S, Kaur J, Nagpal AK, Kaur I (2016) Quantitative assessment of possible human health risk associated with consumption of arsenic contaminated groundwater and wheat grains from Ropar Wetand and its environs. Environ Monit Assess. doi:10.1007/s10661-016-5507-9

Shyamsundar PC, Das M, Maiti SK (2014) Phytostabilization of Mosaboni copper mine tailings: a green step towards waste management. Appl Ecol Environ Res 12:25–32

Singh SK, Ghosh AK (2011) Entry of arsenic into food material – a case study. World Appl Sci J 13:385–390

Singh V, Brar MS, Sharma P, Brar BS (2011) Distribution of arsenic in groundwater and surface soils in south western districts of Punjab. J Indian Soc Soil Sci 59:376–380

Sinha S, Gupta AK, Bhatt K et al (2006) Distribution of metal in the edible plant grown at Jajmau, Kanpur (India) receiving treated tannery wastewater: relation with physico-chemical properties of the soil. Environ Monit Assess 115:1–22

Smedley PL, Zhang M, Zhang G, Luo Z (2003) Mobilisation of arsenic and other trace elements in fluviolacustrine aquifers of the Huhhot Basin, Inner Mongolia. Appl Geochem 18:1453–1477

Smilde KW, Van Luit B (1983) The effect of phosphate fertilizer on cadmium in soils and crops. Rapport 6–8, Inst. voor Bodemvruchtbaarheid, Oosterweg, pp 1–17

Smolders E (2001) Cadmium uptake by plants. Int J Occup Med Environ Health 14:177–183

Somasundaram MV, Ravindran G et al (1993) Ground-water pollution of the madras urban aquifer, India. Ground Water 31:4–11

Someya M, Ohtake M, Kunisue T et al (2010) Persistent organic pollutants in breast milk of mothers residing around an open dumping site in Kolkata, India: specific dioxin-like PCB levels and fish as a potential source. Environ Int 36:27–35

Stalin P, Singh MV, Muthumanickam D et al (2010) Four decades of research on micro and secondary nutrients and pollutant elements in crops and soils of Tamil Nadu. Research Publication No. 8. AICRP Micro- and Secondary- Nutrients and Pollutant Elements in Soils and Plants, IISS, Bhopal

Stracher GB, Taylor TP (2004) Coal fires burning out of control around the world: thermodynamic recipe for environmental catastrophe. Int J Coal Geol 59:7–17

Subramanian A, Tanabe S (2007) Persistent Toxic Substances in India. In: Li A, Tanabe S, Jiang G et al (eds) Developments in environmental science, vol 7. Elsevier Ltd. doi:10.1016/S1474–8177(07)07009-X

Subramanian A, Kunisue T, Tanabe S (2015) Recent status of organohalogens, heavy metals and PAHs pollution in specific locations in India. Chemosphere 137:122–134

Swain BK, Goswami S, Das M (2011) Impact of mining on soil quality: a case study from Hingula open coal mine, Angul district, Orissa. Vistas Geol Res 10:77–81

Tiwari K, Pandey A, Pandey J (2008) Atmospheric deposition of heavy metals in a seasonally dry tropical urban environment (India). J Environ Res Develop 2:605–611

Tiwary RK, Dhakate R, Rao VA, Singh VS (2005) Assessment and prediction of contaminant migration in ground water from chromite waste dump. Environ Geol 48:420–429

Tripathi A, Misra DR (2012) A study of physico-chemical properties and heavy metals in contaminated soils of municipal waste dumpsites at Allahabad, India. Int J Environ Sci 2:2024–2033

Tripathi RM, Ashawa SC, Khandekar RN (1993) Atmospheric depositions of Cd, Pb, Cu and Zn in Bombay, India. Atmos Environ 27:269–273

UNEP (2002) Environmental data report. United Nations Environmental Programme, Nairobi

USEPA (1999) Background report on fertilizer use, contamination and regulations, EPA-747-R-98-003. Office of Pollution Prevention and Toxics, Washington, DC

WHO (1988) Toxicological evaluation of certain food additives and contaminants: Arsenic. The 33rd meeting of the Joint FAO/WHO Expert Committee on Food Additives, 26, 155–162, Geneva

Yellishetty M, Ranjith PG, Kumar DL (2009) Metal concentrations and metal mobility in unsaturated mine wastes in mining areas of Goa, India. Resour Conserv Recycl 53:379–385

Chapter 12
Remediation and Management of Polluted Sites

Abstract Soils perform several important ecosystem functions and therefore polluted land requires remediation and appropriate management for restoration of its life sustaining functions. Several technologies have been developed for their remediation based on clean-up, detoxification and risk minimization approaches. All of these technologies have both advantages and disadvantages in respect of the extent of applicability, side-effects on other components of environment, cost & ease of adoption, speed & effectiveness of remediation etc. While removal of contaminants and the risk minimization are the major approaches for heavy metal polluted soil, degradation to non toxic or less toxic compounds is the most common approach for soils polluted with organic pollutants. Plants, microorganisms, nanotechnology have also been used for remediation of polluted sites with varying degree of success. Waste products from agriculture, industries, city etc. have also exhibited their potential in minimizing risk from pollutants. Growers may minimize risks from polluted land through modification of soil, crop and nutrient managements. This chapter also discusses examples of a remediation approaches followed in case of different polluted sites worldwide.

Keywords Soil pollution • Decontamination • Heavy metals • Organic pollutants • Integrated management • Phytoremediation • Nanotechnology

Contamination of soil, water bodies and sediments with toxic trace elements is the most unfortunate by-product of urbanization, industrialization and mining. The use of contaminated water enhances the chance of occurrences of diseases in plant, animal and human and adversely affects the soil health. As described in previous chapters, occurrence of significant soil and water pollution due to industrial, mining and urban activities have been reported through several studies in India and other countries. In order to restore the natural ecosystem functions, to prevent these from contaminating other components of environment, to improve the quality of human and animal life dependent on the polluted land area, their remediation has been advocated by government organizations, environmentalists, policy makers and land owners. By definition, soil remediation is a set of processes and techniques of reducing the mobile and in consequence, bioavailable fraction of contaminants in soils. The sole objective of soil remediation is to minimize transfer of pollutants into soil organisms, plants, food chains, surface water and groundwater bodies so

J.K. Saha et al., *Soil Pollution - An Emerging Threat to Agriculture*, Environmental Chemistry for a Sustainable World 10, DOI 10.1007/978-981-10-4274-4_12

that expression of toxicity to organisms is either prevented or minimized. Remediation of large rural areas with marginally polluted soils, and agricultural fields should be approached differently than the remediation of heavily contaminated areas such as those around mining and smelting sites. The selection of remediation technologies depends on:

(a) Size of contaminated area and its topographical location
(b) History of contamination process,
(c) Soil mineralogical, physical and chemical characteristics (e.g., structure, texture, pH, oxides, carbonates etc.),
(d) Physical and chemical properties of the contaminants,
(e) Degree of pollution (i.e., contaminants concentration and their vertical and horizontal spread in the profile),
(f) Intended land use after remediation,
(g) Technological options and financial support available,
(h) Possible impact of remediation technologies on environment,
(i) Associated legal and social issues of the region and country.

Objectives of remediation can be achieved either by soil clean-up process through removal and/or inactivation (of toxic metals and metalloids) and degradation (of organic pollutants) of contaminants or by preventing their spread to surrounding soil and groundwater through isolation and/or immobilization. In this section, different available remediation technologies and their appropriate applicable areas have been described.

12.1 Engineering Approaches

12.1.1 Isolation and Containment

Under this method, contaminants are isolated and contained to prevent further movement through reducing their permeability and increasing the strength or bearing capacity for the contaminants. Physical barriers (made of steel, concrete, bentonite, clay and tiled-walls) can be used for vertical and horizontal containment as well as for capping of contaminated zone. Capping with synthetic membranes is also advocated to reduce water infiltration and contamination of groundwater.

Advantages of containment approach are (i) simple and robust; (ii) generally economical compared to contaminant removal technologies; (iii) eliminates contaminant transport to other areas almost completely. This approach of remediation has also limitations like (i) it may not reduce mass, concentration, or toxicity; (ii) slurry/clay walls (if used for containment) are not impermeable and hence provide containment over a finite period; (iii) long-term monitoring of the containment system is required to ensure that contaminants are not migrating.

12.1.2 Excavation and Landfilling

'Soil excavation and landfilling' is the most common remediation method for metal polluted soils. The process consists of two steps: (i) removal of the contaminated soil from the contaminated site to a special land-fill and (ii) restoration of the site which may include backfilling the excavated space by clean soil with subsequent establishment of vegetation. Soil placed in landfills may also require additional treatment to stabilize metals against their leaching to groundwater. Hazardous waste landfills are specially constructed to contain the contaminant metals in effective manner. However, it is extremely environmentally disruptive to the site and can often be extremely expensive. 'Excavation and landfilling' is simple and is most widely used technique for small, heavily contaminated sites. Site managers often prefer this technique because of its low risk of failure, its predictable time frame for decontamination, and because it leaves the site in a relatively pristine condition. In case of contamination over a large area, this technique is simply not feasible due to high costs involved with excavation of large mass of contaminated soil; its replacement with equal quantity of good quality soil and limited availability of landfill space to contain high volume of contaminated soil.

12.1.3 Vitrification

During vitrification, contaminated soil mass is treated through the processes involving melting and refreezing to create a glass-like solid that entraps inorganic contaminants and thereby isolates these from the environment. It can be applied both *in-situ* and elsewhere above ground in a treatment unit (*ex-situ*). The high temperature applied during vitrification process also destroys organic contaminants in soil. Vitrification is thus applicable for treating soil contaminated with both toxic organic chemicals and trace elements. The process involves introduction of electrodes into the wet soil that is capable of carrying current. Heat generated during current flow raises temperature of surrounding soil mass to 1100–1450 °C that is sufficient to liquefy siliceous soil particle. After melting is over, contaminated mass is solidified upon cooling. Toxic gases can also be produced during vitrification of soil contaminated with certain contaminants. Full-scale vitrification technologies exist for soils contaminated with arsenic, lead and chromium. Soil mixed with various toxic wastes can also be treated in this manner.

Limitations Efficiency of vitrification process is reduced by higher contents of clay, moisture and debris. Applicability of solidification/stabilization processes is generally limited to shallow depth and smaller volume of contamination. Vitrification can not be used where soil is contaminated with large amounts of flammable or explosive materials. Heating the soil may cause movement of contaminants into clean areas in the subsurface region. Also, the solidified material may prevent future use of the site. Heating process rapidly volatilizes some organic pollutants and volatile heavy metals including radionuclides (if present). Such vaporized contaminants as well as the high voltage applied during the process pose potential health and safety risks to the human.

12.1.4 In-Place Stabilization/Immobilization

Contaminated soils may be stabilized by mixing with solidifying materials such as cement or other pozzolanic materials (siliceous or alumino-siliceous material possessing cementing properties), thermoplastics, or other suitable agents. Stabilization/immobilization techniques may be applied *in situ* at the site of contamination or *ex situ* on excavated soil. The stabilized matrix may be a solid impermeable mass or a more friable solidified matrix. The solidified soil matrix reduces exposed surface area, resulting in reduced contaminant contact with the surrounding environment and reduced loss by leaching. Cement is the most commonly used solidifying material because of it's low cost, readily available, and is relatively easy to handle. Cement has a high pH at which many metals form insoluble compounds; thereby it also stabilizes the contaminants chemically. The resulting pH of the matrix stabilized with cement, may be more than ten and therefore, the cement stabilized soil mass cannot support plant growth. In some cases, the chemical form of the contaminants, particularly halides and soluble salts, may retard cement setting time or reduce the strength of the soil-cement mixture. Stabilization/immobilization is most effective in low to moderately contaminated soils, especially where contamination is near the surface where excavation or *in situ* mixing is easiest. However, the site may require periodic inspection as most weathering processes may lower pH of the stabilized soil-mass and this may cause the cemented material to break down and liberate the contaminants back into the surrounding environment. Stabilization techniques may not be suitable for some forms of trace element contaminants, such as ionic species that exist as oxyanions (e.g., $Cr_2O_7{}^{2-}$, $AsO_3{}^{-}$) or metals (e.g., Hg) whose solubility as hydroxides is quite significant. Stabilization-immobilization technology may not be suitable at sites contaminated with wastes containing high amount of organic pollutants, especially volatile organics.

12.1.5 Soil Washing

Contaminated soils can also be remediated by soil washing. Two types of soil washing are followed:

12.1.5.1 Particle Separation Techniques

In this case, soil contaminants are associated with one particular range of size fractions of the soil which are subsequently segregated out and disposed separately. Most of the insoluble organic and inorganic toxic contaminants have the affinity to be attached with clay, silt, and organic soil particles through either chemical bonding or physical forces. This verity is utilized for reducing soil contamination through particle size separation. The silt and clay particles are attached to sand and gravels by physical forces like compaction and adhesion. Contaminated organic matter particles, fine clay (along with associated humus fraction) and silt particles are separated from the coarser sand and gravel soil particles by washing processes and thereby contaminants are concentrated effectively into a smaller soil volume for subsequent treatment or disposal. This technique is more suitable for removing high specific gravity particles such as heavy metal-containing compounds (lead, radium oxide, etc.).

Particle separation technique begins with a coarse separation technique such as dry sieving to remove large, relatively uncontaminated materials such as stones and construction material. A variety of engineering methods can subsequently be employed to segregate particles by their density, size and shape. These methods include sieving, use of hydrocyclones (which dissociate the larger particles from the smaller ones via centrifugal force), fluidized-bed separation systems in a vertical column (that carries aways maller particles at the top along with counter-current overflow), or gravimetric settling/flotation (based on the differences in surface-density characteristics of contaminant carrying particles). Addition of certain froth forming chemicals, flotation reagents along with aeration assists contaminate bearing particles to float and thus enhances efficiency of separation. Such washing technology is most effective only in soils with higher proportion of sand and silt, and most of the contaminants being associated with the clay or organic matter fractions.

12.1.5.2 Leaching Technique

In this case, contaminants are chemically solubilized and removed from the soil system. Soil washing as a decontamination technique removes contaminants solubilized using chemical extractants and may be conducted *in situ* or *ex-situ* on excavated soil. Chemical extractants used to enhance heavy metal solubility include inorganic acids (e.g., H_2SO_4, HCl), organic acids (e.g., acetic acid), complexing agents (detergents, EDTA), or combinations of these chemicals. During this process, suitable extracting solution is percolated through the soil. The solution is recovered and treated to remove the contaminants; and extractants are often re-used. This process continues until the remediation goal is achieved or further solubilization/extraction of contaminants is not possible. Soil washing by extraction or leaching has several limitations. It is only effective or possible on recently contaminated soils and where pollutants are relatively soluble. It also requires an extracting agent that can easily be reconditioned. The use of strong acids requires extraction reactors, where tubings and containers are constructed with anti-corrosive material. Furthermore, the chemicals used for extraction may have significant detrimental impacts on the environment (even though may not be strongly acidic), necessitating additional steps in soil remediation process. In case of *in-situ* soil washing, contaminants solubilized during this process may leach into underlying soil and groundwater, if carried out without proper precautions. Some forms of this remediation technique using concentrated acids result in a large amount of decontaminated soil with extremely low pH, as well as a sizable quantity of contaminant-containing solution, both of which require additional attention following remediation. However, this process is only sparingly used due to cost and engineering constraints.

12.1.6 Thermal Treatment

Thermal treatment is rarely used on soils contaminated with heavy metals. It is commonly used for removing organic contaminants through chemical degradation and volatilization by heating the soil to high temperatures. The vaporized contaminants are subsequently collected and treated. Different types of soil heating techniques are available like; electrical resistance heating, radio frequency/electromagnetic heating, hot air/steam injection. Theoretically, this process may also be applied to soil polluted with metals; but it is not usually considered a practical or effective technique unless it is highly contaminated (5–10%). Efficiency of extraction for certain contaminants varies and depends on the maximum temperature achieved in the process selected. Even at temperatures of about 1200 °C the volatility of most of the metals and metal compounds is too low to remove sufficient amount of the contaminants. Among the toxic metals, this method is most suitable for removing mercury, since it is easily transformed to its vapor form (Hg°) at high temperatures even at low concentration (Smith et al. 1995). The high energy demand required to volatilize contaminants makes this a costly remediation technique, and the chance of dispersal of volatilized contaminants to nearby uncontaminated areas may be a concern. This technique of remediation is less efficient for soils with reduced permeability to air due to high compactness or due to high moisture content. Such reduced air permeability hinders the operation of thermally enhanced soil vapor extraction by enhancing energy requirement for increased vacuum and temperature.

12.1.7 Electro-Reclamation

Electro-reclamation is based on an electrokinetic process which occurs during flow of direct current between a cathode and an anode inserted in the soil. The electric current results in the flow of small electrically charged particles and soil-pore water where cations and anions travel to the cathode and anode respectively. In most soils with a predominance of negatively charged surfaces, uncharged molecules migrate with the bulk flow of liquids towards the cathode. The technique is most successfully applied in high clay soils containing organic or relatively mobile contaminants and is applicable to a relatively few metal ions. During the remediation process, electrodes must be equipped with a purge system that helps to maintain pH around the electrodes and removes the contaminants. The effectiveness of electro-reclamation is mostly dependent on the chemical composition of soil pore-water and groundwater. For example in calcareous soils, the process efficiency depends largely on the nature of clay minerals and the calcium and magnesium bearing minerals like calcite, dolomite, gypsum etc. The heavy metal concentration in the soil after remediation depends on their initial concentrations, the electrokinetic mobility and the exchangeability with the other metal ions (Lageman et al. 1990). The efficiency of electroreclamation is less for soils with a high cation ion exchange capacity (causing stronger retention of metals) and having high acid neutralizing capacity (like calcareous soil). If resource-support is available, the efficiency of *in situ* remediation can be increased by irrigating the soil with slightly acidified water (pH 3–4). The level of acidification for irrigation water in such case depends mainly on the cation exchange capacity of the soil. Minimum moisture content of the soil for reclamation should not be less than 10%. Major limitations with electroreclamation process are high cost of equipment installation, substantial amounts of electricity requirement, and frequent maintainance of the system. Due to these constraints and its inability to remove precipitated contaminants, it is not often applicable at site.

12.2 Chemical Approaches

Due to high cost and limited applicability of above engineering based technologies, attention has been focused in recent years on the development of *in situ* contaminants immobilization methods that are relatively cost effective and less disruptive to the surrounding landscape, water bodies, and ecosystems. For severely contaminated barren land, immobilization of contaminants with soil application of suitable amendment materials and subsequent re-vegetation of the area are considered efficient and less expensive alternative remediation method. In this approach, soluble and potentially soluble solid phase forms of contaminants are converted to geochemically more stable solid phases through different soil processes like sorption, ion exchange, and precipitation. However, metals immobilization within soil should be significant and long-lasting so that process reduces contamination potential of the site to plant and water bodies. Subsequently, more vegetation can develop on the reclaimed site to stabilize the soil further and as a result of greater vegetative cover, lateral wind erosion is prevented to a larger extent and heavy metal percolation is greatly minimized. Successful *in situ* immobilization fulfils three main objectives of polluted site remediation: (1) to stabilize soil with vegetation cover and limit trace element transfer to plant and food chain, (2) to alter the trace element speciation in the soil and thus minimize their mobility to contaminate surface and groundwater, and (3) to diminish the direct exposure of soil organisms to contaminants and enhance biodiversity. Different soil amendments used as immobilizing agents and their effectiveness are described below:

12.2.1 Red-mud and Steel Shot Waste

Iron oxides may remove several trace elements from soil solution and the reactions associated have been related to adsorption, coagulation and diffusion of ions into lattice structure of soil minerals. Iron containing waste materials like red mud (a byproduct of aluminum industry) and steel shot waste (an industrial material containing more than 90% iron) has been found effective in decreasing heavy metals uptake by crops through immobilization action in soil (Mench et al. 1998). Application of red mud (at 3 or 5%) had been found to increase soil pH and decrease metal availability in soil (Gray et al. 2006). Perennial grass *Festuca rubra* failed to establish in the unamended contaminated plots, but grew and covered the amended plots almost completely. In another experiment, application of red mud to four different heavy metal contaminated soils reduced plant uptake of Cd, Pb and Zn in *Festuca rubra* and *Amaranthus* hybrids by 38–87%, 50–81% and 66–87% (Friesl et al. 2003). Heavy metals uptake by oilseed rape, pea, and wheat was also reduced with the application of alkaline amendment materials like red mud, lime and beringite (Lombi et al. 2002). In a field trial, Cd concentration in ryegrass decreased by amending the soil with steel shot waste (30%) and with sludge from water treatment plant (20%), Zn concentration fell by about 10% in both cases (Muller and Pluquet 1998). Red mud was found most efficient in decreasing the "mobile" fraction of metals and in promoting bacterial population and enhancing soil biochemical activity (Garau et al. 2007).

12.2.2 Clays and Aluminum Pillared Clays

Due to having high cation exchange capacity and large specific surface area, clay minerals are potential binding agents for various pollutants. Such binding/adsorption capacity of clay minerals toward cationic heavy metals can be enhanced by the addition of aluminium. After addition in soil, acidic aluminium is partially neutralized and is precipitated in the form of its hydroxides at the surfaces and in the interstitial space of clay minerals. Heavy metals can be adsorbed on a medium containing clay and Al-hydroxides present on the surface of clay. Aluminium-pillared smectites can be prepared by adding *in situ* aluminum polymers within clay layers (Bergaya and Barrault 1990). In arsenic contaminated garden soil, addition of 1% of Al pillared smectite decreased soluble As by about 75% and plant As by about 50% (Vangronsveld et al. 1994). Aluminum based binding agents have been recommended for remediation of soils polluted by Ni, Cu, Zn, or Cd (Lothenbach et al. 1997). Nickel, Cu and Zn were adsorbed specifically on both Al-montmorillonite (hydroxy-aluminium montmorillonite) and Al_{13}-montmorillonite (montmorillonite intercalated with polynuclear aluminium complex). These three metals were also incorporated within the interlayers of Al-montmorillonite and Al_{13}-montmorillonite during the aging processes. While Al_{13}-montmorillonite is effective in short pH range (6.1–7.5), Al-montmorillonite was most effective in the wide pH ranges (6–8 for Ni and Zn, 4–6 for Cu, and 7–9 for Cd).

12.2.3 Zeolites

Zeolites may be of both natural and synthetic in origin. These are crystalline, hydrated alumino-silicates of alkali and alkaline earth cations and possess infinite 3-dimensional crystalline structure having internally a series of interconnecting channels and cages with specific dimensions. Ion-exchange capacity is the primary determining factor for the use of zeolites in pollution control. The negative charge density in zeolite depends primarily on the extent of substitution of Al^{+3} for Si^{+4} in silicate tetrahedra. Heavy metal cations are bonded by these negatively charged sites at the surfaces and can also be trapped inside the interconnecting channels and cages. Plethora of research works have dealt with the potential of zeolites to be used as effective and efficient amendments for soil remediation. The application of natural zeolite decreased the available fraction of Pb in the garden soil and restrained the Pb uptake by rape (Li et al. 2009). Application of zeolite @20 g kg $^{-1}$ reduced the Pb concentration in the edible part (shoots) of rape up to 30% in the severely polluted soil (2000 mg kg^{-1}). Copper in the presence of zeolite additives had a much lower availability for leaching out and had lower mobility in the biogeochemical cycle and the stability constant of copper complexes was higher in soils with a zeolite additive (Burlakovs et al. 2012).

12.2.4 Coal Ashes

Large amount of coal ashes are generated from thermal power plants as waste product having tremendous disposal problem. Coal ash is predominantly composed of ferro-aluminosilicate minerals containing Al, Si, Fe, Ca, K and Na as major elements in variable proportions. Increases in both alkalinity and salinity are the most immediate effects of coal ash incorporation into the soil at high dose. Alkaline property of this industrial waste has been used for neutralizing highly acidic soils. Addition of coal ash to acid mine spoils may neutralize acidity and consequently, precipitate soluble heavy metals besides enhancing its physical properties. These chemical processes have made establishment of vegetation possible on metal polluted sites. However, coal-ash application may also result in excessive soluble salt concentrations, excessive boron (B), and in some cases, cause adverse effects on soil properties (e.g., cementation) and increased concentrations of several potentially toxic trace elements depending on their concentrations in original coal.

12.2.5 Phosphate Minerals

The apatite group of minerals (phosphate bearing compounds) have been found effective both in sequestering dissolved metals and in transforming adsorbed metals (on soil surface) to more insoluble phases. Phosphate minerals may immobilize trace metals through surface complexation and co-precipitation. Amendment of soil with apatite diminished availability of Cd and Zn to maize crop at several contaminated sites (Chlopecka and Adriano 1997). In comparison to lime, apatite has been found more effective in decreasing metal content of barley leaves, especially in soil heavily contaminated with Zn (Chlopecka and Adriano 1996). Both synthetic and natural hydroxyapatite had been found to reduce mobile forms (soluble and resin-exchangeable) of Pb in contaminated soils by precipitating soluble Pb as hydroxyl-pyromorphite [$Pb_{10}(PO_4)_6(OH)_2$] (Ma et al. 1993). The use of phosphates to remediate Pb-contaminated soils focuses mainly on *in situ* precipitation by changing the existing soil Pb chemistry through formation of Pb phosphates (like pyromorphites), with low solubility (Hettiarachchi and Pierzynski 2004).

12.2.6 Alkaline Materials

Liming is thought to be the oldest and probably most commonly adopted immobilizing technique for heavy metal cations in the soil solution and is specifically suitable in acid soils. Lime decreases readily plant available forms (water-soluble plus exchangeable fraction) of metals in soils (Saha et al. 1999). Their repeated application at regular intervals (every 2–5 years; @2–10 tha^{-1}) is necessary to keep metal immobile and therefore, larger quantities of lime are required compared to other inorganic alkaline amendments (Knox et al. 2001). High Fe and lime rich composted sludge materials were used successfully in remediation of Zn and Cd contaminated sites at Palmerton (USA). Lime mixed in compost has been found more effective in raising soil pH and hence, inactivation of bioavailable fraction of heavy metals.

Liming is however, not recommended for all types of contaminants. The increase in pH may mobilize toxic anions such as arsenates and chromates (by displacement with OH^-), or organic pollutants by solubilizing humic matter on which these are normally adsorbed. Lime loses its effectiveness significantly with time, and therefore metal may remobilize especially at high levels of soil contamination.

12.2.7 *Biochar*

In recent years, biochar has been recognized as novel carbonaceous material for remediation of heavy metals from soil and water system. A group of researchers found that the soils amended with biochar are having less bioavailable fraction of toxic metals especially Cd, Cr and Pb (Bolan et al. 2014). It acts as a stabilizing agent in soil as well as adsorbs the metals from available pool (soil solution) or from sewage water. Pigeon pea biochar has recently been found to reduce Cd and Cu bioavailability in soil and their uptake by spinach crop (Coumar et al. 2016a, b). Characteristics of biochar are determined by several factors, which include feedstock type, particle size of the feedstock and temperature and conditions of pyrolysis. Such variation in properties makes a particular biochar more suitable for metal stabilization in soil over others (Paz-Ferreiro et al. 2014). Large surface area of biochars is ascribed to be responsible for their higher sorption of heavy metals as demonstrated by scanning electron microscopy (Lu et al. 2012). Mechanisms for sorption may be through complexation of trace elements with different functional groups of biochar, through exchange with cations (such as Ca^{+2} and Mg^{+2}) or through physical adsorption. Biochar is generally alkaline in reaction due to contents of basic oxides. Enhanced pH due to biochar application may result in heavy metals precipitation in soil. Generally, biochar pH value increases with temperature of pyrolysis (Jindo et al. 2014). Studies indicated that biochar can be used along with phytoremediation technology for remediation of trace element polluted soils, though the mechanisms and strategy for remediation may vary with type of pollutant and soil type (Paz-Ferreiro et al. 2014).

12.2.8 *Organic Matter*

Compost is rich in humic matter and acts as a sinks for trace metals due to having several binding sites. Therefore it has been recommended as soil amendment to alleviate heavy metal pollution through a low cost approach (Huang et al. 2016). Compost application can change mobility and bioavailability of heavy metals in soil environment through various physical and chemical processes namely at adsorption, complexation, precipitation and redox reactions and as a result can modify the toxic effects on plants and microorganisms. The consequence of these processes on mobility depends on nature of the metal/metalloid as well as on the nature of composts. For example, cationic metals are immobilized but arsenic is mobilized due to application of compost (Kumpiene et al. 2007; Bissen and Frimmel 2003). Organic matter has been found to immobilize Cu from active fractions (water soluble and exchangeable fraction) into inert residual fraction (Saha et al. 1999). Organic matter added through compost can effectively reduce (electrochemically) soluble Cr(VI) (anionic form) to Cr(III) (cationic form) which is strongly absorbed by soil particles, thereby reducing its bioavailability. Unstable or not-fully decomposed compost may increase bioavailability of heavy metals due to formation of soluble organic complexes (Madrid et al. 2007). Immobilization of heavy metals by compost is a short duration phenomenon and such immobilized metals may be remobilized slowly during microbially controlled C mineralization in the long-run and reach food/fodder parts through plant uptake. Therefore compost application for remediation may be an appropriate low cost technology for lowering down the soluble metal levels in the soil immediately after a pollution/contamination event so that their migration to food and water bodies can be arrested to an acceptable level.

The major source of compost in India is agricultural wastes and animal manures. However due to increased mechanization of agriculture, reduced dependency on livestock population and increased *in situ* residue burning in field, compost production and utilization are becoming less frequent in India. Due to increasing urban population, potential of preparing composts from municipal solid wastes (MSW) has increased, which can be used for remediation of heavy metal polluted soils. However, presence of high amount of heavy metals in the MSW compost is itself an environmental concern (Saha et al. 2010). Though bioavailability of trace elements (Cu, Zn, Fe, Mn, Cr, Ni, Pb, Cd) in soils may increase due to application MSW compost; their contamination in edible part of the plants may not necessarily increase beyond toxicity thresholds (Papafilippaki et al. 2015). Even then, appropriate technology (such as segregation of organic wastes) should be adopted for preparation of such compost for reducing heavy metals in the final product so that risks of contamination can be minimized. Similarly, alum [$Al_2(SO_4)_3$] can be used to pretreat poultry manure for reducing the water-soluble concentration of As, Zn, Cu, and Cd (Rai et al. 2004).

12.3 Biological Approaches

12.3.1 Microorganism Based Techniques

Generally, the practical application of microorganisms in remediation technology is mostly available for soils contaminated with organic components. Microorganism based remediation techniques for metal-contaminated soils are still developing. Such techniques function by modifying bioaccumulation, biological oxidation/

reduction, and biomethylation of heavy metal contaminants. Presently, these techniques have not been applied so far for *on-site* remediation.

Heap leaching has been employed in mining processes to obtain economically desirable metals. Bacteria used to solubilize metals in heap-leaching may include some members of the genus *Thiobacillus*, autotrophic anaerobes that thrive at a pH range of 1–4. These bacteria oxidize reduced form of sulfur compounds (sulfides) and produce sulfuric acid that lowers pH, thereby enhancing metal solubility. Solubilized contaminants may be extracted, which may subsequently be treated by increasing the pH and precipitating these as sulfides under anaerobic conditions. The performance of the bioleaching process is dependent on the speciation of the metal contaminants, as well as the presence of sulfide minerals.

The Bio Metal Slurry Reactor (BMSR) is a technique proposed of removal of heavy metals using microbes from excavated soil (Vangronsveld and Cunningham 1998). This technique operates by accumulating contaminants in bacterial biomass, and thereafter by removing the contaminated biomass. The BMSR is a stirred tank reactor which is continuously fed with contaminated soil to which bacteria, water and nutrients are added. The bacteria, such as the strain of *Alcaligenes eutrophus var. metalloterans,* accumulate metals such as Cd or Zn on cell surfaces and precipitate them. In this process, metals bind to functional groups of the periplasmic peptidoglycan and to polysaccharides and proteins at the outer cell wall. On the cell surface, a crystallization process is induced which leads to very high metal to biomass ratios (between 0.5 to 5.0). Once the bio-precipitation process reaches a certain level, proteins and polysaccharides bearing precipitates are released from the cells and new proteins are synthesized. Bioprecipitation continues on both the released foci and the cell surfaces. In the last stage of the process, the slurry is sedimented in a settler, and a froth floatation process is used to remove the contaminant bearing bacterial biomass which remains in the overstanding suspension.

12.3.2 Phytoremediation of Soils Contaminated with Trace Elements

Two possible approaches can be distinguished for phytoremediation of trace elements-polluted soils, both aiming for a reduction of the bioavailable fraction of metals in the soil: (a) *in situ* metal inactivation by means of revegetation (known as 'phytostabilization' or 'photorestoration'), with or without non-toxic metal-immobilizing soil amendments, and (b) bioextraction by means of hyperaccumulating plants (known as 'phytoextraction'). Different types of phytoremediation technologies are described as below:

12.3.2.1 Phytoextraction

Phytoextraction removes metal contaminants from soil through growing and harvesting of plants that accumulate metals. This technique exploits plants' unique ability to take up, translocate, and accumulate metal ions from the soil. Initially, researchers used pre-existing plants (not genetically modified) which accumulate large quantities of metals and are called as hyperaccumulators. When concentration of metal in the harvestable above ground biomass of the plant reaches 0.1–1% of the dry weight, the uptake is considered hyperaccumulation. Certain hyperaccumulator plants like *Thlaspi caerulenscens, Haumaniastrum robertii, Ipomoea alpina, Macadamia neurophylla, Psychotria douarret, Thlaspi rotundifolium, Cystus ladanifer, Salix* sp. etc. are employed to remove heavy metals from soils. Such plants can tolerate high metal levels in soil and accumulate them at 10 to 500 times higher levels than other plants and crops. Using modern techniques of biotechnology, several high biomass producing phytoaccumulators have been developed by introducing relevant genes (from hyperaccumulator, bacteria, animals) into non-accumulator plants (Singh et al. 2003). Some of high biomass hyperaccumulators for which regeneration protocols have been developed include Indian mustard (*Brassica juncea*), sunflower (*Helianthus annuus*), tomato (*Lycopersicon esculentum*) and yellow poplar (*Liriodendron tulipifera*) (Mello-Farias et al. 2011).

For phytoextraction to be successful, plants should possess the following characteristics: (a) sufficient accumulation of the metal(s) intended to be extracted, preferably at >0.1% plant metal concentrations in the above-ground parts, (b) fast growth with large biomass, (c) suitable plant phenotype for easy harvest, treatment and disposal, and (d) tolerance to site conditions. The phytoextraction of Pb is perhaps the most challenging because of its low soil to plant transfer coefficients. Possible chemical amendments to induce hyperaccumulation include synthetic organic chelates (e.g., EDTA, DTPA, HEDTA) have been developed for agricultural use. Many of these chelates have great affinity for metals and form strong chelate-metal complexes, thereby preventing metals from precipitating once they are taken into plant root, thereby ensuring translocation to the shoots.

Phyto-extraction may not be a practical remediation technique in all sites. Even in the best case, this technique appears to remove only 100–200 μg metal g^{-1} soil $year^{-1}$ to a depth of about 30 cm. Even at this rate this technique may require several centuries to decontaminate deep, heavily contaminated areas. Without multiple cropping, this technique removes at the most 15 μg metal g^{-1} soil per year to the same depth. Phytoextraction may not be applicable to all heavy metal-contaminated sites, particularly those in which contaminants are not located in the rooting zone, or those which contain high level of extremely biotoxic contaminants and therefore not effectively hyper-accumulated. In some cases, careful site management may be necessary to prevent more soluble chelated metals from leaching and off-site migration.

12.3.2.2 Phytovolatilization

During the last decade, research has explored the use of plants to remove metals from soil by converting them to volatile forms. The volatilization process appears to involve both plants and plant-microbe associations, and appears to be primarily associated with the decontamination of mercury and selenium-laden soils. These plants and plant- microbes associations transform mercury and selenium into a more volatile and less toxic form, e.g., Hg^0 and di-methyl selenide. As selenium is an essential element for most animal life, the release of selenium into the atmosphere may not cause an environmental problem. This process may not be applied to other elements such as those which are not readily converted into volatile forms, or those that have potentially toxic volatile forms. The disadvantage with this technology is that the mercury released into the atmosphere is likely to be recycled by precipitation and then re-deposited back into land, lakes and oceans re-exhibiting its toxicity.

12.3.2.3 Rhizofiltration

Rhizofiltration is generally termed as decontamination of polluted groundwater by flowing it through the root mass. The contaminants are either adsorbed onto the root surface or are absorbed within the plant roots and/or transported to stems and leaves. In this method, plants are hydroponically grown in clean water rather than on soil until a sufficient root system has developed. Subsequently the clean water supply is substituted with polluted water supply to acclimatize the plant. After the plants become acclimatized they are planted in the polluted area where the roots uptake the polluted water and the contaminants along with it. As the roots become saturated, these are harvested and disposed of safely. This method has been found effective and cheap in decontaminating polluted aquous ecosystem. Several hydroponically grown plants like Indian mustard, sunflower and various grasses have been found to remove toxic metals effectively from water body. In a study, magnitudes of metals removal by Indian mustard were 131–563 folds as compared to initial solution concentrations (Dushenkov et al. 1995).

12.3.2.4 Rhizodegradation

Rhizodegradation (also known as enhanced biodegradation at rhizosphere, or phytostimulation or plant assisted bioremediation) is the break-down of organic pollutants in the root zone by soil microbes which is enhanced by the rhizodeposition. Compounds present in plant root exudates such as carbohydrates, alcohols, and organic acids act as C and energy sources for the soil micro-flora and enhance microbial growth and activity. The plant roots also make the soil friable and facilitate transport of air and water to the rhizosphere which further enhances microbial activity and consequent biodegradation of pollutants.

12.3.2.5 Phytostabilisation

Phytostabilisation is the process in which plants are used to immobilize soil and water contaminants. Unlike phytoextraction, this process mainly focuses on immobilizing pollutants in soil in the root zone but not sequestration within above-ground plant tissues. Pollutants become less mobile and hence, their exposure to livestock, wildlife, and human is reduced. The contaminants are absorbed and accumulated by roots or precipitated in the rhizosphere. This reduces or even restricts the mobility of the contaminants, preventing their migration into the groundwater or air, and also reduces their contamination of food chain through minimized phytoavailibility. This technique has also been found useful in re-establishing vegetation on sites that have been denuded due to the high levels of trace element contamination. Once a group of tolerant plant species are established, magnitude of wind erosion (and hence spread of soil borne pollutant) and migration of the soil-pollutants to groundwater is reduced. Phytostabilization involves three processes which include: humification, lignification and irreversible binding.

12.4 Degradation of Organic Pollutants in Soil

Growing awareness on the harmful effects of hazardous organic chemicals to the ecosystem and human health has led to a marked increase in framing strategies for remediation and restoration of polluted sites. Originally, the conventional remediation techniques involve excavation of contaminated soil and disposal to landfill (Liedekerke et al. 2014). Though the conventional methods like soil washing, solvent extraction and oxidation process effectively address the issue of contaminants removal from the sites, these have also got a number of limitations such as high cost and complexity of their application, possibility of secondary contamination, and lack of public acceptance which have become major drawbacks for successful implementation of these technologies (Kang 2014; Vidali 2001). Further, with the advancement of *ex situ* and *in situ* remediation techniques along with strict regulatory control of landfill operations, the biological treatments of soil have come into existence as alternative technologies for remediation of contaminated sites (Liedekerke et al. 2014; Singh and Jain 2003).

Organic pollutants are more likely to 'attenuate' and 'break down' (degrade) naturally in the soil environment than the inorganic contaminants in soil. Degradation of such organic contaminants by bacteria and other micro-organisms is well known and documented (Al-Awadhi et al. 1996) (Table 12.1). In this context, the capability of biological agents (i.e., bacteria, fungi, and plants) to degrade and

Table 12.1 Organic pollutant degradation by soil biota

Organisms	Organic pollutant	References
Pesticide		
Clostridium sp.	Dieldrin and Endrin	Maule et al. (1987) and Baczynski et al. (2004)
Bacillus sp.; *Micrococcus* sp.; *Pseudomonas* sp.; *Trichoderma viride*; *Trichoderma harzianum*	Dieldrin, Endrin	Matsumura and Boush (1967), Patil et al. (1970), Matsumura et al. (1971) and Katayama and Matsumura (1993)
Arthrobacter sp.; *Achromobacter*; *Klebsiella pneumoniae*; *Pseudomonas spinose*; *Pseudomonas aeruginosa*; Algae	Endosulfan	Kwon et al. (2002), Hussain et al. (2007), Singh and Singh (2011) and Sethunathan et al. (2004)
Aspergillus niger; *Aspergillus sydoni*; *Fusarium ventricosum; Trichoderma harzianum*	Endosulfan	Katayama and Matsumura (1993), Siddique et al. (2003) and Bhalerao and Puranik (2007)
Alcaligenes sp.; *Bacillus* sp.; *Cyanobacteria*; *Pseudomonas* sp.	DDT	Katayama et al. (1993), Megharaj et al. (2002), Nadeau et al. (1994) and Hay and Focht (2000)
Agrobacterium radiobacter; *Chelatobacterheintzii*	Atrazine	Struthers et al. (1998) and Rousseaux et al. (2003)
Ganoderma sp; Mixed Bacterial Consortium	Lindane	Rigas et al. (2007) and Bhatt et al. (2007)
Serratia; Vibrio proteus, Yersinia,	Tetrachlorvinphos	Ortiz-Hernández and Sánchez-Salinas (2010)
Bacillus sp.; *Brucella*; *Enterobacter Cereus*; *Klebsiella* sp.; *Pseudomonas aeruginosa; Synechocystis* (Cyanobacterium); *Serratia marscecens*	Chlorpyrifos	Lakshmi et al. (2008, 2009) and Singh et al. (2004), (2011)
Burkholderia sp.	Fenitrothion	Qing et al. (2007)
Sphingomonas sp.	2,4,Dichlorophenoxy acetic acid	Karin et al. (2010)
PAH		
Acinetobacter sp.; *Aeromonas* sp.; *A. faecalis*; *A. denitrificans; A.niger; Arthrobacter polychromogenes; Agrocybe* sp.; *Bacillus* sp.; *Beijernickia* sp.; *C. elegans; Curvularia turberculata; Curvularia lunata; Flavobacterium gondwanense*; *Flavobacterium* sp.;	Phenanthrene	Allen et al. (1999), Aitken et al. (1998), Moody et al. (2001), Yamazoe et al. (2004), Miyata et al. (2004), Liu et al. (2004), Zhang et al. (2004a, b), Liu et al. (2004), Habe and Omori (2003), Chauhan et al. (2008), Collins and Dobson (1996), Casillas et al. (1996), Sutherland et al. (1993, 1991), Da-Silva et al.

(continued)

Table 12.1 (continued)

Organisms	Organic pollutant	References
Ganoderma lucidum; Micrococcus sp.; *Mycobacterium* sp.; *Microbacterium* sp.; *Nocardia* sp; *Nocardioides* sp.; *Pseudomonas* sp.; *P. putida; P. aeruginosa*; *P. stutzeri*; *P. chrysosporium*; *P. sordida*; *P. ostreatus; Rhodococcus* sp.; *Rhodotorula glutinis*; *Streptomyces* sp; *S. griseus*; *Sphingomonas* sp.; *T. versicolor; Vibrio* sp.		(2003), Cerniglia and Heitkamp (1989) and Wang et al. (2008)
Aspergillus niger; Acinetobacter calcoaceticus; Bjerkandera adusta; Burkholderia cepacia; Caenorhabditis elegans; Ganoderma lucidum; Gordona sp.; Irpex lacteus; Leclercia adecarboxylata; Mycobacterium sp.; *Mycobacterium gilvum; Mycobacterium pyrenivorans; Pseudomonas putida; P. saccharophilia; P. ostreatus; Rhodococcus* sp.; *Sphingomonas paucimobilis; Stenotrophomonas maltophilia; Sphingomonas* sp.	Pyrene	Chen and Aitken (1999), Boldrin et al. (1993), Sharma et al. (2004), Derz et al. (2005), Kanaly and Harayama (2000), Launen et al. (1995), Bezalel et al. (1996), Schutzendubel et al. (1999), Wunder et al. (1997), Sack and Fritsche (1997), Boonchan (1998), Boonchan et al. (2000), Da-Silva et al. (2003), Vila et al. (2001), Churchill et al. (1999), Rehmann et al. (1998), Miller et al. (2004), Wang et al. (2008), Byss et al. (2008) and Valentin et al. (2007)
Bacillus thermoleovorans; Comamonas testosterone; Pseudomonas putida	Naphthalene	Habe and Omori (2003); Chauhan et al. (2008); Musat et al. (2009); AbuLaban et al. (2010); Bergmann et al. (2011); Hoffman et al. (2005); Annweiler et al. (2000)
Eisenia fetida	Polychlorinated biphenyls; Phenanthrene, anthracene and benzo(a)pyrene	Tharakan et al. (2004) and Contreras-Ramos et al. (2006)
Dyes		
Debaryomyces polymorphus	Reactive Black 5 dye	Yang et al. (2005)
Trichosporon sp.; *Cyberlindera* sp.; *Barnettozyma* sp.; and Candida	Reactive Black 5 dye	Martorell et al. (2012)
Baker's yeast	Astrazone basic dye	Farah et al. (2007)

(continued)

Table 12.1 (continued)

Organisms	Organic pollutant	References
Galactomyces geotrichum	Remazol Red, Golden Yellow HER, Rubine GFL, Scarlet RR, Methyl Red, Brown 3 REL and Brilliant Blue	Waghmode et al. (2011)
Staphylococcus epidermidis	Crystal Violet, Phenol Red, Malachite Green and Methyl Green	Ayed et al. (2010)

detoxify environmental contaminants is considered as a potential alternative technique for remediation of contaminated soil which is often termed as "Bioremediation" (Juwarkar et al. 2010; Megharaj et al. 2011). Bioremediation is considered as a cost-effective remediation approach. Bioremediation process developed from the emerging concept of green engineering, is defined as the design, commercialization, and use of biological processes and products that are feasible and economical while minimizing pollution at the source and risk to human health and the environment (Kirchhoff 2003). Bioremediation technologies are generally classified as either *in situ* or *ex situ*. *In situ* technologies involve treating the polluted soil mass at the site, while *ex situ* bioremediation involves the excavation of the polluted soil material followed by its treatment elsewhere. The potential applicable bioremediation approach/methods are "biostimulation", "bioaugmentation", "phytoremediation" and "rhizoremediation". The appropriateness of a particular bioremediation technology is determined by several factors, such as site characteristics, population of indigenous microorganism, and charcteristics (type, quantity and toxicity) of pollutant chemical species present.

Biostimulation is the technique applicable to the contaminated sites where the abundance of intrinsic/native metabolic activity is low due to less favourable condition and for soils that are not conducive (physical and chemical properties of soil) for biodegradation. The native microbial metabolic activity and the microbial degradation process may be improved by the biostimulants techniques through fertilizer nutrients and organic material addition. Since early 1980, compost or farmyard manure has been applied for remediation of contaminated sites and during the last few decades, it was found to be very efficient in chlorophenol, triazine pesticides and PAH contaminated sites (Häggblom and Valo 1985; Valo and Salkinoja-Salonen 1986; Antizar-Ladislao et al. 2004; Loick et al. 2012; Semple et al. 2001; Abdelhafid et al. 2000). However in most contaminated soils, the inherent metabolic potential is too low to be improved by biostimulation alone. In such scenario, the other bioremediation process "bioaugmentation" plays a major role in degradation and restoration of organic contaminated soil. Bioaugmentation is a process to enhance degradation and to facilitate the transformation rate of xenobiotics through introduction of either native or genetically modified microbes into soil (Kuiper et al. 2004). Bioaugmentation" has proven capability of degrading or transforming the chemical pollutants along with amendments, such as activated sludge or compost with the objective to improve degradation and enhance the transformation rate of organic pollutant in soil (Kuiper et al. 2004). Bioaugmentation is desirable in those locations that need exogenous microbial communities or flora to assist degradation of organic contaminants in a polluted soil (Megharaj et al. 2011). However, the main drawback of bioaugmentation technique is that the number of exogenous microorganisms decreases shortly with the passage of time of inoculation on to the site which might be due to contaminants toxicity, competitiveness and adaptability, relative to the indigenous members of natural communities (Tomei and Daugulis 2013). Further, the whole live cells, those have to be introduced for the purpose of bioremediation through bioaugmentation techniques, needs continuous supply of fresh inoculum, aeration, and nutritional requirements (Rayu et al. 2012). These limitations have boosted the use of formulated enzymes rather than whole cells as bioremediation agents, a process known as "enzymatic bioremediation". In the literature, several enzymes have been reported which can potentially degrade wider range of organic contaminants (Table.12.2) and found to be a relatively better alternative than the use of live microbial cell (Torres et al. 2003).

Another technology that provides promising solution for remediation of soil polluted with organic contaminants is "Phytoremediation". This is a low cost and environment friendly technology involving plants which can extract, sequester, immobilize and/or detoxify contaminants from polluted soil and water (Pilon-Smits 2005). Phytoremediation works through various mechanisms which include phytoextraction, phytodegradation, phytovolatilization, phytostabilization and rhizosphere biodegradation. Phytoremediation takes advantages of the specific extractive and metabolic capabilities of plants (Alagić et al. 2013; van Bohemen and van de Laak 2003; Caliman et al. 2011; Haritash and Kaushik 2009; Macek et al. 2000; Prasad 2007). The choice of plant used for phytoremediation of contaminated sites depends on the physical and chemical characteristics of the contaminants. Several plants including ryegrass (Ahmad et al. 2012), rice (Tao et al. 2006), zucchini (Parrish et al. 2006), cauliflower (Tao et al. 2004), poplar (Skaates et al. 2005) radish (Mikes et al. 2009), pumpkin (Whitfield Åslund et al. 2007), tobacco and nightshade plants (Vrkoslavova et al. 2010) have been reported to absorb and accumulate organic pollutants in biomass (Table 12.3). The added advantage of adopting phytoremediation technology in degraded contaminated sites is increasing the carbon stocks through plant biomass production and soil carbon sequestration. Furthermore, this approach is highly acceptable in local community as it causes minimum site disturbance. However, the technology suffers from some shortcomings/limitations; for example, it may require prolong time (several crop cycles) to achieve the objective and also may be ineffective in the highly contaminated soil (high pollutants concentration) due to restrictive crop growth & biomass.

To overcome this limitation, growing plants in conjunction with contaminant degrading and/or plant growth-promoting rhizobacteria offers much higher potential for the remediation of contaminated soils (McGuinness and Dowling 2009). Plant-associated bacteria, such as rhizobacteria (RB) and endophytic bacteria (EB) have been increasingly investigated for improving remediation of

Table 12.2 Potentially useful enzymes for organic pollutant degradation and their producer

Organisms	Enzymes	Organic pollutants	References
Dehalobacter restrictus; Desulfitobacterium hafniense	Perchloroethene (PCE) reductase	Polychlorinated biphenyls (PCBs)	Maillard et al. (2003)
Rhizobium tropici; Sphingomonas sp.	Carbazole 1,9a–dioxygenase 2-Hydroxy-6-oxo-6-(2′-aminophenyl)-hexadienoate hydrolase	Carbazole, dibenzo-furan and biphenyl	Saiki et al. (2003) and Habe et al. (2002)
Burkholderia sp.	Biphenyl dioxygenase	Polychlorinated biphenyls (PCBs)	Zielinski et al. (2003)
Polaromonas vacuolata; Pseudomonas spp.	Dioxygenase	Naphthalene;	Jeon et al. 2003
Pseudomonas putida	Monooxygenases; Dioxygenases	Herbicides trifluralin; Hexachlorobenzene and pentachlorobenzene	Singh and Walker (2006), Scott et al. (2008) and Riya and Jagatpati (2012)
Burkholderia xenovorans	Glutathione-S-transferase	Organochloride pesticide	McGuinness et al. (2007)
Phanerochaete chrysosporium	laccase, lignin peroxidases and manganese peroxidases	Phenanthrene, fluorene, fluoranthene, anthracene and pyrene	Sanglard et al. (1986) and Gibson and Subramanian (1984)
Escherichia coli	Phosphonatase	Organophosphorus compounds	Singh and Walker (2006), Scott et al. (2008) and Riya and Jagatpati (2012)

contaminated soil (Germaine et al. 2006; McGuinness and Dowling 2009). Plants and their associated bacteria interact synergistically with each other, wherein plant provides carbon source to the bacteria that stimulates their capacity to decompose organic contaminants in the soil. On the other side, plant root associated-bacteria can help their host plant to withstand stress related to contaminant-induced phytotoxicity (through mineralization of organic pollutants to a less toxic compounds/ element), and thereby support plant growth and development. As different enzymes are responsible for effective degradation of various organic pollutants (Table 12.2), consortia of different rhizospheric microorganisms having different enzyme generating capabilities probably work in a sequential and synergistic order to decompose wide spectrum of contaminants more effectively. This indicates that a favourable plant–bacteria association can help in reclamation more effectively of agricultural land contaminated with various types of organic pollutants. Degradation of several aliphatic hydrocarbons in soil was enhanced with growing of

Table 12.3 Potential plant accumulator of organic pollutants

Organic pollutants	Plant type	References
Phenanthrene, Pyrene	Maize	Xu et al. (2006)
Pyrene, PAH mixtures	Alfalfa	Fan et al. (2008) and Pradhan et al. (1998)
Pyrene	White Clover	Xu et al. (2009)
Phenanthrene, Pyrene	Rye grass	Goa et al. (2006) and Kang et al. (2010).
Pyrene Anthracene, Chrysene	Rice	Tao et al. (2006)
Phenanthrene, Pyrene	Amaranthus	Gao and Zhu (2004)
PAH mixtures	Cabbage	Kipopoulou et al. (1999) and Tao et al. (2004)
PAH mixture	Cauliflower	Tao et al. (2004)
Atrazine	Poplar, Rice, Salix	Field and Peel (1971a, b, 1972), Nair et al. (1993), Roche et al. (2009), Shann and Boyle (1994) and Schnabel and White D. M (2001)
Hexachloro-cyclohex-ane(HCH)	Oat	Pereira et al. (2006)
Chlorpyrifos	Ryegrass	Ahmad et al. (2012)
Polybrominated diphenyl ethers	Tobacco	Vrkoslavova et al. (2010)
2,4,6 trinitrotoluene, Phenol	Vetiver grass	Makris et al. (2007)
Dibenzofuran	Bermuda grass, White clover	Wang and Oyaizu (2009)
2,6-dinitrotoluene	Arabidopsis thaliana	Yoon et al. (2007)
POPs Pesticide (DDT and HCH)	Xanthium strumarium	Nurzhanova et al. (2010)
Metolachlor	Rice	Kawahigashi et al. (2006)
2,4-dichlorophenoxyacetic	Pea	Germaine et al. (2006)
Organochlorine pesti-cide (Dicofol)	Water hyacinth	Xia (2008)
Dieldrin; Endrin; Hepta-chlor; DDT; Chlordane	Cucurbits	Otani et al. 2007, Lichtenstein et al. (1965), Hülster et al. (1994), White et al. (2006), Mattina et al. (2004) and Lunney et al. (2004)

ryegrass (*Lolium perenne*) which had been linked to the increase in microbial population and their activities in the rhizosphere (Günther et al. 1996).

In addition to supply of nutrients (biostimulants), substrates, live cells (bioaugmentation), enzymes and plants (phyto and rhizoremediation), abiotic (environmental) factors, including soil physico-chemical properties and seasonal temperature fluctuations are also important for the degradation of organic contaminants in soil. In general, most of the organic contaminants are insoluble in water and therefore it gets strongly adsorbed to lattice structures in clay or on organic matrices within the soil particles (Higginson 1993). As a result, degradation process occurs normally in the organic contaminants-water interface and also with certain dissolved organic contaminants in water (du Plessis et al. 1994). Contaminant bioavailability to micro-organisms is one of major factor that determines its persistence or degradation rates in soil. Its bioavailability is governed by both soil properties (soil pH, cation exchange capacity, organic matter, soil texture, temperature, redox conditions, soil chemical constituents like iron, manganese and aluminium oxide content and the presence of other organic compounds) and contaminant characteristics (chemical structure including size and molecule charge, vapour pressure, polarity, or hydrophobicity and the age of the contaminants in soil as well) (Higginson 1993; Reid et al. 2000; Rao et al. 2014; Wen et al. 2012). A number of studies have indicated that bioavailability for degradation of organic pollutants, particularly PAHs is reduced due to their adsorption onto soil mineral particles and organic matter within the micropores and and hence, these are not readily available for biodegradation. For example, Fu et al. (1994) observed that 30 to 50% of naphthalene, the most water-soluble and bioavailable PAH, resisted desorption from an Oklahoma soil consisting of 92% sand. Recalcitrance of PAHs against degradation increases with prolonged aging time, i.e., extended contact time between the contaminant and the soil particle (Yap et al. 2010). Such aging process causes the transfer of compounds from accessible fraction of soil to less accessible ones resulting in reduction of contaminants' bioavailability for degradation (Reid et al. 2000). Bioremediation technologies for different groups of organic pollutants are discussed below.

12.4.1 Degradation of Polycyclic Aromatic Hydrocarbons (PAHs) in Soil

Generally land contaminated with low to moderate level of PAHs does not pose significant threat to plant. However, direct intake of PAHs may occur through ingestion (by children), inhalation (of dust), or dermal (skin) exposure to contaminated soil/dust during tilling, and grazing (by animals) or through soil → water → animal → human route causing potential health risk. As PAHs are highly toxic to human, these require degradation for minimizing the risk of exposure through above routes. Abiotic degradation of PAHs depends on several natural factors such as climatic conditions and geographic location as well as physical and chemical properties of soil. Some of these factors (e.g., temperature, humidity, pH, organic matter composition) also influence biotic degradation through their effect on growth of microorganisms (Maliszewska-Kordybach 2009; Smreczak and Maliszewska-Kordybach 2003). Photochemical decomposition is an important process of the degradation of PAHs in the surface soil layer (up to 3 mm). Resistance to photochemical decomposition decreases with the solubility as follows: naphthalene > phenanthrene > chrysene > pyrene > anthracene > benzo[a]pyrene > indeno (1,2,3-cd)pyrene. It was observed that abiotic decomposition of hydrocarbons increases with increase in intensity of light and concentration of oxygen (Little et al. 2002). Higher intensity of light may break aromatic ring structure of PAH compounds and causes further degradation of formed derivatives.

In the bioremediation process, microbial communities (including bacteria and fungi) can biologically decompose PAH compounds during metabolic process using these as substrate for carbon and source of energy or by co-metabolism while decomposing another organic compound. Individual microorganism isolates demonstrated a relatively good capacity to degrade PAHs which are more water-soluble (acenaphthene, fluorene, phenanthrene, fluoranthene). On the other hand, less water-soluble PAHs (like anthracene and pyrene) were more efficiently removed by consortia of microbial culture (*Pseudomonas putida*, *Flavobacterium* sp., and *Pseudomonas aeruginosa*) as compared to isolated cultures (Trzesicka-Mlynarz and Ward 1995). Biological degradation is dependent upon several abiotic soil factors, including nutrients, pH, metals, temperature, moisture, salts etc. Degradation of lower molecular weight PAH (upto 3 ring) has been found accelerated by increasing soil temperature; while level of high molecular weight PAHs were unaffected by temperature (Coover and Sims 1987). Due to strong interaction with non-aqueous phases particularly soil organic matter, PAHs become potentially unavailable for microbial degradation since microorganisms are known to degrade chemicals mainly in its soluble form. As hydrophobicity of PAHs increases almost logarithmically with rising molecular mass, biodegradation of high-molecular weight PAHs having five to seven rings becomes difficult (Johnsena et al. 2005). The efficiency of PAHs biodegradation is highly related to their solubility in water, with higher solubility being accounted for their increased accessibility to microorganisms for degradation. Researches have indicated that plants play an important role in the degradation of hydrocarbon compounds derived from crude oil through the release of enzymes (dehalogenase, nitroreductase, peroxides, and lactase) from roots that are able to transform organic pollutants. Several species of grasses such as *Festuca rubra* and *Lolium* planted on soil contaminated with crude oil or diesel oil had indicated significant degradation of pollutants (Haritash and Kaushik 2009). While efficiency of degradation in the presence of *Festuca rubra* and *Lolium* plants were more (77% and 92%), the same in absence of these plants ranged between 60% and 74%. Similarly, Liste and Alexander (2000) observed up to 74% removal of pyrene after 2 months of growing oat, lupin, rape, fennel, parsley, pepper and pine species on clay soil having 2.5% organic carbon as compared to 40% removal in the same soil without any plant. From their experiment, Oleszczuk and Baran (2003) concluded that favourable environment for the growth of microorganisms (neutral soil pH, optimum soil moisture) and the higher content of nutrient elements play an important role in the case of degradation of low molecular weight PAHs (having 3-ring), while the degradation of high molecular weight PAHs (having higher number of 4 or more rings) is determined by their bioavailability. The high status of organic matter and clay in soil decline the rate of degradation process. The goal of biodegradation should be for the process leading to complete mineralization of PAHs to CO_2, water, and other inorganic compounds and C incorporation in microbial biomass carbon. Unfortunately, biodegradation of PAHs often result in the build-up of metabolites (mainly ketones, quinones, dicarboxylic acid anhydrides and coumarins) of the process that can be more toxic and/or more soluble than the parent compound (Lundstedt 2003). Amending soil with biochar is nowadays being advocated as a part of strategy to sequester of C in soil (Powlson et al. 2011) in addition to increase in crop yields and to reduce the leaching of nutrients from soil (Jeffery et al. 2010; Singh et al. 2010). However, as biochar can act both as source and sink for PAHs, its role in remediation of contaminated soil was also investigated (Quiliam et al. 2013). Soil amendment with biochar increased the content of phenanthrene in soil as well as decreased its degradation due to increase in sorption capacity and consequently reduction in bioavailability. Destruction of PAH in coal tar contaminated soil was found to be accelarated after pre-treatment with solvents like acetone and ethanol (Lee et al. 2001).

12.4.2 Degradation of Polychlorinated Biphenyl (PCBs)

Polychlorinated biphenyls are extremely resistant to disintegration because of their very low solubility in water and the lack of a structural site in these molecules for enzymatic attack during biodegradation. Studies have demonstrated that PCBs can be degraded to some extent by augmenting activity of soil microorganisms; however, biodegradation rate declined with increasing degree of their chlorination (Correa et al. 2010). Some bacterial groups like β-Proteobacteria and Acidobacteria have relatively higher capacity for degrading PCBs and these are more abundant in soil contaminated with PCBs, especially the compounds having higher-chlorinated congeners (Correa et al. 2010). Root exudates from some photosynthetic plants species have been found to support the growth of PCB-degrading bacteria in the rhizosphere (Donnelly et al. 1994; Chekol et al. 2004). This indicates scope of phyto-remediation as an effective supplementary strategy for their breakdown. Synergistic interactions between arbuscular mycorrhizal fungus and *Rhizobium* improved phyto-remediation by alfalfa of an agricultural soil contaminated with PCBs (Teng et al. 2010).

12.4.3 Degradation of Dye Pollutants

Microorganism mediated degradation has been found most promising and low cost method of remediation of soil contaminated with dye pollutants (McMullan et al. 2001; Keharia and Madamwar 2003; Khehra et al. 2006; Saratale et al. 2010) and unlike PAHs, their degradation products in most cases has either no or less phytotoxicity. Polycyclic anthraquinone and triphenylmethane groups of azo-dyes however, have been found recalcitrant to biodegradation process (Puvaneswari et al. 2006). Ligninolytic enzymes exuded by white rot fungi have been found to degrade synthetic dye and other organic pollutants effectively (Pointing 2001). In an interesting study, a technology was developed for degrading dye-pollutants while harvesting other benefit to agriculture. Agricultural residues have shown significant potential for removing dye pollutants from industrial effluent, particularly of textile industry (Adegoke and Bello 2015). These residues may subsequently be used as substrate for growing white rot fungi without having any toxicity effect which consequently degrades the dyes. The spent material can be composted for using it as soil conditioner for improving soil quality (Nigam et al. 2000). Carbonaceous adsorbent prepared from carbon slurry waste obtained from fertilizer industry was found to adsorb dyes more as compared to adsorbents prepared from steel industry wastes due to higher porosity and appreciable surface area of the former (Jain et al. 2003).

12.4.4 Degradation of Antibiotics in Soil

Among the abiotic processes of antibiotics degradation, hydrolysis and photodegradation have been identified as major ways of their disintegration in soil. However, rate and quantum of degradation depends on the type of antibiotic compounds as well as on soil chemical condition and management. Hydrolysis caused degradation of chlortetracycline, but not fenbendazole and sulfapyridine (Thiele-Bruhn 2003). Numerous antibiotics are susceptible to microbes mediated (through enzymes) transformation reactions (Thiele-Bruhn 2003; Sarmah et al. 2006). Fixation on the surfaces of soil particles may protect antibiotics from degradation. Aerobic environment and higher temperature enhance degradation kinetics in soil (Halling et al. 2003). Photodegradation is limited to surface and hence, has no significant effect on the overall antibiotics concentration in soils. However, repeated turning of soil may enhance the photodegradation process.

12.5 Application of Nanotechnology for Remediation of Polluted Soil

Conventional polluted site remediation technologies are either very costly or very slow in process which limits the adoptability by their actual users. Recent researches indicate that nanotechnology can generate effective soil remediation solution which can save both cost and time of the process involved. Nanotechnology has been defined as the use of knowledge on characteristic behaviour and manipulation of matter at dimensions between approximately 1 and 100 nanometres, where the particles exhibit unique phenomena enabling their novel applications in different fields. Nanotechnology is being extensively used by industrial sectors involving semiconductors; computer memory and data storage technologies; optical and photonic technologies; energy; biotechnology; and health care. There are increasing efforts to use this technology to protect environment through prevention of pollution, inactivation of pollutants and clean-up of polluted sites. Remediation methods developed through nanotechnology approach (also called nanoremediation) involve the use of reactive nanoparticles for conversion, degradation and detoxification of pollutants. These particles have unique properties that enable these to reduce electrochemically and transform catalytically to degrade/convert the pollutants into their non-toxic or less toxic form. There is substantial reduction in the cost during nanoremediation as this is carried out *in situ* and hence, there is no need to transport the polluted soil to other places for treatment and disposal and also there is no need to pump out groundwater for above-ground treatment.

Several types of nanoscale materials have been explored for remediation, such as nanoscale zeolites, carbon nanotubes and nanofibers, various bimetallic nanoparticles (BNPs), and those prepared from metal oxides and titanium dioxide. Of these, nanoscale zero-valent iron (nZVI) is the most widely used nano-particle for remediation of polluted water and soil. The nZVI (both in its Fe° form and as Fe^{2+}) is able to reduce chemically the redox sensitive contaminants such as Cr(VI), As (V), trichloroethane and many other halogenated compounds; and therefore can immobilise or reduce their availability/mobility (Puls et al. 1999; Kumpiene et al. 2006; Kanel et al. 2006; CLAIRE 2007; Ludwig et al. 2007). These nanoparticles have also been found to remove chlorine from various organic contaminants at both laboratory scale and in field tests and reduce their toxicity (Liang et al. 1996; Kim et al. 2007). In this way, nZVI has been shown to be capable of reducing, transformation and detoxification of many known pollutants like halogenated methanes, halogenated ethanes, chlorinated organic solvents and dyes, organochlorine pesticides, and polychlorinated biphenyls (PCBs) etc. at ambient temperatures (Deng and Hu 2001; Zhang 2003). Due to their smaller size, nanoparticles remain suspended in solution for longer time. When surface is modified (through coating) to reduce surface electrostatic forces, mobility of these particles increases considerably within soil matrix (Yirsaw et al. 2016). In this way, surface modified nanoparticles become a versatile remediation tool suitable for direct injection as a liquid into the subsurface where contaminants are present. Under laboratory condition, iron oxide nanoparticles (12 nm diameter) removed As almost completely (99%) through their strong surface binding and subsequent removal using magnetic field (Rickerby and Morrison 2007). The nZVI has also been found to reduce As (V) electrochemically to As(III) at quite faster rate at neutral pH (Kanel et al. 2006). Most of the field application on use of nanoparticles has been demonstrated for remediation of groundwater contamination with chlorinated organic pollutants. In a field application, nanoparticles was shown to remain reactive in soil and water for up to 8 weeks and can flow with the groundwater for >20 m and as a result, these reduced trichloroethane concentration almost completely within a few days of their injection (Zhang 2003, 2005). Several remediation technologies make use of the magnetic properties of certain nanosized iron minerals (e.g. Fe_3O_4) (Hu et al. 2004). For example during As decontamination from groundwater, magnetized Fe-oxide nanoparticles bind As strongly which can subsequently be separated from water by applying magnetic field (Rickerby and Morrison 2007). Similarly, Cr(VI) was removed from wastewaters using nanosize magnetite crystals (Hu et al. 2004).

Reports of on-site demonstration of soil remediation are few. This is due to the fact of limited mobility of nanoparticles with soil matrix. Effectiveness of nanoparticles in groundwater remediation depends on geologic, hydrogeologic, and subsurface conditions of the contaminated area. These include chemical and mineral constituents, macro & micro porosity, hydraulic conductivity of the soil matrix; groundwater gradient and flow velocity, depth of water table, and geochemical properties of groundwater (e.g., pH, ionic strength, dissolved oxygen, oxidation-reduction potential, and concentrations of nitrate, nitrite, and sulfate). Efficiency of nZVI to decrease the mobility and bioavailability of different toxic elements like Cr, Cu, As and Zn was investigated in a chromated copper arsenate (CCA)-contaminated soil (Kumpiene et al. 2006). Application of this nanoparticle decreased mobile/bioavailable As and Cr concentrations considerably, for example, by 99% and 94% in soil pore water respectively and by 98% and 45%, respectively in soil leachates. Such drastic reduction in soluble fraction of toxic trace elements was reflected as decline in their concentration (As by 84% and Cr by 95%) in plant shoots. This nZVI mediated stabilization technology also reduced microbial toxicity and restored soil enzyme activities. The use of nZVI based soil amendments has shown encouraging results over relatively long treatment periods and in a different types of contaminated soils e.g. As-contaminated agricultural soils (Mench et al. 2006) and CCA-contaminated soils (Kumpiene et al. 2006). Use of nZVI and iron phosphate-based nanomaterials as soil amendments have shown potential to decrease bioavailability of Cr in soil after electrochemical reduction of Cr(VI) to Cr(III) and also to decrease bio-accessibility of Cu through strong bonding (Liu and Zhao 2007; Xu and Zhao 2007). Silicon is known for imparting resistence and for reducing toxicity of metals in plants (Shi et al. 2005; Tripathi et al. 2015). In a hydroponic experiment, both Si and Si-nanoparticle have been found to enhance the As toxicity amelioration in maize by reducing its uptake and oxidative stress (Tripathi et al. 2016). Effectiveness of Si-nanoparticle (SiO_2) was more as compared to Si (applied as Na_2SiO_2) in reducing uptake of As and in minimizing its toxicity on growth of maize (Table 12.4).

Nanotechnology application in soil pollution remediation also suffers from several limitations. Nanoparticles have the tendency to aggregate and settle, or form mobile nanoclusters within soil matrix, thereby reducing their efficiency of remediation. Beside self-aggregation, nanoparticles may get attached with suspended solid depending on the properties of the nanoparticle and chemical condition of the soil matrix, making them immobile and inactive (Kern et al. 2009). Thus challenge in the successful nanoremediation technology is to maintain nanoparticles in completely dispersed form within soil pore water.

Table 12.4 Effect of Si-nanoparticle on uptake of As and growth of 15-day old maize seedling under hydroponic experiment

	As in plant (μg g^{-1})			Growth reduction (%)		
As level (μM)	Control	Si	SiNP	Control	Si	SiNP
25	124	110	104	−14	−7	−4
50	272	226	212	−29	−19	−12

Tripathi et al. (2016)

12.6 Integrated management of Polluted Soil

Normally, engineering methods of decontamination and immobilization of pollutants as described above in section 12.1 are recommended for land containing very high levels of toxic metals, specifically with the immediate objective of preventing their spread in unpolluted area and in living organisms, and bringing down the threat within an acceptable level in a short span of time. Further, successful implementation of these technologies requires involvement/support of government departments as these involve high cost which may not be affordable by farmers/owners. On the other hand, agricultural land affected with low to moderate level of pollutants may require modifications/interventions in the existing soil, crop and input managements in order to improve and sustain soil quality, crop productivity and produce quality (Saha et al. 2014). For sustainable agriculture in unpolluted area, agriculturists advocate combined strategy of managing nutrient, crop, water, soil and land effectively for achieving the objectives of improving soil fertility, water use efficiency, conservation of soil and water and increasing cropping intensity; and such combined strategy is commonly termed as 'Integrated soil management' (NAAS 2012). In the context of polluted agricultural land, however, integrated soil management should additionally address the issues like protection from pollutant build-up, its remediation and improving soil biological environment. Most often, such management strategies depend on the following parameters (Saha et al. 2014):

(a) Type of pollutants and their mode of toxicity expression
(b) Contamination level
(c) Purpose of land use during and after remediation process
(d) Soil type (that determine their interaction with pollutants), depth of profile and topography
(e) Climate of the area
(f) Cropping pattern
(g) Availability of resources
(h) Economics

Inter-relationships among the above soil, crop, and climatic factors as well as agricultural management technologies are often multifaceted (Fig. 12.1). This results in complexicity in devising appropriate strategy for combating threats in polluted soil. The task becomes complicated further because of two reasons: firstly, ownership of land (and consequently decision making process) lies with farmers'

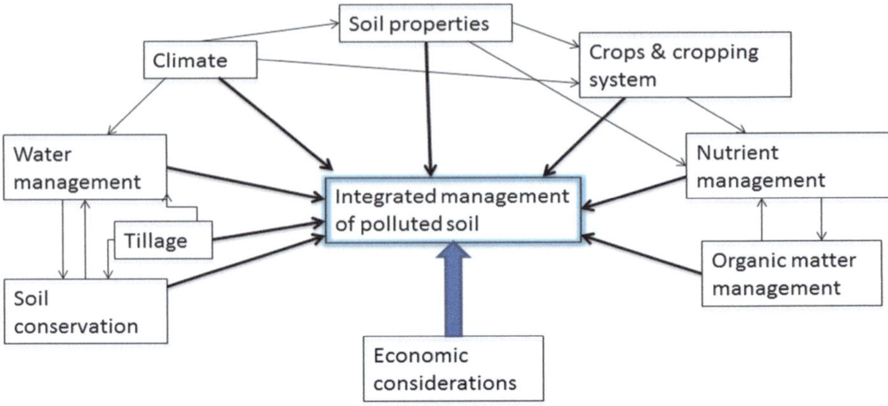

Fig. 12.1 Different controllable and uncontrollable factors affecting management of polluted soils and their inter-relationships (Saha et al. 2014)

and secondly, threat from pollutants is occasionally more to consumers rather than to the growers of food due to contamination (Saha et al. 2014).

Soils of India vary widely in respect of physical (depth, texture, structure, compactness etc.), chemical (mineralogical composition, clay type, Fe & Al oxides, organic matter contents, pH, base-saturation, CEC, soil solution composition, nutrient availability etc.) and biological properties (microbial population and their diversity and activity etc.). Such variations in soil properties result in effectiveness and adoptability of some of the soil management options over others during remediation process. While *in situ* soil washing for removal of metal ions may be expedient in light textured soils (like inceptisol and entisol), these may not be easily achieved in heavy textured soils (like Vertisol) due to compactness and lower macroporosity. On the contrary, manipulation of redox potential for reducing metal toxicity (through precipitation as sulphides) may be easier in latter soil type due to its slower infiltration capacity. Climatic factors (particularly magnitude and distribution pattern of rainfall and temperature) also control selection of appropriate management strategy due to their influence on transformation, mobility and degradation of pollutants in the soil profile. Roles of some of predominant soil and climatic factors on the processes and management for remediation of polluted soils are summarized below mentioned in Table 12.5.

12.6.1 Agricultural Operations Influencing Remediation of Polluted Land

Fertilization Fertilizers play two essential role in the management of polluted agricultural land: (1) enhancing production of food, fodder and other economic crops, and (2) enhancing removal of metals & metalloids by phytoremediating

Table 12.5 Important soil and climatic factors influencing different pedogenic processes and management strategy for remediation

Properties	Process influenced	Effect on pollution process and soil management
Climate	Infiltration and evapotranspiration	Accumulation of Na, Cl and other mobile pollutants on the surface. Leaching of salts and metals down the profile during rainy season.
	Available soil moisture	Solubility and bioavailability of pollutants and their uptake by plants
	Soil temperature	Activity of soil microorganisms, degradation of organic pollutants, inactivation of inorganic pollutants (e.g., Cr^{6+} to Cr^{3+}) (Weber 1972)
	Length of growing period of crops	Phytoremediation potential under rainfed situations
Geology of underground rocks	Groundwater recharge	Fractured underground rock formation enhances rate and magnitude of groundwater contamination
Land topography	Runoff, soil erosion	Spread of hazards to nearby land and water bodies
Texture and structure of soil	Infiltration	Contamination of groundwater, removal of salts from rhizosphere through subsurface drainage
	Aeration	Biodegradation of organic pollutants
	Compactness	Soil pulverization and photochemical degradation of organic pollutants during land farming.
Depth of soil profile	Vertical and horizontal mobility of water	Spread of polluted area due to industrial effluents
Clay types and contents	CEC	Sink capacity of soils for cationic pollutants, removal efficiency during decontamination
Organic matter content	CEC, coordinate complex formation	Sorption and chelation of heavy metals and sorption of organic pollutants influencing their bioavailability
	Microbe population and diversity	Degradation of organic pollutants
Soil reaction	Solubility of minerals and humus, strength of sorption	Expression of heavy metal toxicity, selection of amendment materials for immobilization and phytoextraction
Exchangeable Na	Dispersion of clay, solubilization of humus	Suspended particles carrying contaminated soil particles during runoff and contaminate water bodies
Oxides of Fe, Al	Sorption of anions, soluble Fe and Al ions,	Sorption and solubility of anionic pollutants like As, Se, Cr, organic pollutants etc.
	Hardness of soil upon drying (lateritic)	Ease of land farming to degrade organic pollutants

Saha et al. (2014)

crops through higher biomass production. While nutrient management through fertilizer is desired to enhance mobilization of pollutant elements in case of phytoremediation, it should not facilitate pollutants transfer to edible part during growing of food and fodder crops. Application of chloride containing fertilizers like NH_4Cl and muriate of potash (KCl) enhances mobility of the Cd through formation of soluble $CdCl_n^{2-n}$ complex and promote crop contamination. Therefore, these fertilizers should preferably be avoided in Cd contaminated soil for growing food and fodder crops (Sparrow et al. 1994: Lopez-Chuken et al. 2012). Fertilizers like ammonium sulfate and monoammonium phosphate lowers the soil pH during their continuous application and consequently, may increase the availability of metal pollutants (Levi-Minzi and Petruzzelli 1984; Willaert and Verloo 1992); and hence, may be avoided in metal/metalloid polluted neutral and calcareous soils for growing food & fodder crops. Monoammonium phosphate has been found to decrease soil suspension pH and to increase adsorbed fraction of Cd, while diammonium phosphate precipitated and fixed the metal particularly in low organic matter containing soil (Levi-Minzi and Petruzzelli 1984). As adsorbed Cd is considered more plant available as compared to its precipitated & fixed fractions, the differential effect of monoammonium phosphate fertilizers on food chain contamination can be anticipated. Contamination of food crops with arsenic (As) is widely observed in lower Gangetic plains of India (as discussed elaborately in previous chapter) and Bangladesh due to irrigation with contaminated groundwater (Meharg and Rahman 2003, Huq et al. 2006). In such soils, phosphatic fertilizer application has been found to reduce As contamination in wheat grain (Pigna et al. 2010). Under puddled cultivation condition, however, P addition suppressed uptake of arsenate, not arsenite by rice plant which indicate different uptake mechanism operating for different forms of As species (Abedin et al. 2002). High available calcium in soils has been found to inhibit accumulation of several heavy metals by crop (Kurtyka et al. 2008: Suzuki 2005) and hence, frequent application of Ca containing fertilizers (like SSP, CAN, gypsum) and lime may be advisable for growng food and fodder crops in non-calcareous soils in high rainfall areas. On the contrary, acid forming (and Ca lacking) fertilizers application can be an appropriate strategy during soil clean-up through phytoremediation.

Soil Organic Matter Management Due to its role in maintaining soil fertility and crop productivity, augmentation and maintenance of organic matter status in agricultural soil through periodic application of manure is essentially recommended, particularly in tropical and sub-tropical region of the globe where rate of C mineralization is quite high (Saha 2002). Application of organic matter has been recommended for ameliorating polluted soils due to its role in several chemical, biochemical and biological activities. Availability of heavy metals in contaminated soil has been found reduced with the application of organic matter due to their complexation by insoluble humic acids (Walker et al. 2004; Sklodowski et al. 2006). However, strength of metal complexes with humic acid increases with increasing pH and decreasing ionic strength (McBride 1994) and therefore, combined application of lime and organic matter may be most effective strategy in reducing/eliminating the adverse impacts (i.e., phytotoxicity and food contamination) in heavy metal polluted acidic soils. However, in another study, combined application of organic matter and lime also transformed metals from unavailable to potentially available pool (organically complexed and amorphous oxides bound); besides decreasing readily plant available pool (soluble plus exchangeable fraction) (Saha et al. 1999).

Due to their availability, composts prepared from sewage-sludge and MSW are widely used as organic matter amendment in agricultural land in nearby areas of generation. As contents of toxic elements and organic pollutants are generally quite high in these organic manures, their repeated application is likely to build-up toxicants in soil beyond permissible levels (Santodonato et al. 1981; Grossi et al. 1998; Saha et al. 2010). However, as contents of Fe-, Al-, Ca-oxides are significantly higher in sludge biosolids, expression of toxicity due to their repeated application may appear later as compared to MSW compost having similar level of toxic elements (McBride 1995). In order to avoid toxic element build-up in soil beyond the unsafe level for ecosystem, several countries have put forward regulatory guidelines on the use of such contaminated organic amendments in agricultural land in the form of 'maximum concentrations of metals allowed in agricultural soils', 'maximum metal concentrations in sewage sludge and MSW compost', 'annual metal loading limits for agricultural soils' etc. (McGrath et al. 1994; Düring and Gath 2002). In order to protect agricultural land as well as to harvest benefits from organic amendment, following scheme has been proposed for utilization of different quality of MSW composts with variable plant nutrients and heavy metals contents for different land use systems (Table 12.6).

Organic matter application has also been recommended for amelioration of agricultural land contaminated with organic pollutants. An integrated soil and crop management approach involving addition of stable compost, soil pH correction to near neutral and intensive cropping has been found to enhance the clean-up of PAH-contaminated soils during *in situ* remediation (Reilley et al. 1996; Oleszczuk and Baran 2003; Sayara et al. 2010). Improved biological activity due to these interventions in the rhizosphere appears to be the reason for such accelerated degradation. High pH and salinity reduces pace of PAHs degradation in saline-alkali soil and such constraints in amelioration can be reduced by application of organic matter (Luqueño et al. 2008).

Tillage Operations Methods of tillage (or no tillage) control root proliferation (hence volume and depth of rhizosphere), aeration, run-off of rainwater and soil erosion. As all of these processes influence degradation/inactivation of pollutants, transfer to crop biomass as well as their spread to other areas, type of tillage practices can be important for the management of polluted soils. Few studies have indicated role of tillage methods on remediation and management of polluted soil. Conservation tillage (reduced or no tillage with retention of residue from previous crop) has been recommended in case of polluted land as it reduces runoff and soil erosion during high intensity rainfall and thereby, prevents spread of pollutants to nearby unpolluted area and aquatic life (Holland 2004). Reduced

Table 12.6 Classification of MSW composts for their marketability and use in different area

Class		Organic matter and nutrient content	Heavy metal content	Quality control compli-ance	Overall quality and area of application
Marketable classes	A	High	Very low	Complying for all heavy metal parameters	**Best quality.** High manurial value and low heavy metal content. Can be used for high value crops and in organic farming
	B	Medium	Very low		**Very good quality.** Medium fertilizing potential and low heavy metal content
	C	High	Low		**Good quality.** High fertilizing potential and medium heavy metal content.
	D	Medium	Low		**Medium quality.** Medium fertilizing potential and medium heavy metal content.
Restricted use classes	RU-1	Low	–		Low fertilizing potential but safe for environment. Can be used as soil conditioner.
	RU-2	High	Low	Not complying for all heavy metal parameters	Can be used for growing non-food crops. Requires periodic monitoring of soil quality if used repeatedly.
	RU-3	Medium	Low		Can be used only for developing lawns/gardens, tree plantation (with single application).

Saha et al. (2010)

tillage or no-tillage promoted higher metal uptake from surface horizon due to restricted root proliferation in the deeper layer (Oliver et al. 1993) and therefore may be advisable during phytoextraction of surface contaminated land. On the other side, intensive tillage operations are recommended for land contaminated with organic pollutants as it enhances aeration and exposure of below-surface soils to sunlight and thereby, accelerates their biodegradation (Rhykerd et al. 1999).

Soil and Water Conservation and Management: Highly contaminated agricultural land also acts as a source of contamination for nearby land and surface water bodies and therefore, management strategies necessitate taking suitable measures to check soil erosion. Suitable measures like tillage of agricultural land across the slope, contour farming, cover cropping, mulching etc. may be adopted in high rainfall area to reduce impact of raindrops and consequent soil loss through runoff water. Raising deep rooted grasses and cover crops possesing significant phytoremediation potential can be better strategy in such situation as these can reduce runoff associated soil loss as well (Cook et al. 2009). Additionally, trees having phytoremediation potential (like *Populus* spp. and *Salix* spp.) may be grown along with grasses in order to phytostabilize metals in deeper soil layer and prevent downward metal flux besides preventing soil erosion (Quinn et al. 2001). In arid region, wind erosion may blow away contaminated soil particles from polluted land mass and have severe health implications on human and animal populations when enter the body through respiratory route. Capping of such land with fertile soils and growing grasses can prevent air pollution with contaminated dust (Vamerali et al. 2010).

Groundwater of the area in and around industrial clusters often records elevated salinity and use of such water for irrigation is likely to increase osmotic potential in the rhizosphere causing physiological water stress to crop (Saha 2005; Saha et al. 2014). Moreover, elevated Cl^- (normally associated with salinity of the groundwater) in soil solution due to use of saline water as irrigation may enhance the mobility and uptake of heavy metals (particularly Cd) in contaminated land (Lo´pez-Chuken et al. 2012). Therefore, agronomic management approach for cultivated lands around the industrial area should stress on enhancing water use efficiency through efficient irrigation methods (like drip, sprinkler) and also on reducing dependency on groundwater through conservation of rainwater and soil moisture.

Selection of Crops and Cropping Systems Crops differ widely on their ability to take up metals and also on ability to combat toxicity from heavy metals & metalloids through various mechanisms like exclusion, sequestration, metal homeostasis (Manara 2012). Quantum of threat(s) perceived in the farming of polluted agricultural land determines the crops species that can be grown. A relatively lower level of pollution generally poses threat only to the quality of food and fodder due to contamination with toxic elements/compounds and may not have adverse impact on crop growth and economic yield (Saha et al. 2013). Therefore, minimizing their movement to edible part appears to be the appropriate strategy in such situation. Researches indicate that bio-concentration of toxic elements is generally the highest in leafy vegetables followed by root vegetables and minimum in grain crops (Page et al. 1987). Therefore, cultivation of leafy and root vegetables (like spinach, lettuce, cabbage, potato, radish, beet, coriander etc.) may not be recommended in soils having elevated levels of metals/metalloids and cultivation of other crops should only be taken up after ensuring that legislative requirement for food and fodder quality in respect of toxic elements contents is complied. In an experiment under controlled condition, various cereals, oilseeds, sugarcane, and fiber crops showed tolerance to high soil Cd levels (Wang 2002). While grain of cereal crops got contaminated, edible parts of other crops remained free from Cd contamination. Even level of metal contamination in grains varied significantly among crop varieties due to differences in uptake pattern, upward and basipetal translocation in plant parts (Chan and Hale 2004). This emphasizes the importance of crop and genotype selection in the contaminated soil and need for further researches in this direction in view of large pollutants types. Chemical composition of root exudates and its role on mobilization of nutrients and metals vary widely with crop species (Gleba et al. 1999). As result, crop rotation has also been found to influence heavy metal mobilization and uptake by plants caused by residual effect of several organic acids in the root exudates from previous crops (Nigam et al. 2001). Concentration of Cd in grain was more in wheat grown after lupins and was the lowest in wheat grown after cereal, particularly under zero-tillage (Oliver et al. 1993).

Agricultural land severely contaminated with toxic elements requires decontamination as it affects biomass yield of common food crops as well as contaminate the food-chain. Appropriate cost effective strategy for such land is avoidance of food and fodder crops cultivation and removal of metals & metalloids by phytoextracting plants. Several oilseed crops (particularly *Brassica* sp.), flowering plants, fibre crops, short duration trees (like eucalyptus, poplar, willow etc.) have exhibited phyto-remediation potential while generating economic produce during the process (Broadley et al. 1999; Angelova et al. 2004; Su and Wong 2004; Indoria and Poonia 2006; Lal et al. 2008; Ruttens et al. 2011). Use of metal contaminated land for growing of crops generating bio-fuel like sugar (sugarcane, sugar beet), starch (maize, wheat, rice), oil (rapeseed, sunflower, jatropha) can be a feasible option. Contaminated biomass can be incinerated to reduce the volume for appropriate disposal or for metal-recovery while generating energy for useful purpose. It has been estimated that cultivation of maize could produce 33,000–46,000 kWh of renewable energy (electrical and thermal) per hectare per year (Meers et al. 2010). Hence this plant can be grown in moderately contaminated soil as its phytoremediation potential is low. Hemp (*Cannabis sativa*), flax (*Linum usitatissimum*) and peanut (*Arachis hypogaea*) were also found appropriate for both biodiesel production as well as for phytoextraction of Cd from contaminated land (Shi and Cai 2009). Biomass residues after biodiesel/bioethanol production can further be incinerated for energy production with consequent reduction in volume for disposal of hazardous waste (ash) (Saha et al. 2014).

Management of Soil Contaminated with Organic Pollutants Degradability of organic pollutants depends primarily on their physical and chemical properties, concentration level, bioavailability (mostly solubility in aqueous phase) and activity of microorganisms (both general and specific) in soil. Although cost effective and environment friendly, degradation technologies of organic pollutants depend mostly on microorganisms. Indigenous microflora may not be effective for degradation of toxic and recalcitrant types of pollutants. Under such situation, bio-augmentation with single microorganism or consortia of several microorganisms capable of degrading target pollutant is found more effective (Diez 2010). As bioavailability of organic pollutants is desirable for their degradation, application of organic manure may be ineffective during degradation of hydrophobic compounds due to their sorption on SOM causing diminished solubility (Neumann et al. 2014). On the other hand, compost or manure application may be a highly effective in soils with low organic matter contents for degrading relatively soluble simple hydrocarbons and 2–3 ring aromatic hydrocarbons as it provides substrate for microbial proliferation and hence accelerates degradation rate (Bosma et al. 1997). Neutral to slightly alkaline pH, abundance of electron acceptors (like O_2, nitrate, Mn and Fe oxides), optimum level aeration and moisture, and moderate temperature usually favors microbial growth, and consequently contaminant degradation. An integrated remediation strategy involving physical (volatilization), photochemical and microbial methods of remediation, along with phyto-remediation has been found superior in degrading 16 priority PAHs in soil over individual techniques (Huang et al. 2004). The techniques applied in this integrated approach for soil remediation were land-farming (promoting aeration and exposure to sunlight through repeated tillage), bioaugmentation (inoculating soil with contaminant degrading bacteria and plant growth promoting rhizobacteria, PGPR), and growing contaminant tolerant plant (*Festuca arundinacea*). The investigators used strains of PGPR to enhance plant tolerance to contaminants and to accelerate plant growth in heavily contaminated soils. Fertilization and clipping of above-ground biomass were also found to enhance degradation of PAHs due to enhanced root turn-over and rhizosphere activity (Olson et al. 2008).

12.7 Examples of Major Soil Remediation Projects

12.7.1 Homebush Bay, New South Wales, Australia

Homebush Bay and the adjacent land area were heavily contaminated due to decades of industrial activity with various organic contaminants, including Persistent Organic Pollutants (POPs), such as DDT and dioxins. Other contaminants included phthalates, polycyclic aromatic hydrocarbons, and heavy metals. As a result, fishes from this area were found unsafe for human consumption and subsequently fishing was banned. Remediation from 2008–2010 removed about 75% of the dioxin from the bay. Due to persistent nature of the contaminants and their capability to transfer in the marine environment by natural pathways, removal of soil and sediments and their treatment were selected as part of remediation strategy. The remediation method involved dredging of about upper 50 cm layer of sediment from the bay and subjecting the removed material to thermal treatment to decompose the POPs entirely. A clean shale cap was then placed into the dredged area.

12.7.2 Zinc Smelter-Contaminated Sites at Palmerton, USA

Zinc smelting operations for about a century dispersed heavy metals throughout the adjoining area in Palmerton (Pennsylvania). This caused build-up of high metal concentrations in soil which resulted in ecosystem collapse and prevented any vegetation to grow across a 3000 acre area. Such loss of vegetation increased erosion of the contaminated soil particles to further contaminate surface and groundwater bodies (Anderson et al. 1980). Because extensive land area had been so contaminated, it had become difficult and impossible to approach for removal and replacement of contaminated soils, especially when high slopes are involved. Major approach in the remediation process was *in situ* immobilization of heavy metals in the soil so that contaminants do not spread in the surrounding land and water bodies and such heavy metals inactivation/immobilization was done by adding amendment materials and through phytostabilization. The main objectives of *in situ* inactivation and phytostablization were: (a) to transform the trace element speciation for the purpose of reducing its mobility in soil; (b) to minimize soil erosion through establishing vegetation cover; and (c) to limit uptake of trace element by food & fodder crops; (c) to minimize the direct exposure of living organisms to the contaminated soil; and (d) to improve biodiversity. Biosolid compost was used as amendment which reduced soluble Zn and Cd concentrations in soil considerably and ameliorated phytotoxicity. Significant increase pH and levels of organic matter, Fe-oxides, and phosphate in soil due to application of biosolid-compost were ascribed as the reasons for decreased bioavailability of metals which led to the overall success of vegetation establishment, particularly turf-grasses in the highly contaminated area (Li et al. 2000). The combined strategy of the metal inactivation and subsequent phytorestoration not only resulted in the restoring of normal ecosystem functions but also subsequently minimized lateral spread of trace elements through wind erosion, as well as contamination of surface and groundwater. Vegetation establishment was not aimed to clean-up of contaminated soil, but for stabilizing (inactivating) trace elements which are potentially toxic. Mushroom compost, agricultural lime, coal fly/bottom ash from thermal power plant, fertilizers and seed mixtures were the other soil amendments used for metal immobilization and phytostabilization during reclamation of highly contaminated land of Palmerton area.

12.7.3 Love Canal at New York, USA

Love Canal is one of the most widely discussed environmental tragedies in American history, where several toxic organic pollutants (dioxin was major compound identified) released into environment due corrosion of containers containing wastes from chemical factory buried in the landfill site. After diagnosing the release of toxic chemicals in the land and air, the area was evacuated for remediation purpose. The landfill area was covered and fenced; and a leachate treatment plant was constructed. A system was installed to collect leachate from the site. In order to contain hazards from landfill site, a barrier drain was constructed; a leachate collection system was installed and the site was covered with a synthetic material to prevent rain from coming into contact with the buried wastes. Contaminated sediments and soils were removed from sewers and creeks and were subjected to thermal treatment to remove the pollutants. The waste materials and leachates collected were also subjected to thermal treatment/incineration and non-hazardous residues from the incineration process was subsequently disposed on-site.

12.7.4 Phytoremediation of Metal Mined Soils

Mining activities generate a large volume of waste rocks and tailings, and result in huge loss of land area under forest, cultivation and grazing, and induce trace element pollution of soils. In heavily contaminated soils of a large area, removal of these elements through phyextraction would require an unrealistic time period. The recommended practice is to revegetate the land with drought resistant, fast-growing plants, which can reduce metal bioavailability and mobility so as to protect food chain and underground water from contamination. Vetiver grass (*Vetiveria zizanioides*) was successfully used for revegetating tailings from Pb/Zn mine in South China (Shu et al. 2000). Vetiver grass performed better as compared to other three grass species, namely *Paspalum notatum*, *Cynodon dactylon* and *Imperata cyclindrica* in terms of biomass production and land coverage. This plant had also been found suitable for the treatment of landfill leachate as well as for the rehabilitation of metal contaminated soils in Queens land, Australia (Truong and Baker 1988). Suitability of vetiver grass for phytoremediation has been due to its several unique characteristics. It has been found much tolerant to extreme ecological conditions including prolonged drought, submergence, flood, extreme variations of atmospheric temperature (−10 to 48 °C) and soil chemical environment (soil pH 3 to 10.5). Also the plant has massive, finely structured and deep root system capable of reaching 3–4 m within short time. The plant is also highly tolerant to heavy metal (such as As, Cd, Cr, Ni, Pb, Zn, Hg, Se and Cu) toxicities in the soil. Due to such unique morphological and physiological characteristics, it has generally been known for its effectiveness in erosion and sediment control which is highly desirable in rehabilitating polluted mining area.

12.7.5 Phytoremediation of Soil Cd Contamination in Japan

A survey conducted by the Ministry of Agriculture, Forestry, and Fisheries of Japan Government on Cd contamination in rice indicated that 3.3% and 0.3% of rice cultivated in Japan exceeded the limits of 0.2 mg kg^{-1} or 0.4 mg kg^{-1} respectively, normally identified as international standards for polished rice (MAFF 2002). As rice is the largest source of dietary intake of Cd for the people in Japan, pressing need to reduce the Cd concentration in rice was felt through phytoremediation. Cadmium contamination level were low to moderate in paddy fields of major targeted areas which were suspected to be producing paddy rice or upland rice contaminated with Cd. Researchers found that some of the indica-japonica cross varieties cultivated under upland conditions accumulated a substantially high concentration of Cd in grains and shoots (Ishikawa 2005).

Uptake of Cd by several rice varieties was compared with *Brassica juncea* as it is the most popular cultivated plant considered for phytoremediation purpose. Shoot Cd concentration in rice varieties was 2 to 4 times higher than in *B. juncea* grown in two types of Cd-polluted soils; whereas, Cd uptake in the shoots of rice varieties was approximately 6 times higher than those of *B. juncea*. A field experiment was conducted in a paddy field with moderate level of Cd contamination (@2.5 mg kg^{-1}), the results of which showed high removal efficiency of Cd (about 190 g ha^{-1}) by high yielding rice cultivar Milyang 23. A notable decrease in Cd content in the rhizosphere was found after harvesting this high Cd-accumulating rice variety. A computational analysis indicated that repeated cultivation of 'Milyang 23' rice cultivar for 3–4 years can reduce Cd concentration of contaminated soil (having 0.7 mg kg^{-1}) to a level of uncontaminated soil (containing 0.3 mg kg^{-1}).

12.7.6 Phytoremediation of Arsenic (As) Contaminated Paddy Soils

Accumulation of As in paddy soils is an extensive problem in southeast Asia, most predominantly in India and Bangladesh due to irrigation of lowland rice field with As-laden (geogenic source) groundwater. Phytoremediation with *Pteris vittata* (Chinese brake fern) has been found to markedly reduce (18–83% decrease) As accumulation in rice (Ye et al. 2011). Total As removal from the soils by this plant was relatively small (only 3.5–11.4%) as compared to the total amount of As in the soils. However, highly mobile pools like phosphate-extractable As and soil pore-water soluble As (mainly as As[V]) decreased considerably by 11–38% and 18–77% respectively. Therefore *P. vittata* removed mainly bioavailable fraction of As from the soil, which resulted in considerable decreases in its accumulation by subsequent rice crop. Phytoremediation with *P. vittata* also influenced As speciation in rice grain (Ye et al. 2011). Rice grains produced from four of unremediated soils contained As predominantly as dimethylarsinic acid (DMA), which happened to be a minor species or undetectable after phytoremediation. Growing *P. vittata* decreased inorganic As concentration in the bioavailable pool (in soil solution) which subsequently decreased methylation of As by soil microbes resulting in decreased DMA in rice grain. Methylated pentavalent form of arsenic (DMA) may be reduced to trivalent oxidation state (methyloxoarsine and dimethylarsine) and such trivalent methylated As species are extremely toxic. Therefore the study indicated high potential of phytoextraction technology using *P. vittata* in reducing both magnitude of contamination and toxicity of As in food chain in affected areas of southeast Asia.

12.7.7 Remediation of Radionuclides Contaminated Land Near Chernobyl, Russia

Chernobyl disaster is considered as worst nuclear power plant disaster in history in which large amount of radionuclides were released into environment. In the first weeks after the accident at Chernobyl in erstwhile USSR in April 26, 1986, major concern was lowering ^{131}I-activity concentrations in milk, or preventing contaminated milk entering into the food-chain (IAEA 2005). In this direction, main effort was to feed animals with "clean" fodder, because this had the potential to reduce ^{137}Cs in cattle (and hence in milk) to acceptable levels within 1–2 months. Restriction was imposed on agricultural activity and keeping dairy cattle in the severely affected regions (Schimmack et al. 1989). Severely contaminated land was banned for cultivation of food and fodder crops during the growing period in the year 1986. For less heavily contaminated land, appropriate recommendations on soil and crop management practices were developed as suitable countermeasures against radionuclide transfer in the food chain through on-going activity (Fesenko et al. 2007). Harvesting of forage and food crops was delayed in order to lessen the impact of contamination levels in food chain and human health. During survey on agricultural products for radionuclides, samples from four provinces of Russia, five provinces of Ukraine and three provinces of Belarus exceeded action levels for radioactive caesium by the end of 1986. Further, 101,285 ha of agricultural lands in Ukraine were also withdrawn from cultivation in the early 1990s as about 30% of this area showed ^{137}Cs-contamination level more than 555 kBq m^{-2} (Kimura et al. 2015).

Ex-situ decontamination of the top soil layer was not found feasible for agricultural lands due to high cost of the technology, vast area of contamination, anticipated damage to the soil fertility and severe anticipated ecological troubles related to disposal of the contaminated soil. Therefore, the focus during the first few years after the accident was on modification in the soil and agronomic management practices including greatly increased fertilization rates. Generally, high value legume and cereal grasses were grown on the treated land. As most plant roots absorb contaminants from upper soil layers along with their nutrients, deep ploughing was recommended for the purpose of diluting the degree radioactive contamination in the rhizosphere (IAEA 2005). The use of fertilizers increased crop biomass production, thereby diluting the radioactivity in the edible part. Further, the use of fertilizers reduced plant-root uptake of ^{137}Cs by decreasing the Cs:K ratio in the soil solution (Alexakhin 1993). Acidic soils were limed to reduce phytoavailability of radioactive metals. Marshy lands were converted into grassland after draining, deep ploughing, amending the land to improve fertility (IAEA 2005). Traditional surface improvement involving soil discing, fertilization and surface liming was however, less effective in this case. During nineties, focus was more on site-specific technology based on soil characteristics to ascertain that the soil treatment and management used be the most effective under the prevailing conditions. Several investigations have shown that the reduction factors attained for soil-plant transfer of radioactive caesium following alteration in soil tillage, liming, fertilization and crop management were in the range of 2–4 for less fertile sandy soils and 3–6 for more organic soils (IAEA 2005). Despite these countermeasures, radioactive caesium contamination in 20% of grasses and hay samples from farms during 1997–2000 exceeded the permissible levels in the highly contaminated south-western zone. Fodder crops, such as lupine, peas, buckwheat and clover were completely or partly debarred from cultivation as these crops were found to accumulate high amounts of radioactive caesium. In Belarus, less radionuclide accumulating varieties of rape seed were grown with additional fertilizers and lime in contaminated areas with the aim of producing less contaminated (with radio-caesium and radio-strontium) edible oil and protein cake (for animal).

References

Abdelhafid R, Houot S, Barriuso E (2000) Dependence of atrazine degradationon C and N availability in adapted and non-adapted soils. Soil Biol Biochem 32:389–401

Abedin MJ, Feldmann J, Meharg AA (2002) Uptake kinetics of arsenic species in rice plants. Plant Physiol 128:1120–1128

Abu Laban N, Selesi D, Rattei T, Tischler P, Meckenstock RU (2010) Identification of enzymes involved in anaerobic benzene degradation by a strictly anaerobic iron-reducing enrichment culture. Environ Microbiol 12:2783–2796

Adegoke KA, Bello OS (2015) Dye sequestration using agricultural wastes as adsorbents. Water Resour Ind 12:8–24

Ahmad F, Iqbal S, Anwar S et al (2012) Enhanced remediation of chlorpyrifos from soil using ryegrass (*Lollium multiflorum*) and chlorpyrifos-degrading bacterium *Bacillus pumilus* C2A1. J Hazard Mater 237–238:110–115

Aitken MD, Stringfellow WT, Nagel RD et al (1998) Characteristics of phenanthrene-degrading bacteria isolated from soils contaminated with polycyclic aromatic hydrocarbons. Can J Microbiol 44:743–752

Alagić SČ, Šerbula SS, Tošić SB et al (2013) Bioaccumulation of arsenic and cadmium in birch and lime from the Bor Region. Arch Environ Contam Toxicol 65(4):671–682

Al-Awadhi N, Al-Daher R, El-Nawawy A, Balba MT (1996) Bioremediation of oil-contaminated soil in Kuwait. I. Landfarming to remediate oil-contaminated soil. J Soil Contam 5:243–260

Alexakhin RM (1993) Countermeasures in agricultural production as an effective means of mitigating the radiological consequences of the Chernobyl accident. Sci Total Environ 137:9–20

Allen CCR, Boyd DR, Hempenstall F et al (1999) Contrasting effects of nonionic surfactant on the biotransformation of polycyclic aromatic hydrocarbons to cis-dihydrodiols by soil bacteria. Appl Environ Microbiol 65:1335–1339

Anderson AM, Johnson AH, Siccama TG (1980) Levels of lead copper, and zinc in the forest floor of the northeastern United Stated. J Environ Qual 9:293–296

Angelova V, Ivanova R et al (2004) Bio-accumulation and distribution of heavy metals in fibre crops (flax, cotton and hemp). Ind Crop Prod 19:197–205

Annweiler E, Richnow HH, Antranikian G et al (2000) Naphthalene degradation and incorporation of naphthalene-derived carbon into biomass by the thermophile *Bacillus thermoleovorans*. Appl Environ Microbiol 66:518–523

Antizar-Ladislao B, Lopez-Real JM, Beck AJ (2004) Bioremediation of polycyclic aromatic hydrocarbons (PAH)-contaminated waste using Composting strategies. Crit Rev Environ Sci Technol 34:249–289

Ayed L, Chaieb K, Cheref A, Bakhrouf A (2010) Biodegradation and decolorization of triphenyl methane dyes by *Staphylococcus epidermidis*. Desalination 260:137–146

Baczynski TP, Grotenhuis T, Knipscheer P (2004) The dechlorination of cyclodiene pesticides by methanogenic granular sludge. Chemosphere 55:653–659

Bergaya F, Barrault J (1990) Mixed Al-Fe pillard laponites preparation, characterization and their catalytic properties in syngas conversion. In: Mitchell LV (ed) Pillard layered structures. current trends and applications. Elsevier Applied Science, London, pp 167–184

Bergmann F, Selesi D, Weinmaier T et al (2011) Genomic insights into the metabolic potential of the polycyclic aromatic hydrocarbon degrading sulfate-reducing Delta proteobacterium N47. Environ Microbiol 13:1125–1137

Bezalel Y, Hadar P, Fu P et al (1996) Initial oxidation products in the metabolism of pyrene, anthracene, fluorene, and dibenzothiophene by the white rot fungus *Pleurotus ostreatus*. Appl Environ Microbiol 62(7):2554–2559

Bhalerao TS, Puranik PR (2007) Biodegradation of organochlorine pesticide, endosulfan, by a fungal soil isolate, *Aspergillus niger*. Int Biodeterior Biodegrad 59:315–321

Bhatt P, Suresh Kumar M, Chakrabarti T (2007) Assessment of bioremediation possibilities of technical grade hexachlorocyclohexane (tech-HCH) contaminated soils. J Hazard Mat 143:349–353

Bissen M, Frimmel FH (2003) Arsenic – a review. Part I: occurrence, toxicity, speciation, mobility. Acta Hydrochim Hydrobiol 31(1):9–18

Bolan N, Kunhikrishnan A, Thangarajan R et al (2014) Remediation of heavy metal(loid)s contaminated soils – to mobilize or to immobilize? J Hazard Mater 266:141–166

Boldrin B, Andreas T, Fritzche C (1993) Degradation of phenanthrene, fluorene, fluoranthene and pyrene by a *Mycobacterium* spp. Appl Environ Microbiol 59:1927–1930

Boonchan S (1998) Biodegradation of polycyclic aromatic hydrocarbons: application of fungal–bacterial coculture sand surfactants. Dissertation, Victoria University of Technology, Melbourne Victoria

Boonchan S, Britz ML, Stanley GA (2000) Degradation and mineralization of high-molecular-weight polycyclic aromatic hydrocarbons by defined fungal bacterial cocultures. Appl Environ Microbiol 66:1007–1019

Bosma TNP, Middeldorp PJM, Schraa G et al (1997) Mass transfer limitation of biotransformation: quantifying bioavailability. Environ Sci Technol 31:248–252

Broadley MR, Willey NJ et al (1999) A method to assess taxonomic variation in shoot caesium concentration among flowering plants. Environ Pollut 106:341–349

Burlakovs J, Klavins M, Karklina A (2012) Remediation of soil contamination with heavy metals by using zeolite and humic acid additives. Latv J Chem 4:336–341

Byss M, Elhottov´ac D, Tˇr´ıskab J, Baldriand P (2008) Fungal bioremediation of the creosote-contaminated soil: Influence of *Pleurotus ostreatus* and *Irpex lacteus* on polycyclic aromatic hydrocarbons removal and soil microbial community composition in the laboratory-scale study. Chemosphere 73:1518–1523

Caliman FA, Robu BM, Smaranda C et al (2011) Soil and groundwater cleanup: benefits and limits of emerging technologies. Clean Techn Environ Policy 13:241–268

Casillas RP, Crow SA, Heinze TM et al (1996) Initial oxidation and subsequent conjugative metabolites produced during the metabolism of phenanthrene by fungi. J Ind Microbiol 16:205–215

Cerniglia CE, Heitkamp MA (1989) Microbial degradation of polycyclic aromatic hydrocarbons (PAH) in the aquatic environment. In: Varanasi U (ed) Metabolism of polycyclic aromatic hydrocarbon in aquatic environment. CRC Press Inc., Boca Raton, pp 41–68

Chan DY, Hale BA (2004) Differential accumulation of Cd in durum wheat cultivars: uptake and retranslocation as sources of variation. J Exp Bot 55:2571–2579

Chauhan A, Fazlurrahman, Oakeshott JG, Jain RK (2008) Bacterial metabolism of polycyclic aromatic hydrocarbons: strategies for bioremediation. J Ind Microbiol 48:95–113

Chekol T, Vough LR et al (2004) Phytoremediation of polychlorinated biphenyl-contaminated soils: the rhizosphere effect. Environ Int 30:799–804

Chen SH, Aitken MD (1999) Salicylate stimulates the degradation of high molecular weight polycyclic aromatic hydrocarbons by *Pseudomonas saccharophila* P15. Environ Sci Technol 33:435–439

Chlopecka A, Adriano DC (1996) Mimicked *in situ* stabilization of metals in a cropped soil: bioavailability and chemical form of zinc. Environ Sci Technol 30:3294–3303

Chlopecka A, Adriano DC (1997) Inactivation of metals in polluted soils using natural zeolite and apatite. In: Iskandar IK, Hardy SE, Chang AC, Pierzynski GM (eds) Proceeding of extended abstracts 4th international conference biogeochemistry of trace elements. Berkeley, USA, pp 415–416

Churchill SA, Harper JP, Churchill PF (1999) Isolation and characterization of a *Mycobacterium* species capable of degrading three- and four-ring aromatic and aliphatic hydrocarbons. Appl Environ Microbiol 65:549–552

CLAIRE (2007) Treatment of chromium contamination and chromium ore processing residue. Technical Bulletin TB14. Available at: http://www.claire.co.uk/component/phocadownload/category/17-technical-bulletins?download=55:technicalbulletin14. Accessed 17 Sept 2016

Collins PJ, Dobson ADW (1996) Oxidation of fluorene and phenanthrene by Mn(II) dependent peroxidase activity in whole cultures of Trametes (coriolus) versicolor. Biotechnol Lett 18:801–804

Contreras-Ramos SM, Alvarez-Bernal D, Dendooven L (2006) *Eisenia foetida* increased removal of polycyclic aromatic hydrocarbons (PAHs) from soil. Environ Pollut 141:396–401

Cook LL, Inouye RS et al (2009) Evaluation of four grasses for use in phytoremediation of Cs-contaminated arid land soil. Plant Soil 324:169–184

Coover MP, Sims RC (1987) The effect of temperature on polycyclic aromatic hydrocarbon persistence in an unacclimated agricultural soil. J Hazard Mater 4:69–82

Correa PA, Lin L et al (2010) The Effects of Individual PCB Congeners on the soil bacterial community structure and the abundance of biphenyl dioxygenase genes. Environ Int 36:901–906

Coumar MV, Parihar RS, Dwivedi AK et al (2016a) Pigeon pea biochar as a soil amendment to repress copper mobility in soil and its uptake by spinach. Bioresources 11:1585–1595

Coumar MV, Parihar RS, Dwivedi AK et al (2016b) Impact of pigeon pea biochar on cadmium mobility in soil and transfer rate to leafy vegetable spinach. Environ Monit Assess 188:31–31

Da-Silva M, Cerniglia CE, Pothuluri JV et al (2003) Screening filamentous fungi isolated from estuarine sediments for the ability to oxidise polycyclic aromatic hydrocarbons. World J Microbiol Biotechnol 19:399–405

Deng B, Hu S (2001) Reductive dechlorination of chlorinated solvents on zerovalent iron surfaces. In: Smith JA, Burns SE (eds) Physicochemical groundwater remediation. Kluwer Academic, New York, pp 139–159

Derz K, Klinner U, Schupan I et al (2005) *Mycobacterium pyrenivorans* sp. nov., a novel polycyclic-aromatic hydrocarbon-degrading species. Int J Syst Evol Microbiol 54:2313–2317

Diez MC (2010) Biological aspects involved in the degradation of organic pollutants. J Soil Sci Plant Nutr 10:244–267

Donnelly PK, Hegde RS, Fletcher JS (1994) Growth of PCB-degrading bacteria on compounds from photosynthetic plants. Chemosphere 28:981–988

du Plessis CA, Senior E, Hughes JC (1994) The physical-chemical approach to organic pollutant attenuation in soil. In: Hinchee R, Anderson DB, Blaine Metting F Jr., Sayles GD (eds) Applied biotechnology for site remediation. Second international symposium on *In situ* and on-site bioreclamation, San Diego, April 1993

Düring R, Gath S (2002) Utilization of municipal organic wastes in agriculture: where do we stand, where will we go? J Plant Nutr Soil Sci 165:544–556

Dushenkov V, Kumar PBAN, Motto H, Raskin I (1995) Rhizofiltration: the use of plants to remove heavy metals from aqueous streams. Environ Sci Technol 29(5):1239–1245

Fan S, Li P, Gong Z et al (2008) Promotion of pyrene degradation in rhizosphere of alfalfa (*Medicago sativa L.*) Chemosphere 71:1593–1598

Farah JY, El-Gendy NS, Farahat LA (2007) Biosorption of Astrazone Blue basic dye from an aqueous solution using dried biomass of Baker's yeast. J Hazard Mater 148:402–408

Fesenko SV, Alexakhin RM, Balonov MI et al (2007) An extended critical review of twenty years of countermeasures used in agriculture after the Chernobyl accident. Sci Total Environ 383:1–24

Field RJ, Peel AJ (1971a) The metabolism and radial movement of growth regulators and herbicides in willow stems. New Phytol 70:743–749

Field RJ, Peel AJ (1971b) The movement of growth regulators and herbicides into the sieve elements of willow. New Phytol 70:997–1003

Field RJ, Peel AJ (1972) The longitudinal mobility of growth regulators and herbicides in sieve tubes of willow. New Phytol 71:249–254

Friesl W, Lombi E, Horak O, Wenzel WW (2003) Immobilization of heavy metals in soils using inorganic amendments in a greenhouse study. J Plant Nutr Soil Sci 166:191–196

Fu G, Kan AT, Tomson M (1994) Adsorption and desorption hysteresis of PAHs in surface sediment. Environ Toxicol Chem 13:1559–1567

Gao Y, Zhu L (2004) Plant uptake, accumulation and translocation of phenanthrene and pyrene in soils. Chemosphere 55:1169–1178

Gao Y, Ling W, Wong MH (2006) Plant-accelerated dissipation of phenanthrene and pyrene from water in the presence of a nonionic-surfactant. Chemosphere 63:1560–1567

Garau G, Castaldi P, Santona L et al (2007) Influence of red mud, zeolite and lime on heavy metal immobilization, culturable heterotrophic microbial populations and enzyme activities in a contaminated soil. Geoderma 142:47–57

Germaine KJ, Liu X, Cabellos GG et al (2006) Bacterial endophyte-enhanced phytoremediation of the organochlorine herbicide 2, 4-dichlorophenoxyacetic acid. FEMS Microbiol Ecol 57:302–310

Gibson DT, Subramanian V (1984) Microbial degradation of aromatic hydrocarbons. In: Gibson DT (ed) Microbial degradation of organic compounds. Marcel Dekker, New York, pp 181–252

Gleba D, Borisjuk NV, Borisjuk LG et al (1999) Use of plant roots for phytoremediation and molecular farming. PNAS 96:5973–5977

Gray CW, Dunham SJ, Dennis PG et al (2006) Field evaluation of in situ remediation of a heavy metal contaminated soil using lime and red-mud. Environ Pollut 142:530–539

Grossi G, Lichtig J et al (1998) PCDD/F, PCB, and PAH content of Brazilian compost. Chemosphere 37:2153–2160

Günther T, Dornberger U, Fritsche W (1996) Effects of ryegrass on biodegradation of hydrocarbons in soil. Chemosphere 33:203–215

Habe H, Omori T (2003) Genetics of polycyclic aromatic hydrocarbon degradation by diverse aerobic bacteria. Biosci Biotechnol Biochem 67:225–243

Habe H, Ashikawa Y, Saiki Y et al (2002) *Sphingomonas* sp. strain KA1, carrying a carbazole dioxygenase gene homologue, degrades chlorinated dibenzo-p-dioxins in soil. FEMS Microbiol Lett 211:43–49

Häggblom M, Valo RJ (1985) Bioremediation of chlorophenol wastes. In: Young LY, Cerniglia CE (eds) Microbial transformation and degradation of Toxic Organic Chemicals. Wiley, New York, pp 389–434

Halling SB, Sengelov G, Ingerslev F, Jensen LB (2003) Reduced antimicrobial potencies of oxytetracycline, tylosin, sulfadiazin, streptomycin, ciprofloxacin, and olaquindox due to environmental processes. Arch Environ Contam Toxicol 44:7–16

Haritash AK, Kaushik CP (2009) Biodegradation aspects of polycyclic aromatic hydrocarbons (PAHs): a review. J Hazard Mater 169:1–15

Hay AG, Focht DD (2000) Transformation of 1,1-dichrolo-2,2-(4-chrolophenyl)ethane (DDD) by Ralstonia eutropha strain A5. FEMS Microb Ecol 31:249–253

Hettiarachchi GM, Pierzynski GM (2004) Soil lead bioavailability and *in situ* remediation of lead-contaminated soils: a review. Environ Prog 23:78–93
Higginson FR (1993) The composition of mineral soil, physico-chemical and ion exchange properties. In: Hazelton PA, Koppi AJ (eds) Soil Technology – Applied Soil Science. A course of lectures. 2nd edition. ASSSI NSW Branch and Dept. Agric. Chem. & Soil Science, University of Sydney. pp 39–59
Hoffman DR, Okon JL, Sandrin TR (2005) Medium composition affects the degree and pattern of cadmium inhibition of naphthalene biodegradation. Chemosphere 59:919–927
Holland JM (2004) The environmental consequences of adopting conservation tillage in Europe: reviewing the evidence. Agric Ecosyst Environ 103:1–25
Hu J, Lo IM, Chen G (2004) Removal of Cr(VI) by magnetite nanoparticle. Water Sci Technol 50:139–146
Huang X, El-Alawi Y et al (2004) A multi-process phytoremediation system for removal of polycyclic aromatic hydrocarbons from contaminated soils. Environ Pollut 130:465–476
Huang M, Zhu Y, Li Z et al (2016) Compost as a soil amendment to remediate heavy metal-contaminated agricultural soil: mechanisms, efficacy, problems, and strategies. Water Air Soil Pollut. doi:10.1007/s11270-016-3068-8
Hülster A, Muller JF, Marschner H (1994) Soil-plant transfer of polychlorinated dibenzo-p-dioxins and dibenzofurans to vegetables of the cucumber family (*Cucurbitaceae*). Environ Sci Technol 28:1110–1115
Huq SMI, Joardar JC et al (2006) Arsenic contamination in food-chain: transfer of arsenic into food materials through groundwater irrigation. J Health Popul Nutr 24:305–316
Hussain S, Arshad M, Saleem M, Khalid A (2007) Biodegradation of α- and β-endosulfan by soil bacteria. Biodegradation 18:731–740
IAEA (2005) Environmental consequences of the Chernobyl accident and their remediation: twenty years of experience. Report of the UN Chernobyl Forum Expert Group "Environment" (EGE). International Atomic Energy Agency, Vienna, p 246
Indoria AK, Poonia SR (2006) Phytoextractability of lead from soil by some oilseed crops as affected by sewage sludge and farmyard manure. Arch Agron Soil Sci 52:667–677
Ishikawa S (2005) Promising technologies for reducing cadmium contamination in rice. In: Toriyama K, Heong KL, Hardy B (eds) Rice is life: scientific perspectives for the 21st century. Proceedings of the World Rice Research Conference held in Tokyo and Tsukuba, Japan, 4–7 November 2004. International Rice Research Institute, Los Baños (Philippines) and Japan International Research Center for Agricultural Sciences, Tsukuba (Japan). pp 381–384
Jain AK, Suhas Jain S, Bhatnagar A (2003) Utilization of industrial wastes for the removal of anionic dyes. Toxicol Environ Chem 84:41–52
Jeffery S, Verheijen FGA, van der Velde M, Bastos AC (2010) A quantitative review of the effects of biochar application to soils on crop productivity using meta- analysis. Agric Ecosyst Environ 144:175–187
Jeon CO, Park W, Padmanabhan P et al (2003) Discovery of a bacterium, with distinctive dioxygenase, that is responsible for in situ biodegradation in contaminated sediment. Proc Natl Acad Sci U S A 100:13591–13596
Jindo K, Mizumoto H, Sawada Y et al (2014) Physical and chemical characterization of biochars derived from different agricultural residues. Biogeosciences 11:6613–6621
Johnsena AR, Wickb LY et al (2005) Principles of microbial PAH-degradation in soil. Environ Pollut 133:71–84
Juwarkar A, Singh S, Mudhoo A (2010) A comprehensive overview of elements in bioremediation. Rev Environ Sci Biotechnol 9(3):215–288
Kanaly RA, Harayama S (2000) Biodegradation of high molecular weight polycyclic aromatic hydrocarbons by bacteria. J Bacteriol 182(8):2059–2067
Kanel SR, Greneche JM, Choi H (2006) Arsenic(V) removal from groundwater using nano scale zero-valent iron as a colloidal reactive barrier material. Environ Sci Technol 40:2045–2050

Kang J (2014) Removing environmental organic pollutants with bioremediation and phytoremediation. Biotechnol Lett 36(6):1129–1139

Kang F, Chen D, Gao Y, Zhang Y (2010) Distribution of polycyclic aromatic hydrocarbons in subcellular root tissues of ryegrass (*Lolium multiflorum* Lam.) BMC Plant Biol 10:210

Karin O, Jonsson A, Stenstro J (2010) A new concept for reduction of diffuse contamination by simultaneous application of pesticide and pesticide degrading microorganisms. Biodegradation 21:21–29

Karn B, Kuiken T, Otto M (2009) Nanotechnology and *in situ* remediation: a review of the benefits and potential risks. Environ Heal Perspect 117:1823–1831

Katayama A, Matsumura F (1993) Degradation of organochlorine pesticides, particularly endosulfan, by *Trichoderma harzianum*. Environ Toxicol Chem 12:1059–1065

Katayama A, Fujimura Y, Kuwatsuka S (1993) Microbial degradation of DDT at extremely low concentrations. J Pestic Sci 18:353–359

Kawahigashi H, Hirose S, Ohkawa H, Ohkawa Y (2006) Phytoremediation of the herbicides atrazine and metalachlor by transgenic rice plants expressing human CYP1A1, CYP2B6 and CYP2C19. J Agric Food Chem 54:2985–2991

Keharia H, Madamwar D (2003) Bioremediation concepts for treatment of dye containing wastewater: a review. Indian J Exp Biol 41:1068–1075

Khehra MS, Saini HS et al (2006) Biodegradation of azo dye C.I. Acid Red 88 by an anoxic aerobic sequential bioreactor. Dyes Pigments 70:1–7

Kim JS, Shea PJ, Yang JE, Kim JE (2007) Halide salts accelerate degradation of high explosives by zerovalent iron. Environ Pollut 147:634–641

Kimura Y, Okubo Y, Hayashida N et al (2015) Evaluation of the relationship between current internal ^{137}Cs exposure in residents and soil contamination west of Chernobyl in Northern Ukraine. PLoS One 10(9):e0139007. doi:10.1371/journal.pone.0139007

Kipopoulou AM, Manoli E, Samara C (1999) Bioconcentration of polycyclic aromatic hydrocarbons in vegetables grown in an industrial area. Environ Pollut 106:369–380

Kirchhoff MM (2003) Promoting green engineering through green chemistry. Environ Sci Technol 37(23):5349–5353

Knox AS, Seaman JC, Mench MJ, Vangronsveld J (2001) Remediation of metal and radionuclides-contaminated soils by in situ stabilization techniques. In: Iskandar IK (ed) Environmental restauration of metals-contaminated soils. CRC Press LLC, Boca Raton, pp 21–60

Kuiper EL, Lagendijk GV, Lugtenberg B (2004) Rhizoremediation: A beneficial plant microbe interaction. Mol Plant-Microbe Interact 17:6–15

Kumpiene J, Ore S, Renella G et al (2006) Assessment of zerovalent iron for stabilization of chromium, copper, and arsenic in soil. Environ Pollut 144:62–69

Kumpiene J, Montesinos IC, Lagerkvist A, Maurice C (2007) Evaluation of the critical factors controlling stability of chromium, copper, arsenic and zinc in iron-treated soil. Chemosphere 67(2):410–417

Kurtyka R, Małkowski E et al (2008) Effect of calcium and cadmium on growth and accumulation of cadmium, calcium, potassium and sodium in maize seedlings. Pol J Environ Stud 17:51–56

Kwon GS, Kim JE, Kim TK et al (2002) *Klebsiella pneumoniae* KE-1 degrades endosulfan without formation of the toxic metabolite, endosulfan sulfate. FEMS Microbiol Lett 215:255–259

Lageman R, Pool W, Seffinga G (1990) Electro-reclamation: state-of-the-art and future developments. In: Arendt F, Hinsenveld M, Van Den Brink WJ (eds) Contaminated soil '90. Springer, Dordrecht, pp 1071–1078

Lakshmi CV, Kumar M, Khanna S (2008) Biotransformation of chlorpyrifos and bioremediation of contaminated soil. Int Biodeterior Biodegrad 62(2):204–209

Lakshmi CV, Kumar M, Khanna S (2009) Biodegradation of chlorpyrifos in soil by enriched cultures. Curr Microbiol 58:35–38

Lal K, Minhas PS, Shipra et al (2008) Extraction of cadmium and tolerance of three annual cut flowers on Cd-contaminated soils. Bioresour Technol 99:1006–1011

Launen L, Pinto LJ, Wiebe C et al (1995) The oxidation of pyrene and benzo[a]pyrene by non-basidiomycete soil fungi. Can J Microbiol 41:477–488

Lee PH, Ong SK et al (2001) Use of solvents to enhance PAH biodegradation of coal tar. Water Res 35:3941–3949

Levi-Minzi R, Petruzzelli G (1984) The influence of phosphate fertilizers on Cd solubility in soil. Water Air Soil Pollut 23:423–429

Li YM, Chaney RL, Siebielec G, Kerschner BA (2000) Response of four turfgrass cultivars to limestone and biosolids-compost amendment of a zinc and cadmium contaminated soil at Palmerton, Pennsylvania. J Environ Qual 29:1440

Li H, Shi W, Shao H, Shao M (2009) The remediation of the lead-polluted garden soil by natural zeolite. J Hazard Mater 169:1106–1111

Liang L, Gu B, Yin X (1996) Removal of technetium-99 from contaminated groundwater with sorbents and reductive materials. Sep Sci Technol 6:111–122

Lichtenstein EP, Schulz KR, Skrentny RF, Stitt PA (1965) Insecticidal residues in cucumbers and alfalfa grown on aldrin- or heptachlor treated soils. J Econ Entomol 58:742–746

Liedekerke MV, Prokop G, Rabl-Berger S et al (2014) Progress in the management of contaminated sites in Europe. EUR 26376 – Joint Research Centre – Institute for Environment and Sustainability, European Commission. pp 1–68

Liste HH, Alexander M (2000) Plant promoted pyrene degradation in soil. Chemosphere 40:7–10

Little C, Hepher MJ, El-Sharif M (2002) The sono-degradation of phenanthrene in an aqueous environment. Ultrasonics 40:667–674

Liu RQ, Zhao DY (2007) In situ immobilization of Cu(II) in soils using a new class of iron phosphate nanoparticles. Chemosphere 68:1867–1876

Liu Y, Zhang J, Zhang Z (2004) Isolation and characterisation of polycyclic aromatic hydrocarbons- degrading *Sphingomonas* sp. Strain ZL5. Biodegradation 15:205–212

Loick N, Hobbs PJ, Hale MCD, Jones DL (2012) Bioremediation of poly-aromatic hyrdocarbon (PAH)-contaminated soil by composting. Crit Rev Environ Sci Technol 39(4):271–332

Lombi E, Zhao FJ, Wieshammer G, Zhang G, McGrath SP (2002) In situ fixation of metals in soil using bauxite residue: biological effects. Environ Pollut 118:445–452

Lopez-Chuken UJ, Lopez-Domınguez U et al (2012) Implications of chloride-enhanced cadmium uptake in saline agriculture: modeling cadmium uptake by maize and tobacco. Int J Environ Sci Technol 9:69–77

Lothenbach B, Furrer G, Schulin R (1997) Immobilization of heavy metals by polynuclear aluminium and montmorillonite compounds. Environ Sci Technol 31:1452–1462

Lu H, Zhang YY, Huang X, Wang S, Qiu R (2012) Relative distribution of Pb^{2+} sorption mechanisms by sludge-derived biochar. Water Res 46:854–862

Ludwig RD, Su C, Lee TR et al (2007) In situ chemical reduction of Cr(VI) in groundwater using a combination of ferrous sulphate and sodium dithionite: a field investigation. Int J Environ Sci Technol 41:5299–5305

Lundstedt S (2003) Analysis of PAHs and their transformation products in contaminated soil and remedial processes. Dissertation, Umea University, Umea, Sweden

Lunney AI, Zeeb BA, Reimer KJ (2004) Uptake of weathered DDT in vascular plants: potential for phytoremediation. Environ Sci Technol 38:6147–6154

Luqueño FF, Marsch R et al (2008) Remediation of PAHs in a saline–alkaline soil amended with wastewater sludge and the effect on dynamics of C and N. Sci Total Environ 402:18–28

Ma QY, Traina SJ, Logan TJ (1993) In situ lead immobilization by apatite. Environ Sci Technol 27:1803–1810

Macek T, Mackova M, Kas J (2000) Exploitation of plants for the removal of organics in environmental remediation. Biotechnol Adv 18:23–34

Madrid F, López R, Cabrera F (2007) Metal accumulation in soil after application of municipal solid waste compost under intensive farming conditions. Agric Ecosyst Environ 119 (3–4):249–256

MAFF (2002) Survey of the cadmium contained in agricultural products. Ministry of Agriculture, Forestry, and Fisheries, Gov. of Japan, pp 1–45

Maillard J, Schumacher W, Vazquez F et al (2003) Characterization of the corrinoid iron-sulfur protein tetrachloroethene reductive dehalogenase of *Dehalobacter restrictus*. Appl Environ Microbiol 69:4628–4638

Makris KC, Shakya KM, Datta R, Sarkar D, Pachanoor D (2007) High uptake of 2, 4, 6-trinitrotoluene by vetiver grass – potential for phytoremediation? Environ Pollut 146:1–4

Maliszewska-Kordybach B (2009) Concentrations, sources and spatial distribution of individual polycyclic aromatic hydrocarbons (PAHs) in agricultural soils in the Eastern part of the EU: Poland as a case study. Sci Total Environ 407:3746–3753

Manara A (2012) Plant responses to heavy metal toxicity. In: Furini A (ed) Plants and heavy metals, briefs in molecular science. Springer, Dordrecht, pp 27–53

Martorell MM, Pajot HF, de Figueroa LIC (2012) Dye-decolourizing yeasts isolated from Las Yungas rainforest. Dye assimilation and removal used as selection criteria. Int Biodeter Biodegr 66:25–32

Matsumura F, Boush GM (1967) Dieldrin: degradation by soil microorganisms. Science 156:959–961

Matsumura F, Khanvilkar VG, Patil KC, Boush GM (1971) Metabolism of endrin by certain soil microorganisms. J Agric Food Chem 19:27–31

Mattina MJI, Eitzer BD, Iannucci-Berger W et al (2004) Plant uptake and translocation of highly weathered, soil bound technical chlordane residues: data from field and rhizotron studies. Environ Toxicol Chem 23:2756–2762

Maule A, Plyte S, Quirk AV (1987) Dehalogenation of organochlorine insecticides by mixed anaerobic microbial populations. Pestic Biochem Physiol 27:229–236

McBride MB (1994) Environmental chemistry of soils. Oxford University Press Inc, New York

McBride MB (1995) Toxic metal accumulation from agricultural use of sludge: are U.S. EPA regulations protective? J Environ Qual 24:5–18

McGrath SP, Chang AC, Page AL, Witter E (1994) Land application of sewage sludge: scientific perspectives of heavy metal loading limits in Europe and the United States. Environ Rev 2:108–118

McGuinness M, Dowling D (2009) Plant-associated bacterial degradation of toxic organic compounds in soil. Int J Environ Res Public Health 6:2226–2247

McGuinness M, Mazurkiewicz V, Brennan E, Dowling DN (2007) Dechlorination of pesticides by a specific bacterial glutathione S-transferase, BphKLB400: Potential for bioremediation. Eng Life Sci 7:611–615

McMullan G, Meehan C et al (2001) Microbial decolourisation and degradation of textile dyes. Appl Microbiol Biotechnol 56:81–87

Meers E, Slycken SV et al (2010) The use of bio-energy crops (*Zea mays*) for 'phytoattenuation' of heavy metals on moderately contaminated soils: a field experiment. Chemosphere 78:35–41

Megharaj M, Kantachote D, Singleton I, Naidu R (2002) Effects of long-term contamination of DDT on soil microflora with special reference to soil algae and algal transformation of DDT. Environ Pollut 109:35–42

Megharaj M, Ramakrishnan B, Venkateswarlu K et al (2011) Bioremediation approaches for organic pollutants: a critical perspective. Environ Int 37(8):1362–1375

Meharg AA, Rahman MM (2003) Arsenic contamination of Bangladesh paddy field soils: implications for rice contribution to arsenic consumption. Environ Sci Technol 37:229–234

Mello-Farias PC, Chaves ALS, Lencina CL (2011) Transgenic plants for enhanced phytoremediation – physiological studies, genetic transformation. In: Alvarez M (ed) InTech. doi:10.5772/24355, 2011. Available at: http://www.intechopen.com/books/

genetic-transformation/transgenic-plants-for-enhanced-phytoremediation-physiological-studies, Accessed 18 Sept 2016

Mench M, Vangronsveld J, Lepp NW, Edwards R (1998) Physico-chemical aspects and effiency of trace element immobilization by soil amendments. In: Vangronsveld J, Cunningham SD (eds) Metal-contaminated soils: In Situ inactivation and phytorestoration. Springer, Heidelberg/New York

Mench M, Vangronsveld J, Beckx C, Ruttens A (2006) Progress in assisted natural remediation of an arsenic contaminated agricultural soil. Environ Pollut 144:51–61

Mikes O, Cupr P, Trapp S, Klanova J (2009) Uptake of polychlorinated biphenyls and organochlorine pesticides from soil and air into radishes (*Raphanus sativus*). Environ Pollut 157 (2):488–496

Miller CD, Hall K, Liang YN et al (2004) Isolation and characterization of polycyclic aromatic hydrocarbon-degrading *Mycobacterium* isolates from soil. Microb Ecol 48:230–238

Miyata N, Iwahori K, Foght JM, Gray MR (2004) Saturable, energy dependent uptake of phenanthrene in aqueous phase by Mycobacterium sp. strain RJGII-135. Appl Environ Microbiol 70(1):363–369

Moody J, Freeman J, Doerge D, Cerniglia C (2001) Degradation of phenanthrene and anthracene by cell suspensions of *Mycobacterium* sp. PYR-1. Appl Environ Microbiol 67(4):1476–1483

Muller I, Pluquet E (1998) Immobilization of heavy metals in sediment dredged from a seaport by iron bearing materials. Water Sci Technol 37:379–386

Musat F, Galushko A, Jacob J et al (2009) Anaerobic degradation of naphthalene and 2-methylnaphthalene by strains of marine sulfate-reducing bacteria. Environ Microbiol 11:209–219

NAAS (2012) Sustaining agricultural productivity through integrated soil management. Policy Paper No. 56, National Academy of Agricultural Sciences, New Delhi

Nadeau LJ, Menn FM, Breen A, Sayler GS (1994) Aerobic degradation of 1,1,1-trichloro-2,2-bis (4-chlorophenyl)ethane (DDT) by *Alcaligenes eutrophus* A5. Appl Environ Microbiol 60:51–55

Nair DR, Burken JG, Licht LA, Schnoor JL (1993) Mineralization and uptake of triazine pesticide in soil-plant systems. J Environ Eng 119:842–854

Neumann D, Heuer A, Hemkemeyer M et al (2014) Importance of soil organic matter for the diversity of microorganisms involved in the degradation of organic pollutants. ISME J 8:1289–1300

Nigam P, Armour G et al (2000) Physical removal of textile dyes from effluents and solid-state fermentation of dye-adsorbed agricultural residues. Bioresour Technol 72:219–226

Nigam R, Srivastava S et al (2001) Cadmium mobilization and plant availability – the impact of organic acids commonly exuded from roots. Plant Soil 230:107–113

Nurzhanova A, Kulakow P, Rubin E et al (2010) Obsolete pesticides pollution and phytoremediation of contaminated soil in Kazakhstan. In: Kulakow PA, Pidlisnyuk VV (eds) Application of phytotechnologies for cleanup of industrial, agricultural, and wastewater contamination. Springer, Dordrecht, pp 87–111

Oleszczuk P, Baran S (2003) Degradation of individual polycyclic aromatic hydrocarbons (PAHs) in soil polluted with aircraft fuel. Pol J Environ Stud 12:431–437

Oliver DP, Schultz JE, Tiller KG, Merry RH (1993) The effect of crop rotations and tillage practices on cadmium concentration in wheat grain. Aust J Agric Res 44:1221–1234

Olson PE, Castro A et al (2008) Effects of agronomic practices on phytoremediation of an aged PAH-contaminated soil. J Environ Qual 37:1439–1446

Ortiz-Hernández ML, Sánchez-Salinas E (2010) Biodegradation of the organophosphate pesticide tetrachlorvinphos by bacteria isolated from agricultural soils in México. Rev Int Contam Ambient 26(1):27–38

Otani T, Seike N, Sakata Y (2007) Differential uptake of dieldrin and endrin from soil by several plant families and Cucurbita genera. Soil Sci Plant Nutr 53:86–94

Page AL, Logan TJ et al (1987) Land application of sludge: Food chain implications. Lewis Publications, Chelsea

Papafilippaki A, Paranychianakis N, Nikolaidis NP (2015) Effects of soil type and municipal solid waste compost as soil amendment on *Cichorium spinosum* (spiny chicory) growth. Sci Hortic-Amsterdam 195:195–205

Parrish Z, White J, Isleyen M et al (2006) Accumulation of weathered polycyclic aromatic hydrocarbons (PAHs) by plant and earthworm species. Chemosphere 64:609–618

Patil KC, Matsumura F, Boush GM (1970) Degradation of endrin, aldrin, and DDT by soil microorganisms. Appl Microbiol 19:879–881

Paz-Ferreiro J, Lu H, Fu S et al (2014) Use of phytoremediation and biochar to remediate heavy metal polluted soils: a review. Solid Earth 5:65–75

Pereira CR, Camps-Arbestain M, Rodríguez Garrido B et al (2006) Behaviour of a-, b-, c-, and d-hexachlorocyclohexane in the soil–plant system of a contaminated site. Environ Pollut 144:210–217

Pigna M, Cozzolino V et al (2010) Effects of phosphorus fertilization on arsenic uptake by wheat grown in polluted soils. J Soil Sci Plant Nutr 10:428–442

Pilon-Smits E (2005) Phytoremediation. Annu Rev Plant Biol 56:15–39

Pointing SB (2001) Feasibility of bioremediation by white rot fungi. Appl Microbiol Biotechnol 57:20–33

Powlson DS, Whitmore AP, Goulding KWT (2011) Soil carbon sequestration to mitigate climate change: a critical re-examination to identify the true and the false. Eur J Soil Sci 62:42–55

Pradhan SP, Conrad JR, Paterek JR, Srivastava VJ (1998) Potential of phytoremediation for treatment of PAHs in soil at MGP sites. Soil Sediment Contam 7:467–480

Prasad MNV (2007) Aquatic plants for phytotechnology. In: Singh SN, Tripathi RD (eds) Environmental bioremediation technologies. Springer, Berlin, pp 259–274

Puls RW, Paul CJ, Powell RM (1999) The application of *in situ* permeable reactive (zero-valent iron) barrier technology for the remediation of chromate-contaminated groundwater: a field test. Appl Geochem 14:989–1000

Puvaneswari N, Muthukrishnan J, Gunasekaran P (2006) Toxicity assessment and microbial degradation of azo-dyes. Indian J Exp Biol 44:618–626

Qing H, Zhang Z, Hong Y, Li S (2007) A microcosm study on bioremediation of fenitrothion-contaminated soil using *Burkholderia* sp. FDS-1. Int Biodeterior Biodegrad 59:55–61

Quilliam RS, Rangecroft S, Emmett BA et al (2013) Is biochar a source or sink for polycyclic aromatic hydrocarbon (PAH) compounds in agricultural soils? GCB Bioenergy 5:96–103

Quinn JJ, Negri CM et al (2001) Predicting the effect of deep-rooted hybrid poplars on the groundwater flow system at a large scale phytoremediation site. Int J Phytoremed 3:41–60

Rai UN, Pandey S, Sinha S et al (2004) Revegetating fly ash landfills with *Prosopis juliflora* L: impact of different amendments and Rhizobium inoculation. Environ Int 30:293–300

Rao MA, Scelza R, Acevedo F et al (2014) Enzymes as useful tools for environmental purposes. Chemosphere 107:145–162

Rayu S, Karpouzas D, Singh B (2012) Emerging technologies in bioremediation: constraints and opportunities. Biodegradation 23(6):917–926

Rehmann K, Noll HP, Steinberg CEW, Kettrup AA (1998) Pyrene Degradation by *Mycobacterium* Sp. Strain KR2. Chemosphere 36:2977–2992

Reid BJ, Jones KC, Semple KT (2000) Bioavailability of persistent organic pollutants in soils and sediments – a perspective on mechanisms, consequences and assessment. Environ Pollut 108(1):103–112

Reilley KA, Banks MK et al (1996) Dissipation of polycyclic aromatic hydrocarbons in the rhizosphere. J Environ Qual 25:212–219

Rhykerd RL, Crews B et al (1999) Impact of bulking agents, forced aeration, and tillage on remediation of oil-contaminated soil. Bioresour Technol 67:279–285

Rickerby DG, Morrison M (2007) Nanotechnology and the environment: a European perspective. Sci Technol Adv Mater 8:19–24

Rigas F, Papadopoulou K, Dritsa V, Doulia D (2007) Bioremediation of a soil contaminated by lindane utilizing the fungus *Ganoderma australe* via response surface methodology. J Hazard Mater 140:325–332

Riya P, Jagatpati T (2012) Biodegradation and bioremediation of pesticides in soil: its objectives, classification of pesticides, factors and recent developments. World J Sci Technol 2:36–41

Roche H, Vollaire Y, Martin E et al (2009) Rice fields regulate organochlorine pesticides and PCBs in lagoons of the Nature Reserve of Camargue. Chemosphere 75:526–533

Rousseaux S, Hartmann A, Lagacherie B et al (2003) Inoculation of an atrazine-degrading strain, Chelatobacter heintzii Cit1, in four different soils: effects of different inoculum densities. Chemosphere 51:569–576

Ruttens A, Boulet J et al (2011) Short rotation coppice culture of willows and poplars as energy crops on metal contaminated agricultural soils. Int J Phytoremediation 13:194–207

Sack U, Fritsche W (1997) Enhancement of pyrene mineralization in soil by wood-decaying fungi. FEMS Microbiol Ecol 22(1):77–83

Saha JK (2002) Role of farmyard manure in integrated nutrient management in Indian soils for sustainable crop production. In: Dris R, Abdelaziz FH, Jain M (eds) Plant nutrition: growth and diagnosis. Oxford & IBH Publishing Co. Pvt. Ltd, New Delhi, pp 277–288

Saha JK (2005) Changes in salinity and sodicity of soils with continuous application of contaminated water near industrial area. J Indian Soc Soil Sci 53(4):612–617

Saha JK, Adhikari T, Mandal B (1999) Effect of lime and organic matter on distribution of zinc, copper, iron, and manganese in acid soils. Commun Soil Sci Plant Anal 30:1819–1829

Saha JK, Panwar N et al (2010) An assessment of municipal solid waste compost quality produced in different cities of India in the perspective of developing quality control indices. Waste Manag 30:192–201

Saha JK, Panwar N et al (2013) Risk assessment of heavy metals in soil of a susceptible agro-ecological system amended with municipal solid waste compost. J Indian Soc Soil Sci 61:15–22

Saha JK, Sharma AK, Srivastava A (2014) Impact of different types of polluted irrigation water on soil fertility and wheat grain yield in clayey black soils of central India. Environ Monit Assess 186:2349–2356

Saiki Y, Habe H, Yuuki T et al (2003) Rhizoremediation of dioxin-like compounds by a recombinant Rhizobium tropici strain expressing carbazole 1,9a- dioxygenase constitutively. Biosci Biotechnol Biochem 67:1144–1148

Sanglard D, Leisola MSA, Fiechter A (1986) Role of extracellular liginases in biodegradation of benzo[a]pyrene by *Phanerochaete chrysoporium*. Enzym Microb Technol 8:209–212

Santodonato J, Howard P et al (1981) Health and ecological assessment of polynuclear aromatic hydrocarbons. J Environ Pathol Toxicol Oncol 5:1–364

Saratale RG, Saratale GD et al (2010) Decolorization and biodegradation of reactive dyes and dye wastewater by a developed bacterial consortium. Biodegradation 21:999–1015

Sarmah AK, Meyer MT, Boxall ABA (2006) A global perspective on the use, sales, exposure pathways, occurrence, fate and effects of veterinary antibiotics (VAs) in the environment. Chemosphere 65:725–759

Sayara T, Pognani M et al (2010) Anaerobic degradation of PAHs in soil: Impacts of concentration and amendment stability on the PAHs degradation and biogas production. Int Biodeterior Biodegrad 64:286–292

Schimmack W, Bunzl K, Zelles L (1989) Initial rates of migration of radionuclides from the Chernobyl fallout in undisturbed soils. Geoderma 44:211–218

Schnabel WE, White DM (2001) The effect of mycorrhizal fungi on the fate of aldrin: phytoremediation potential. Int J Phytoremediat 3(2):221–241

Schutzendubel A, Majcherczyk A, Johannes C, Huttermann A (1999) Degradation of fluorene, anthracene, phenanthrene, fluoranthene and pyrene lacks connection to the production of extracellular enzymes by *Pleurotus ostreatus* and *Bjerkandera adjusta*. Int Biodeter Biodegr 43:93–100

Scott C, Pandey G, Hartley CJ et al (2008) The enzymatic basis for pesticide bioremediation. Indian J Microbiol 48:65–79

Semple KT, Reid BJ, Fermor TR (2001) Impact of composting strategies on the treatment of soils contaminated with organic pollutants. Environ Pollut 112:269–283

Sethunathan N, Megharaj M, Chen ZL et al (2004) Algal degradation of a known endocrine disrupting insecticide, a-endosulfan, and its metabolite, endosulfan sulfate, in liquid medium and soil. J Agric Food Chem 52(10):3030–3035

Shann JR, Boyle JJ (1994) Influence of plant species on *in situ* rhizosphere degradation. In: Anderson TA, Coats JR (eds) Bioremediation Through Rhizosphere Technology, ACS Symposium Series, American Chemical Society, Washington, DC 563:70–81

Sharma PM, Bhattacharya D, Krishnan S, Lal B (2004) Degradation of polycyclic aromatic hydrocarbons by a newly discovered enteric bacterium *Leclercia adecarboxylata*. Appl Environ Microbiol 70(5):3163–3166

Shi G, Cai Q (2009) Cadmium tolerance and accumulation in eight potential energy crops. Biotechnol Adv 27:555–561

Shi Q, Bao Z, Zhu Z et al (2005) Silicon-mediated alleviation of Mn toxicity in *Cucumis sativus* in relation to activities of superoxide dismutase and ascorbate peroxidase. Phytochemistry 66:1551–1559

Shu WS, Lan CY, Zhang ZQ, Wong MH (2000) Use of vetiver and other three grasses for revegetation of Pb/Zn mine tailings at Lechang, Guangdong Province: field Experiment. Second international vetiver conference, Bangkok, Thailand

Siddique T, Okeke BC, Arshad M, Frankenberger WT Jr (2003) Enrichment and isolation of endosulfan-degrading microorganisms. J Environ Qual 32:47–54

Singh OV, Jain RK (2003) Phytoremediation of toxic aromatic pollutants from soil. Appl Microbiol Biotechnol 63:128–135

Singh NS, Singh DK (2011) Biodegradation of endosulfan and endosulfan sulfate by *Achromobacter xylosoxidans* strain C8B in broth medium. Biodegradation 22:845–857

Singh BK, Walker A (2006) Microbial degradation of organophosphorus compounds. FEMS Microbiol Rev 30:428–471

Singh OV, Labana S et al (2003) Phytoremediation: an overview of metallic ion decontamination from soil. Appl Microbiol Biotechnol 61:405–412

Singh KP, Dinesh M, Sinha S, Dalwani R (2004) Impact assessment of treated/untreated wastewater toxicants discharged by sewage treatment plants on health, agricultural, and environmental quality in the wastewater disposal area. Chemosphere 55:227–255

Singh B, Singh BP, Cowie AL (2010) Characterisation and evaluation of biochars for their applications a soil amendment. Aust J Soil Res 48:516–525

Singh V, Brar MS, Sharma P, Brar BS (2011) Distribution of arsenic in groundwater and surface soils in south western districts of Punjab. J Indian Soc Soil Sci 59:376–380

Skaates SV, Ramaswami A, Anderson LG (2005) Transport and fate of dieldrin in poplar and willow trees analyzed by SPME. Chemosphere 61(1):85–91

Skłodowski P, Maciejewska A et al (2006) The effect of organic matter from brown coal on bioavailability of heavy metals in contaminated soils. In: Twardowska I, Allen HE (eds) Soil and water pollution monitoring, protection and remediation. Springer, pp 3–23

Smith DJT, Edelhauser EC, Harrison RM (1995) Polycyclic aromatic hydrocarbon concentrations in road dust and soil samples collected in the United Kingdom and Pakistan. Environ Technol 16:45–53

Smreczak B, Maliszewska-Kordybach B (2003) Primary investigations into determination of potentially bioavailable fractions of PAHs in soils contaminated with those compounds. Arch Environ Prot 29(4):41–50

Sparrow LA, Salardini AA et al (1994) Field studies of cadmium in potatoes (*Solanum tuberosum* L.). III. Response of cv. Russet Burbank to sources of banded potassium. Aust J Agric Res 45:243–249

Struthers JK, Jayachandran K, Moorman TB (1998) Biodegradation of atrazine by *Agrobacterium radiobacter* J14a and use of this strain in bioremediation of contaminated soil. Appl Environ Microbiol 64(9):3368–3375
Su DC, Wong JWC (2004) Selection of mustard oilseed rape (*Brassica juncea* L.) for phytoremediation of cadmium contaminated soil. Bull Environ Contam Toxicol 72:991–998
Sutherland JB, Selby AL, Freeman JP et al (1991) Metabolism of phenanthrene by *Phanerochaete chrysosporium*. Appl Environ Microbiol 57:3310–3316
Sutherland JB, Fu PP, Yang SK et al (1993) Enantiomeric composition of the trans-dihydrodiols produced from phenanthrene by fungi. Appl Environ Microbiol 59:2145–2149
Suzuki N (2005) Alleviation by calcium of cadmium-induced root growth inhibition in Arabidopsis seedlings. Plant Biotechnol 22:19–25
Tao S, Cui YH, Xu FL et al (2004) Polycyclic aromatic hydrocarbons (PAHs) in agricultural soil and vegetables from Tianjin. Sci Total Environ 320:11–24
Tao S, Jiao XC, Chen SH et al (2006) Uptake of vapor and particulate polycyclic aromatic hydrocarbons by cabbage. Environ Pollut 140:13–15
Teng Y, Luo Y et al (2010) Influence of arbuscular mycorrhiza and *Rhizobium* on phytoremediation by alfalfa of an agricultural soil contaminated with weathered PCBs: a field study. Int J Phytoremedition 12:516–533
Tharakan J, Addagada A, Tomlinson D, Shafagati A (2004) Vermicomposting for the bioremediation of PCB congeners in SUPERFUND site media. In: Waste management and the environment II: International Conference on Waste Management and the Environment No. 2, Rhodes, pp 117–124
Thiele-Bruhn S (2003) Pharmaceutical antibiotic compounds in soils-a review. J Plant Nutr Soil Sci 166:145–167
Tomei MC, Daugulis AJ (2013) Ex situ bioremediation of contaminated soils: an overview of conventional and innovative technologies. Crit Rev Environ Sci Technol 43(20):2107–2139
Torres E, Bustos-Jaimes I, Le Borgne S (2003) Potential use of oxidative enzymes for the detoxification of organic pollutants. Appl Catal B-Environ 46(1):1–15
Tripathi DK, Singh VP, Prasad SM et al (2015) Silicon-mediated alleviation of Cr(VI) toxicity in wheat seedlings as evidenced by chlorophyll florescence, laser induced breakdown spectroscopy and anatomical changes. Ecotoxicol Environ Saf 113:133–144
Tripathi DK, Singh S, Singh VP (2016) Silicon nanoparticles more efficiently alleviate arsenate toxicity than silicon in maize cultivar and hybrid differing in arsenate tolerance. Front Environ Sci 4:46
Truong PN, Baker D (1988) Vetiver grass system for environmental protection. Technol Bull No 1. Pacific Rim Vetiver Network. Office of the Royal Development Projects Board, Bangkok, Thailand
Trzesicka-Mlynarz D, Ward OP (1995) Degradation of polycyclic aromatic hydrocarbons (PAHs) by a mixed culture and its component pure cultures, obtained from PAH-contaminated soil. Can J Microbiol 41:470–476
Valentíın L, Lu-Chau TA, L´opez C et al (2007) Biodegradation of dibenzothiophene, fluoranthene, pyrene and chrysenein a soil slurry reactor by the white-rot fungus *Bjerkandera* sp. BOS55. Process Biochem 42:641–648
Valo R, Salkinoja-Salonen M (1986) Bioreclamation of chlorophenol contaminated soil by composting. Appl Microbiol Biotechnol 25:68–75
Vamerali T, Bandiera M, Mosca G (2010) Field crops for phytoremediation of metal-contaminated land. A review. Environ Chem Lett 8:1–17
van Bohemen HD, van de Laak WHJ (2003) The influence of road infrastructure and traffic on soil, water, and air quality. Environ Manag 31(1):50–68
Vangronsveld J, Cunningham SD (1998) Metal contaminated soils: In situ inactivation and phytorestoration. Springer and R G Landes Company, Georgetown
Vangronsveld J, Carleer R, Clijsters H (1994) Transfer of metals and metalloids from soil to man through vegetables cultivated in polluted gardens: risk assessment and methods from

immobilizations of these elements in soils. In: Varnavas SP (ed) Environmental contamination. CEP Consultants, Edinburgh, pp 142–145
Vidali M (2001) Bioremediation. An overview. Pure Appl Chem 73:1163–1172
Vila J, Lopez Z, Sabate J et al (2001) Identification of a novel metabolite in the degradation of pyrene by *Mycobacterium* sp. strain AP1: actions of the isolate on two- and three-ring polycyclic aromatic hydrocarbons. Appl Environ Microbiol 67:5497–5505
Vrkoslavova J, Demnerova K, Mackova M et al (2010) Absorption and translocation of polybrominated diphenyl ethers (PBDEs) by plants from contaminated sewage sludge. Chemosphere 81:381–386
Waghmode TR, Kurade MB, Govindwar SP (2011) Time dependent degradation of mixture of structurally different azo and non azo dyes by using *Galactomyces geotrichum* MTCC 1360. Int Biodeter Biodegr 65:479–486
Walker DJ, Clemente R et al (2004) Contrasting effects of manure and compost on soil pH, heavy metal availability and growth of *Chenopodium album* L. in a soil contaminated by pyritic mine waste. Chemosphere 57:215–224
Wang KR (2002) Tolerance of cultivated plants to cadmium and their utilization in polluted farmland soils. Acta Biotechnol 22:189–198
Wang Y, Oyaizu H (2009) Evaluation of the phytoremediation potential of four plant species for dibenzofuran-contaminated soil. J Hazard Mater 168:760–764
Wang X, Gong Z, Li P, Zhang L, Hu X (2008) Degradation of pyrene and benzo[a]pyrene in contaminated soil by immobilized fungi. Environ Eng Sci 25:677–684
Weber WJ (1972) Physicochemical processes for water quality control. Wiley, New York, pp 34–36
Wen Y, Ehsan S, Marshall WD (2012) Simultaneous mobilization of macro- and trace elements (MTEs) and polycyclic aromatic hydrocarbon (PAH) compounds from soil with a non-ionic surfactant and [S, S]-ethylenediamine-disuccinic acid (EDDS) in admixture: PAH compounds. J Hazard Mater 199–200:240–246
White JC, Parrish ZD, Isleyen M et al (2006) Influence of citric acid amendments on the availability of weathered PCBs to plant and earthworm species. Int J Phytorem 8:63–79
Whitfield Aslund ML, Zeeb BA, Rutter A, Reimer KJ (2007) *In situ* phytoextraction of polychlorinated biphenyl - (PCB) contaminated soil. Sci Total Environ 374:1–12
Willaert G, Verloo M (1992) Effect of various nitrogen fertilizers on the chemical and biological activity of major and trace elements in a cadmium contaminated soil. Pedology 43:83–91
Wunder T, Marr J, Kremer S et al (1997) 1-Methoxypyrene and 1,6-dimethoxypyrene: two novel metabolites in fungal metabolism of polycyclic aromatic hydrocarbons. Arch Microbiol 167:310–316
Xia H (2008) Enhanced disappearance of dicofol by water hyacinth in water. Environ Technol 29:297–302
Xu YH, Zhao DY (2007) Reductive immobilization of chromate in water and soil using stabilized iron nanoparticles. Water Res 41:2101–2108
Xu SY, Chen YX, Wu WX et al (2006) Enhanced dissipation of phenanthrene and pyrene in spiked soils by combined plants cultivation. Sci Total Environ 363:206–215
Xu SY, Chen YX, Lin KF et al (2009) Removal of pyrene from contaminated soils by white clover. Pedosphere 19:265–272
Yamazoe A, Yagi O, Oyaizu H (2004) Degradation of polycyclic aromatic hydrocarbon by a newly isolated dibenzofuran utilizing *Janibacter* sp strain YY-1. Appl Microbiol Biotechnol 65:211–218
Yang Q, Yediler A, Yang M, Kettrup A (2005) Decolorization of an azo dye, reactive black 5 and MnP production by yeast isolate: *debaryomyces polymorphus*. Biochem Eng J 24:249–253
Yap CL, Gan S, Ng HK (2010) Application of vegetable oils in the treatment of polycyclic aromatic hydrocarbons-contaminated soils. J Hazard Mater 177(1–3):28–41

Ye WL, Khan MA, McGrath SP, Zhao FJ (2011) Phytoremediation of arsenic contaminated paddy soils with *Pteris vittata* markedly reduces arsenic uptake by rice. Environ Pollut 159:3739–3743

Yirsaw BD, Megharaj M, Chen Z, Naidu R (2016) Environmental application and ecological significance of nano-zero valent iron. J Environ Sci 44:88–98

Yoon JM, Oliver DJ, Shanks JV (2007) Phytotoxicity and phytoremediation of 2, 6-dinitrotoluene using a model plant, *Arabidopsis thaliana*. Chemosphere 68:1050–1057

Zhang WX (2003) Nanoscale iron particles for environmental remediation: an overview. J Nanopart Res 5:323–332

Zhang WX (2005) Nanotechnology for water purification and waste treatment. Frontiers in nanotechnology, U.S. EPA millennium lecture series. Available at http://www.epa.gov/ncer/nano/lectures/zhang0705.pdf. Accessed 14 Mar 2016

Zhang H, Kallimanis A, Koukkou AI, Drainas C (2004a) Isolation and characterization of novel bacteria degrading polycyclic aromatic hydrocarbons from polluted Greek soils. Appl Microbiol Biotechnol 65:124–131

Zhang W, Wang H, Zhang R et al (2004b) Bacterial communities in PAH contaminated soils at an electronic-waste processing center in China. Ecotoxicology 19:96–104

Zielinski M, Kahl S, Hecht HJ, Hofer B (2003) Pinpointing biphenyl dioxygenase residues that are crucial for substrate interaction. J Bacteriol 185:6976–6980

Chapter 13
Soil Protection Policy

Abstract Since independence, agriculture and food security has received considerable importance in policy documents of Planning Commission of Indian government. Although issue of land protection has been raised and addressed in the Five Yearly Plan documents, focus remained mainly on rational use of land by different sectors of economic activity and on prevention of soil erosion from agricultural land. Protection of land from chemicals & radionuclides pollution and prevention of toxic contaminants entry into soil have received less reference in environmental acts and rules as compared to that received by air and water pollution. Therefore, the need for soil protection policy has been felt by environmentalists that can regulate anthropogenic activities responsible for land pollution and put forward guidelines on remedial measures of already contaminated land. This chapter identifies key considerations for soil protection policy after analyzing basic functions of soil and types of threats perceived from human activities.

Keywords Soil • Protection • Environmental acts • Rules • Policy

Soil an important natural resources, supports several ecosystem services by providing anchorage for plants; habitat for diverse group of organisms including human, animals, insects, microorganisms; producing food, fuel and fibre; maintaining or improving air quality in favour of sustaining life; regulating water cycling processes for sustaining its quality; sequestering atmospheric carbon and recycling of nutrient. In this context, soil quality is defined as the capacity of specific kind of soil to function within natural ecosystem and/or managed land-use boundaries, to sustain plant and animal productivity, promote air and water quality, and maintain plant, animal and human health (Karlen et al. 2001). In brief, soil quality has been related to "fitness of soil for desired use" or "capacity of the soil to function as desired". Soil also controls the pattern of human settlement and land use; determines local characteristics of landscapes, and conserves cultural and archaeological remnants and other aspects of our geo-diversity heritage.

J.K. Saha et al., *Soil Pollution - An Emerging Threat to Agriculture*, Environmental Chemistry for a Sustainable World 10, DOI 10.1007/978-981-10-4274-4_13

13.1 Land Related National Policies Towards Food Security

Land is an important source of livelihood for the people in India and State is entitled to collect and share the income derived from the land. As per the Seventh Schedule (entry no. 18 and 45 in II-State list) in the Constitution of India, land revenues including their assessment and collection, maintenance of land records, land management and alienation of revenue etc. fall within the powers of the State Governments. As "Land" is a State subject, only state (provincial) legislatures have the power to enact and implement land-reform laws. Consequently, such laws differ from State to State as these are formulated and enacted in furtherance to the cause of people under its jurisdiction. However, the central government as a whole plays an obligatory advisory and financial role in land related policy as per the provision in constitution in respect of social and economic planning (a responsibility held concurrently with the states). To fulfil the role of social and economic planning, the Government of India established a National Planning Commission immediately after Independence. The Commission prepared a series of Five-Yearly Plans since 1951 which also stressed upon the action plans on various land related policies (Table 13.1). In 1976, the National Commission on Agriculture stress upon on scientific land use planning for attaining food security, self reliance and enhanced livelihood security for the people of country. The National Policy for Farmers, in 2007 had recommended for reviving the existing Land Use Boards and linking them to district-level land-use committees, so that they can provide effective and proactive advice to farmers on land use. The committee on "State Agrarian Relations and the Unfinished Task in Land Reforms" (2009) has also put emphasis on the need for land use planning in the country.

13.2 Focus of National Policies on Protection of Land Resources

Department of Land Resources under Ministry of Rural Development (Government of India) prepared 'National Land Utilisation Policy: framework for land use planning and management' (published in September 18, 2013), which seeks to order and regulate land use in an efficient and rational way, thus taking care of the needs of the community (livelihood, food and water security) while safeguarding natural resources and minimizing land use conflicts or negative environmental impacts. Following issues have been addressed in the document:

- Unregulated land use shifts
- Reducing per capita land resource
- Meeting the demands of all developmental sectors (agriculture, urban, industrial and mining)

Table 13.1 Land related policy formulation throughout Five yearly Plan period

Plan period	Major issues identified	Policy thrust
First Plan (1951–1956)	Huge uncultivated lands sealed under large size holdings. Need to expand area under cultivation and community development (CD) network	Land reforms to bring vast fallow area under cultivation and to increase land use efficiency. Abolition of intermediaries. Tenant to be given the rights to cultivate land
Second Plan (1956–1961)	Concern about vast rainfed agriculture and low land productivity. Less thrust on irrigated agriculture. Need for land related institutional framework	Soil conservation and irrigation infrastructure development for the rainfed areas. First phase of land reform implementation. Training and extension effort for the technology through community development
Third Plan (1961–1966)	Concern for Food security. To bring cultivable waste land under cultivation through reclamation. To bring backward regions under mainstream growth	Emphasised on area development programme and developing integrated land policy. Soil surveys and Intensive area development programme (IADP) adopted for selected districts
Fourth Plan (1969–1974)	Emphasis on food security to be continued for meeting minimum dietary requirements of people. Incentives for land diversion towards food crops cultivation and enhancing the productivity of such land. Domination of large holding sizes and low allocation and technical efficiency	More thrust on irrigation and soil conservation in dryland areas. Concern expressed for increasing cropping intensity. Second phase of land reforms with land ceiling acts and consolidation of holding introduced. Institutional and technological changes brought in
Fifth Plan (1974–1979)	Problems of land degradation highlighted. Drought-prone areas and land management in irrigated command areas attracted attention	Soil conservation, Drought-prone area development and Desert area development programmes started and enhanced further. New impetus to dry farming
Sixth Plan (1980–1985)	Under-utilization of land resources and Drought-prone areas continued to draw attention. Increased focus felt for green revolution in economically backward regions	Land and water management programme initiated in selected areas under drought-prone area programme
Seventh Plan (1985–1990)	Soil erosion and land degradation, Diversion of cultivated land to non-agricultural sector, and Deforestation and degradation of forest lands surfaced as major issues	Conservation of soil and water, and protection of land from degradation were emphasized. Specific attention to degraded lands like Wastelands Development programmes and land management in long-term perspective
Eighth Plan (1992–1997)	Degradation of land in irrigated command areas. Dryland and rainfed areas needed more attention. Lack of peoples' participation in land management at village level surfaced as major issue	Emphasis on land management through watershed approach. Soil conservation was brought under watershed programmes. Agro-climatic zone based regional planning approach was introduced
Ninth Plan (1997–2002)	Considerable increase in area under land degradation. Integrating Watershed Development Programme with	Focus on bringing more under-utilized land under agriculture. Management of wastelands. Maintenance of village

(continued)

Table 13.1 (continued)

Plan period	Major issues identified	Policy thrust
	various components of other developmental programmes. Relook on on-going land reforms. Narrowing gap between potentials and actual crop yields. Need for a long-term policy document	community land. Management of village community lands through Panchayat Raj institutions. Decentralization of land management system. Revisit on existing land legislations
Tenth Plan (2002–2007)	Degradation of land caused by biotic and abiotic pressures. Macro environmental contamination of air, land, water, and food	Bringing wasteland and degraded lands into productive use. Creation of comprehensive regional land use database
Eleventh Plan (2007–2012)	Land reforms for promoting agricultural productivity. Measures and guidelines in respect of land going out of cultivation. Micronutrient deficiencies and depleting C in soil	A major thrust at watershed development through expansion of the accelerated irrigation benefit programme. Better water management in rainfed areas
Twelfth Plan (2012–2017)	Enhanced soil erosion, soil quality deterioration from imbalanced use of fertilizers	Land reform for equitable and rational use of land. Land records modernization programme. Promotion of best practices of soil fertility management. Recycling of organic matter in soil

- Protecting lands under natural environment and its resources that provides ecosystem services
- Meeting urbanization demands
- Meeting industrialization demands
- Meeting mining sector demands
- Meeting transport sector demands
- Development vs. sustainable development
- Protecting social interests
- Protecting heritage
- Inadequate land use planning capacities

However, this policy document does not mention about the guiding principles related to soil protection from polluting activities and remediation of anthropogenically degraded land.

In order to protect environment, India has enacted different laws which provided the instruments for controlling or regulating the acquisition, use, and conservation of the natural resources. These aim to facilitate a sustained use of resources and allow their need-based management. Majority of the laws/rules implemented were directed towards prevention and control of different environmental components and pollutants, like, water resources, air, hazardous wastes, biomedical wastes, plastics, fly ash, municipal solid wastes, batteries etc.; and adequately addressed the issues on protection of air, water bodies, forests and wild animals, fragile ecosystems. Although, the land Acquisition Act, 1894 provides greater control of the government on lands and subsequently on its planning, development and building of housing etc., it does not address the issue of protecting soil and its valuable functions. Following different statutes/legislations have been enacted for the purpose of protecting environment in India.

- The Water (Prevention and Control of Pollution) Act, 1974
- The Water (Prevention and Control of Pollution) Rules, 1975
- The Water (Prevention and Control of Pollution) Cess Act, 1977
- The Water (Prevention and Control of Pollution) Cess Rules, 1978
- The Air (Prevention and Control of Pollution) Act, 1981
- The Air (Prevention and Control of Pollution) Rules, 1982
- The Environment (Protection) Act, 1986
- The Environment (Protection) Rules, 1986
- Hazardous Wastes (Management and Handling) Rules, 1989
- Manufacture, Storage and Import of Hazardous Chemical Rules, 1989
- The Forest (Conservation) Act, 1980
- The Forest (Conservation) Rules, 1981
- The Wildlife Protection Act, 1972
- The Wildlife (Transactions and Taxidermy) Rules, 1973
- The Wildlife (Stock Declaration) Central Rules, 1973
- The Wildlife (Protection) Licensing (Additional Matters for Consideration) Rules, 1983
- The Wildlife (Protection) Rules, 1995
- The Wildlife (Specified Plants - Conditions for Possession by Licensee) Rules, 1995
- The Public Liability Insurance Act, 1991
- The Public Liability Insurance Rules, 1991
- The National Environment Tribunal Act, 1995
- The National Environment Appellate Authority Act, 1997

13.3 Environment Protection Acts and Rules

The Government of India enacted Environment Protection Act (EPA) in 1986 with the objective of providing protection and improvement of the environment. It empowers the Central Government to set up authorities with the assigned mandate of preventing environmental pollution in all forms from any source and to combat specific environment related problems that are unusual to different parts of the country. The Act was last amended in 1991.

Under 'The Environment (Protection) Rules, 1986', following clauses mention on the protection of natural resources from industrial activities which may also be applicable on land/soil.

Clause 3. "Standards for emissions or discharge of environmental pollutants".

Clause 5. "Prohibitions and restrictions on the location of industries and the carrying on processes and operations in different areas. The Central government may take into consideration the following factors while prohibiting or restricting the location of industries and carrying on of processes and operations in different areas.

(i) Standards for quality of environment in its various aspects laid down for an area;
(ii) The maximum allowable limits of concentration of various environmental pollutants (including noise) for an area;
(iii) The likely emission or discharge of environmental pollutants from an industry, process or operation proposed to be prohibited or restricted
(iv) Environmentally compatible land use;
(v) Net adverse environmental impact likely to be caused by an industry, process or operation proposed to be prohibited or restricted".

13.4 National Environmental Policy 2006 (NEP 2006)

Our national commitment to a clean environment is mandated in Articles 48 A and 51 of Constitution of India. In response to these Articles, National Environmental Policy 2006 was framed with objective of maintaining and enhancing the quality of the environment. The policy also seeks to strengthen the management aspect of environmental quality through building partnerships linkage among different stakeholders, like local agencies, public institutions, scientific and academic communities, and international development partners. Section 5.3.1 of National Environmental Policy 2006 (approved by the Union Cabinet on 18 May, 2006) emphasizes the need to lay down different Environmental Standards; wherein 'Environmental Standards' refer both to the permissible levels of discharges of specified waste streams by different classes of activities ("emission standards"), as well as acceptable levels of particular environmental quality parameters at different categories of locations ("ambient standards"). It also explicitly indicates that environmental standards cannot be same and uniformly applicable for all countries; and therefore, **each country should set standards in accordance with national priorities, objectives of national policies and available resources**. In general, these standards may vary, and become more stringent as country develops with greater access to technologies and higher fund availability for environmental management. Environmental standards need to relate with other measures for risk alleviation in the country, so that a given societal obligation for resource allocation delivers the maximum aggregate reduction in risk to environment.

In order to control soil pollution, NEP 2006 stressed following in their action plans:

1. Developing appropriate techniques for treatment of toxic waste and hazardous material and their safe disposal through secured landfills and incineration.

2. Developing a national inventory and setting up an online monitoring system to keep record of toxic and hazardous waste movements and dumps
3. Addressing emergencies during chemical accidents arising out of transportation, handling, and disposal of hazardous wastes.
4. Strengthening the capacities of local bodies and outsourcing of services like waste segregation, recycling, and reuse.
5. Strengthening the waste collection system and recycling of various materials through institutional finance and scientific technologies.
6. Reclamation of chemical contaminated land and promoting traditional crop varieties through organic farming.
7. Promoting strategies for recycle and reuse concept through technological advancement and substituting non-biodegradable materials with environmentally safe biodegradable and recyclable resources
8. Developing and implementing guidelines and strategies for management of e-waste.
9. Creating awareness and promoting beneficial utilisation of nonhazardous waste materials such as red mud, fly ash, bottom ash, and slag in different non-agricultural sectors.

In the action plan, however, nothing has been mentioned on the other important issues like need for maintaining database on reference soil quality; combating stresses created on soil through setting-up of industries and Special Economic Zone (SEZs); policy on remediation of polluted soil etc.

13.5 Need for Soil Protection Policy

Soil, the uppermost layer of the earth crust, evolved through weathering action during thousands to millions of years and provides basis for life on earth. Existence and sustenance of quality of life depends on the quality of soil; and hence, there is a need for a national policy towards protection and improvement of soil quality. Above list of legislations indicate that no specific rule/policy has been enacted for the purpose of protecting land/soil from human interferences and use of this non-renewable resource. This may probably be because soil resources not being direct major route of toxic pollutants entry into human beings. However, need for 'Soil protection policy' is urgently being felt because of two important reasons: (1) soil is precious non-renewable resources for healthy life on earth and (2) remediation of polluted land/soil (being an immobile matrix) is more difficult as compared to other natural resources air and water. In order to move forward in the direction of formulating 'Soil protection policy' for the country, there is a need to understand the important ecological functions performed by soil and the major anthropogenic pressures acting on it.

13.5.1 Basic Functions of Soil

- **Soil is the storehouse of plant macro and micronutrients**: Soil serves as medium for plant growth and also habitat for soil micro flora and macro fauna. It serves as a store house and moderates the release of nutrients and other elements through chemical, biochemical and microbial processes. Various nutrient cycling processes have greater impact on plant growth (for food, fodder and timber production) soil fertility and agricultural sustainability as well as on other components of environment.
- **Soils to perform partitioning, filtering and regulation of water flow:** Soil acts as a filter before water moves into the underground aquifers, and flows into lakes and streams before eventually ending up in ocean. Therefore, capacity of soil to filter determines the quality of surface and ground water bodies. With proper functioning, soil acts as hydrologic buffer and partitions water for groundwater recharge and for use by plants and soil animals.
- **Soil harbours biodiversity:** A significant part of valuable soil functions is carried out by the thousands of millions of organisms that inhabit within it. These include bacteria, algae, fungi, nematodes, earthworms, insects etc. Soil micro-organisms play a major role in soil formation; and therefore, soil performs to sustains biological activity and diversity. Microbial decomposition of existing soil organic matter or organic residue releases nutrients that are more available to the plant. Microorganisms are considered a primary decomposer of soil and perform its greatest functions in the global ecosystem. Microbial decomposition of dead plants, animals, and organisms by soil biota converts complex organic form of nutrient to simpler inorganic mineral forms, which are then easily utilized by other living plants, animals, and microorganisms.
- **Soils act as a sink for carbon:** Due to burning of fossil fuel, deforestation and large-scale urbanization, carbon dioxide emission post industrialization has been constantly increasing and has reached to dangerous mark of imparting serious environmental problems. Under natural process, carbon cycle is dynamic and complex where carbon is released to the atmosphere from "carbon sources" and again stored back in plants, animals, rocks, fossil fuel, soil and water as "carbon sinks". During soil carbon sequestration process, CO_2 is removed from the atmosphere and stored in the soil carbon pool. This process is mediated by natural forest vegetation and agricultural crops that have the ability to fix atmospheric carbon dioxide in leaves, shoots, and roots using energy from sun through the process called "photosynthesis". The carbon in the dead remains of plant and microbial cells are temporarily locked in the soil profile against their decay and thereby, help in reducing carbon di-oxide (CO_2) concentration in the atmosphere and global warming. Carbon addition and loss rates from soil depend on several factors like soil type and condition, landscape topography, rainfall pattern and temperature.
- **Soils of the land also act as a sink for several toxic pollutants generated due to anthropogenic activities:** Metals which play significant role in our daily life,

are continuously being mined out from different natural geological deposits in order to meet human and industrial needs. The wastes arising out from metal mining and industrial processes are discharged profusely on the land surface in the form industrial effluents and solid waste. Soil thus acts as a natural sink for contaminants from various sources. However during their continuous entry over a period time, the contaminants in soil reach to a level that poses risks to environment and ecosystem service. Though each soil has a definite tolerable limit (loading limit) for pollutants, these contaminate food chain and ground water bodies once these limits are exceeded.

13.5.2 Threats to Soil and Issues of Soil Protection

Soil degradation, a result of natural process or anthropogenic activities is expanding worldwide. It generally refers to the decline in soil's inherent capacity to produce economic goods and perform ecologic functions. Anthropogenic activities have not only accelerated degradation processes, but also created new types of threats in addition to natural degrading processes like desertification, erosion and salinization. These new threats can be, broadly categorized into four areas:

(a) Land consumption

  (i) Removal of upper fertile soil layer for brick making and for utilizing in other developmental/constructional activities.
  (ii) Sealing of land surface during development of infrastructure like housing, commercial buildings, road and rail networks.

(b) Chemical pressure

  (i) Entry of hazardous and toxic contaminants in soil through polluted water, atmospheric deposition and precipitation.
  (ii) Passage of contaminants into soil through amendments, chemical fertilizers and waste derived composts.
  (iii) Illegal disposal of solid and liquid wastes.
  (iv) Entry of contaminants via polluted groundwater like arsenic.
  (v) Salinization from canal water seepage.
  (vi) Nutrient imbalance in soil caused by mismatch between nutrient removal from and addition to soil through fertilizers, resulting in loss crop productivity and produce quality.

(c) Physical pressure

  (i) Soil compaction by excessive use of heavy machinery for agricultural activities.
  (ii) Loss of top-soil through erosion as a result improper soil management like tilling along the slope and excessive tillage operation, faulty land use

adoption, lack of surface mulching, crop cover and other soil conservation measures during heavy rainfall etc.

(d) Biological pressure

(i) Introduction of genetically modified pathogenic or alien organisms, which can adversely affect soil fertility and disrupt the ecosystem.
(ii) Decline in soil biodiversity

13.5.3 Key Considerations in the Formulation of Soil Protection Policy

1. **Establishment of framework for monitoring soil quality**
   Scientific information on parameters representing qualitative state of soil needs to be made available to create soil protection measures useful and acceptable to those concerned. Minimum dataset for evaluation and their quantification to indicate soil quality should be established for different agroecosystems. Critical level of soil quality index for each agroecosystem requiring urgent attention should be determined. Establishing a network of permanent soil quality measuring stations can be helpful for follow-up monitoring which will form a basis for timely corrective actions at national or state level.
2. **Creation of reliable database on status of soil quality in different agro-ecological regions of the country**
   A mechanism to monitor soil health may be created by constituting a high power body like National Authority on Soil Health. A national level data repository may be set up for effective monitoring and initiating protective measures against degrading forces. Identification of land contaminated with pollutants due to industrial or other anthropogenic activities is very important for the policy makers so as to take necessary precautions and control measures well in advance. Hence there is a need to develop database on background concentration of pollutants which will help in detecting land in area that has been contaminated due to the industrial, mining or other anthropogenic activities.
3. **Popularization of soil testing programme for polluted land areas**
   Although soil testing laboratories, both permanent and mobile type have been established across the country to cater the need of farmers, the facilities are grossly unequipped to measure parameters related to soil pollution. There is need to revamp some of the existing soil test laboratories and/or establish new soil testing laboratories with appropriately trained staff and modern equipments for monitoring soil health of the land areas facing pressure from high non-agricultural anthropogenic activities. Protocols for assessing risks from polluted soils along with recommendations for farmers on appropriate soil and

crop management practices need to be developed for different agroecosystems. Appropriate soil extractant and critical limits of different pollutants in soil should be established through network research.

4. **Appropriation of prudent steps to shield fertile soil from physical, chemical and biological pressures**

   Priority objective of efficient soil protection measures should be to act at the source. So better strategy is to prevent generation of the risks to the soil in the first place, or at least, to avoid them as far as possible. All the individual, organization and entrepreneurs associated with an activity that can directly or indirectly impact soil quality, should be bound by the principle of precautionary soil protection (i.e. exploring alternatives to possibly harmful action). This policy should specifically be implied to authorities responsible for passing regulations pertaining to the (i) manufacture or use of processes, materials, organisms, equipment, machinery and automobiles that can harm the soil ecosystem functioning; (ii) industries and related professions that produce, sell or use such substances, organisms or devices; (iii) authorities that set standards for emissions of particulate matters relevant to the soil quality (e.g. air pollution control) or that plan, approve and monitor technological and construction-related interventions affecting soil resource and its quality; (iv) any individual who cultivates the soil as a farmer, gardener or forester, or who damages the soil for construction purposes. As adequate soil organic matter supports soil biodiversity that plays crucial role in combating the adverse effects of soil pollution, policies need to be oriented towards popularizing recycling of aboveground organic wastes in the land. Conservation of rainwater and soil moisture should be promoted through extension of technologies and supply of subsidized equipments (wherever required) to reduce dependency on polluted water bodies for irrigation to crops nearby wastewater generating activities.

5. **Generation of awareness among all sections of society about the vital functions of soil in the ecosystem and its vulnerability**

   Institutions and organizations at all levels, especially those imparting training and education to people in soil-related occupations (for example, farmers, gardeners, construction managers and workers) may be associated to create awareness on soil functions, threats perceived and its implications. A national level nodal organization may be assigned the responsibility for devising a framework and resource materials for such awareness program.

6. **Sustainable, careful and cautious use of soil should only be permitted**

   Several vital functions are performed in the ecosystem by soil mass that has been created over centuries of pedogenic processes. This non-renewable resource is not available at will; and hence, must be used sparingly and judiciously. Every time a new surface area of soil is sealed (for example during housing, road construction, etc.), it represents a reduction in the finite resource. Also, excessive use of agrochemicals (pesticides, herbicides, fungicides etc.), application of soil amendments containing pollutants (municipal solid waste composts, sewage-sludge composts, fly ash, basic slag etc.) degrade the functional quality of soils. Hence, soil protection policy must address these issues while formulating rules/acts/guidelines. Maximum concentration limits for each of such contaminated soil amendments need to be determined through experiments. A special fund may be created to support research on soil protection issues with the help of compulsory contributions from industrial and mining sectors as a part of corporate social responsibility.

7. **Conservation of fertile soils during any developmental activities**

   As a state policy, soil should be declared as precious non-renewable natural resource. Owner and user of land should have the duty to use and conserve this precious resource properly. Mining and construction activities remove large volume of top layer of fertile soil and generally dumped on nearby fertile land along with stones and gravels dug from bottom layer damaging both productive land and fertile dug-out soil. Appropriate guidelines are required for separately conserving fertile top soil layers removed during mining & construction activities and use these appropriately for growing vegetation wherever required and for rehabilitating degraded land.

8. **Remediation of contaminated soils must be accomplished to ascertain whether all risks have been eliminated**

   Contaminated soils cause dangers to all strata of living beings including humans, animals and crops and therefore, must not be left untreated. Financial aspects, particularly high cost of technologies, make it difficult to accomplish the remediation of soils contaminated with chemicals to the point where soil multi-functionality is totally restored. Soil remediation therefore must focus on averting risks and supporting minimum but safe farming operations. Soil remediation objectives should incorporate changing or limiting farm land use as well as provide technical support for contaminants removal or soil replacement.

9. **Responsibility soil protection should also lie on the owners of land**

   Like other natural resources like water, air and forests, soil is a common national property, which may be utilized but not misused or damaged by its owner. Farmers should adopt good agricultural practices like adoption of soil conservation measures, *in-situ* rainwater harvesting, balanced use of nutrients, adoption of integrated nutrient management system, incorporating leguminous crops in the cropping system, improvement and maintenance of soil structure though application of manures, following appropriate tillage practices etc. Government and non-government organizations must be involved in creating awareness among the farmers.

10. **Enforcement of legal standards to ensure protection of soil resources in terms of quantity and quality**

    A well devised set of legal instruments is needed to implement remedial measures that are essential but not undertaken intentionally. This applies especially to the interventions related to soil use methods that can bring about irremediable harm to the society. Therefore, due consideration must be given to ensure that soil protection is incorporated in the provisions regulating permissible levels of emissions/release/addition in the legislation on environmental protection like maximum permissible loading limits of heavy metals and other pollutants in different soils types, maximum concentration of heavy metals that can be allowed to receive materials derived from solid wastes, sewage sludge etc., and also in regulations relating to utilization of the soil (e.g. agriculture, construction, industry etc.).

11. **Institutional framework for soil protection needs to be enhanced and endowed with more budgetary and human resources**

    Presently, only a very small fraction of the funding is available for soil protection as compared to other fields of environmental protection like water, air, forests and landscapes. Significance of soil protection needs to be recognized and supports from the environmental authorities, research and teaching societies need to be increased.

12. **Cooperation among the departments related to land use planning, agriculture and forestry must be established for protection of soil resources**

    Without the supports from land use planning and forestry sector, it is impossible to protect the soil. The agricultural sector on one hand damages (to a limited extent) the soil ecosystem via its activities; while on the other hand, it has an interest in ensuring that the soil remains healthy. As a policy, efforts should be to direct agriculture towards more natural, environmentally sound farming practices for crop diversification, nutrient, pest and weed management those support soil protection. Intensive utilization of forest resources is putting forest soils at risk due to increased physical pressures. Land use planning is responsible for assigning land to different uses. It should thus be able, while drawing up structural plans and spatial plans, to promote types of use that are in line with soil properties and its functional capacities. It should also develop instruments/mechanisms to ensure more economical and careful use of the land. A genuinely effective land use policy needs to be evolved which should earmark areas not suitable for farming on economic grounds, and therefore can be used for other purposes.

13. **Environmental impact assessment of any industrial, infrastructural, mining projects should compulsorily include soil health evaluation**

    Environmental Impact Assessment is pre-requisite for statutory clearance of any project related to infrastructure, energy, industry, mining etc. which have direct impact on environment. Very few experts involved in EIA are presently having the academic background related to soil science, soil pollution and remediation. Even NABET accredited training programmes don't have training modules related soil pollution-protection. Therefore, appropriate emphasis should be given for involving soil scientists in EIA as well as in training of experts involved in EIA on the subject of soil science, soil pollution and remediation.

14. **Creation and strengthening of infrastructural facility for soil pollution assessment**

    As analysis of pollutants requires sophisticated and costly instruments and reagents, few laboratories in the country are appropriately equipped with instruments and soil experts to conduct soil pollution related monitoring and research. This is more relevant in case of National Agricultural Research System of the country where emphasis is more on increasing agricultural productivity. However, since non-agricultural anthropogenic (urban, industrial, mining, transportation etc.) activities are increasing exponentially in the quest of increasing GDP, instances of soil pollution is also expected to increase as a side effect. This necessitates setting up of more research and monitoring facilities and laboratories to address abovementioned researchable issues.

A 'Soil Protection Policy' if formulated and enacted at national level based on above considerations, will certainly help to fulfill the objectives of recently implemented India's 'National Mission for Sustainable Agriculture'. For this, consultation process needs to be initiated involving different stakeholders including concerned research organizations, industry, farmers', policy makers as well as various concerned Government departments.

Reference

Karlen DL, Andrews SS, Doran JW (2001) Soil quality: current concepts and applications. Adv Agron 74:1–40

MIX
Papier aus verantwortungsvollen Quellen
Paper from responsible sources
FSC® C105338
FSC
www.fsc.org

If you have any concerns about our products,
you can contact us on
ProductSafety@springernature.com

In case Publisher is established outside the EU,
the EU authorized representative is:
Springer Nature Customer Service Center GmbH
Europaplatz 3, 69115 Heidelberg, Germany

Printed by Libri Plureos GmbH
in Hamburg, Germany